Functional Ecology of Woodlands and Forests

J. R. Packham, D. J. L. Harding, G. M. Hilton and
R. A. Stuttard

Wolverhampton Woodland Research Group
University of Wolverhampton
Wolverhampton, UK

CHAPMAN & HALL
London · Glasgow · New York · Tokyo · Melbourne · Madras

Published by Chapman & Hall, 2–6 Boundary Row, London SE1 8HN

Chapman & Hall, 2–6 Boundary Row, London SE1 8HN, UK

Blackie Academic & Professional, Wester Cleddans Road, Bishopbriggs, Glasgow G64 2NE, UK

Chapman & Hall, 29 West 35th Street, New York NY10001, USA

Chapman & Hall Japan, Thomson Publishing Japan, Hirakawacho Nemoto Building, 6F, 1-7-11 Hirakawa-cho, Chiyoda-ku, Tokyo 102, Japan

Chapman & Hall Australia, Thomas Nelson Australia, 102 Dodds Street, South Melbourne, Victoria 3205, Australia

Chapman & Hall India, R. Seshadri, 32 Second Main Road, CIT East, Madras 600 035, India

First edition 1992

© 1992 J. R. Packham, D. J. L. Harding, G. M. Hilton and R. A. Stuttard

Typeset in 10/12 pt Meridien by Graphicraft Typesetters Ltd, Hong Kong
Printed in Great Britain at the University Press, Cambridge

ISBN 0 412 44390 2 (HB)   0 412 43950 6 (PB)

A catalogue record for this book is available from the British Library

Library of Congress Cataloging-in-Publication data
Functional ecology of woodlands and forests/J. R. Packham . . . [et al.].
       p.   cm.
    Includes bibliographical references and index.
    ISBN 0-412-44390-2. — ISBN 0-412-43950-6 (pbk.)
    1. Forest ecology. 2. Plant communities. 3. Animal–plant relationships.
QK938.F6F86   1992   91–47028
581.5'2642—dc20   CIP

To all who strive to protect and maintain our woods and forests

I will put in the wilderness the cedar, the acacia, the myrtle, and the olive;
I will set in the desert the cypress, the plane and the pine together;
that men may see and know, may consider and understand together,
that the hand of the Lord has done this.

Isaiah 41, v. 19–20

Contents

Preface

The ecosystem approach to the understanding of woodlands and forests adopted in this book begins with the study of the plant, herbivore and decomposition subsystems, and of the essential processes of nutrient cycling and energy flow. The first chapter shows how organisms react with each other and with human beings, as well as with their physical environment. It also describes the main types of forest community, from the cold boreal forests of the northern taiga to the steamy rainforests of the tropics, and the ways in which so many are being damaged or destroyed.

Succeeding chapters deal with various aspects of forest ecosystems in more detail, presenting observations and inferences from world-wide research, with direct references to the original papers in which this research was first described. It has been the deliberate policy of the authors to choose many of these references from periodicals which are readily available in academic libraries, e.g. *Journal of Ecology, Journal of Animal Ecology, Quarterly Journal of Forestry, Ecology, New Phytologist*. Where direct reference is made to a particular paper or book it will appear in the References; particularly important work may also be listed under Further Reading.

Key ideas and concept words are emphasized by **bold type** when first explained, and references to definitions are printed in bold in the index. The organisms most commonly referred to in the text are listed after the index.

The final two chapters consider forests as a resource, dealing with aspects of management, contemporary problems, and future prospects. Woodlands and forests are unique communities of enormous value, but all too many are being destroyed before they have been described, let alone understood. That is why it is so important to study them directly – as well as by reading and visiting botanic gardens – and then to join the fight for their survival.

This book evolved from *Ecology of Woodland Processes* (Packham and Harding, 1982) which was written to provide a standard text in woodland ecology for undergraduates. The present volume has been enlarged to deal in much greater detail with aspects of woodland management and forestry, and also thoroughly scrutinized and brought up to date in those areas where there have been significant advances. Competitive foraging by plants, succession, patch dynamics and regeneration, root productivity, population dynamics of forest insects, and a rethink of conventional ideas on nutrient cycling in tropical forests are amongst the areas which have received particular attention.

We remain indebted to all those who assisted with the original volume, particularly to Professor A. J. Willis who was its editor. Dr John Barkham made several useful suggestions regarding the structure of the present book and also commented on the final draft. We have again had numerous discussions and communications with other woodland ecologists; we are very grateful to them and to the authors and publishers who gave permission for the reproduction of many of our illustrations. In a few cases it proved impossible to obtain a reply to copyright enquiries, but the origins of all non-original illustrations are given. The book owes much to the support afforded by the School of Applied Sciences, Wolverhampton Polytechnic, and to the stimulating queries of our students. We have benefited from discussion with other members of the Wolverhampton Woodland Research Group and owe a particular debt to Mrs E. V. J. Cohn (help with sections 3.1 and 11.5) and Mr P. R. Hobson (original drawings). On the publishing side, Dr C. Earle and Dr R. C. J. Carling have been a consistent source of help and encouragement. Finally, the two original authors thank their new colleagues for the zest and fresh insights they have brought to this new book: all four of us are grateful to our wives and families for their patience and understanding.

J. R. P.
D. J. L. H.
G. M. H.
R. A. S.
Woodland Research Group,
University of Wolverhampton

1

Introduction

1.1 Ecology and trees

Woodlands are long established, complex ecosystems dominated by trees. Ecologists seek to discover how they are organized and how they function. In such studies of pattern and process, attempts are made to unravel the interrelationships and roles of various species in particular communities, while function at the ecosystem level involves studying such processes as the flow of energy and the cycling of nutrients. Knowledge of ecosystem processes was markedly increased by the International Biological Programme (IBP, 1964–74). Since its completion there have been many more field studies and also a great increase in the number of predictive models attempting to simulate such phenomena as pest outbreaks, succession, regeneration, and the relationship between stability and species diversity.

The importance of woods and forests in helping to stabilize world climates through their role in the carbon cycle is increasingly recognized, as is their function as a reservoir of natural drug precursors and the only habitat of uncounted species of organisms. In a changing world it is important to identify the structure and processes characteristic of particular woodlands, and be able to predict the likely outcome of disturbance to such systems. To a considerable extent ecologists such as Peterken (1981) now seek to use the past as a key to the future, employing a variety of techniques to ascertain the causes of previous woodland changes. At the same time interest in population dynamics, and in the physiology and biochemistry of woodland organisms has never been greater.

Trees vary greatly in shape, size and many other attributes, but all are large woody plants with a comparatively long life-span. Typically, though not invariably, they are single-stemmed whereas many shrubs are multi-stemmed. Their woody nature enables trees to grow far higher than herbaceous forms, allowing their leaves to receive more light than shorter plants, effectively displaying flowers to pollinating animals or to the wind, and increasing the distance over which the seeds can be shed. There are, however, concomitant disadvantages to this life form in which a mature tree possesses a dead, though essential, skeleton (largely of heartwood) as well as great amounts of living cambia, phloem and wood parenchyma, which together form a major respiratory burden. The enormous accumulations of dead tissue within living trees may provide food for heterotrophs, including certain pathogens, while great height itself increases the possibility of damage by fire, lightning, wind or hurricane. In addition, as it grows older, the form and physiology of a tree must allow it to function under the changing conditions of life associated first with the ground vegetation, then the shrub layer, and finally the tree canopy.

The **size**, **form** and **longevity of a tree** all influence the development of the community to which it belongs, as do the shade which it casts, the plant litter which it produces, and its ability to resist disease, water loss and fire. The characteristics of various species, especially with regard to initial establishment, competition and edaphic requirements, fit them for a wide variety of habitats which they occupy for different periods of time. Many British spoil tips abandoned soon after the beginning of the present century were colonized by birch, most of which are now mature or dying, and frequently a process of succession to oak, sycamore or mixed woodland can be observed. In the English Midlands, 60 or 70 years is a typical life-span for native birch (*Betula* spp.) which casts a light shade, spreads widely and rapidly by winged seeds and often acts as a pioneer species. In contrast, the big tree (*Sequoiadendron giganteum*), a redwood from the Sierra Nevada, is so huge, long-lived and has such a dense shade that it may exclude competitors for a period equal to several generations of smaller conifers. The **heaviest** living tree is an individual of this species known as 'General Sherman' whose estimated weight is 2030 tonnes, while the **tallest** is a coastal redwood (*Sequoia sempervirens*) estimated at 366 ft (111.6 m) in 1970. Records of even taller trees of Douglas fir (*Pseudotsuga menziesii*) and *Eucalyptus regnans* remain unverified. 'General Sherman' is known to be between 3500 and 4000 years old from a ring count made on a core drilled in 1931. The **oldest** recorded living tree, however, is a bristlecone pine (*Pinus aristata*) which grows at an altitude of 10 750 ft (3277 m) under bleak conditions on Wheeler Peak in California, USA. Dendrochronologists found this to be about 4900 years old.

Ring count dating, which employs the characteristic patterns of stem rings which develop in response to the climatic patterns of succeeding seasons, was extended to earlier than 6200 BC by the use of fallen bristlecone pine wood. Superlong chronologies extending back over 10 000 years from the present have been constructed by employing fossil tree trunks, especially from riverine deposits, which can be dated by means of their radioisotope contents. Annual growth rings are particularly prominent in such trees as elm (*Ulmus*), whose ring-porous wood contains large vessels formed in the spring and much smaller ones formed later in the season. In contrast, the wood of birch (*Betula*), in which the xylem vessels are essentially uniform and fairly evenly distributed throughout, is diffuse-porous though the annual rings in this tree are also quite distinct. Examples of radial increment diagrams, which show variation in the width of the annual rings during the life of the tree, are shown in Figure 5.2.

Though the structures of all tree trunks are basically similar, there are many variations. The bark of the fire-resistant redwoods is up to 0.3 m thick, while that of many birches is little more than a skin. The form and texture of the bark are determined by the activities of the **cork cambium**. When the cork cambium forms a series of shallow overlapping arcs the bark eventually cracks at their edges; the London plane (*Platanus hybrida*) owes much of its resistance to air pollution to the ease with which large flakes of bark are discarded, often after frost. The more deep-seated **cambium** continues to produce new cells throughout the life of the tree, phloem on the outside and xylem on the inside. The living **phloem** transports photosynthetic products, while the functional **xylem** contains dead, hollow elements through which water is transported upwards by mass flow, a process which also involves mineral salts and a variety of other compounds including aminoacids, hormones and traces of pesticides which may have entered the tree. As xylem produced in spring contains large cells and that formed in late summer small cells, a series of annual rings is developed. In most trees conduction is restricted to the young sapwood which surrounds the non-functional heartwood. The latter usually becomes impregnated with various materials including tannins, resins and pigments, many of which are antiseptic and help prevent decay by inhibiting the growth of fungi and bacteria. The heartwood of western red cedar (*Thuja plicata*), whose timber is exceptionally resistant to decay, contains terpenoid substances known as thujaplicins, while teak (*Tectona grandis*) has an oil which makes the wood immune to insect attack. At the other extreme many willows (*Salix*), which lack antiseptic substances in the older wood, develop hollow trunks at an early stage. Beech is unusual in that most of the xylem remains functional, little heartwood being formed.

Almost all conifers and some angiosperm trees are **evergreen**, their leaves lasting for several seasons so that the tree canopy casts a heavy shade throughout the year. Beneath **deciduous** trees, however, a far higher proportion of the available light reaches the forest floor in winter than in summer. The structure and chemical nature of leaves influences woodland processes when they are dead as well as when they are alive. Many conifers have needle-like leaves reinforced with sclerenchyma that keep their shape in droughts, possess thick cuticles and have stomata lying in grooves. Such leaves often take several years to decay, forming a thick litter which impedes the growth of herbs and the devel-

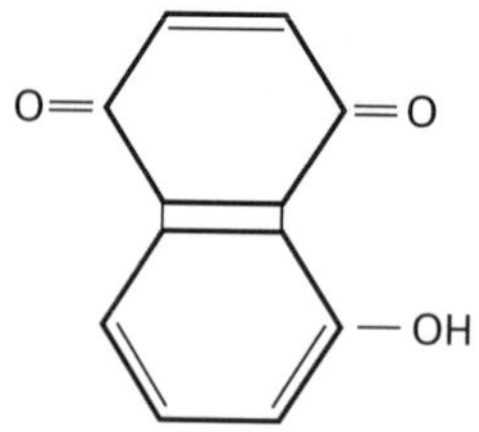

Scheme 1.1   Structure of juglone, an allelopathic chemical leached in the bound form by rain water from the leaves and stems of the black walnut (*Juglans nigra*) and then released as the active toxin by hydrolysis and oxidation in the soil. **Allelopathy**, in which chemical compounds produced by plants inhibit or depress growth in other plants, is increasingly recognized as a significant interaction amongst forest species (Spurr and Barnes, 1980).

opment of tree seedlings. They usually yield relatively little in the way of mineral nutrients and give rise to a discrete layer of humus. In contrast, the large pinnate leaves of elder (*Sambucus nigra*), a deciduous shrub or small tree common on soils rich in nitrogen and phosphorus, are much less rigid, rich in minerals and are decomposed and incorporated in the soil within a few weeks. The chemical content of leaves also affects their palatability to herbivores. The sparseness of herbivorous species living on elder leaves is presumably due to distasteful or toxic chemicals. 'Defensive compounds' such as certain phenolics, terpenoids and alkaloids, are widespread in the plant kingdom, and display their greatest diversity in the tropics. Toxic chemicals in trees may also lead to reduced competition from other plants. A number of hardwood species, including black walnut (*Juglans nigra*) and butternut (*J. cinerea*), produce biochemical exudates with a herbicidal action (Scheme 1.1). Both these species have rather sparse foliage, but few saplings and little understorey grow beneath it. The needles of pines and many other conifers contain phenolic compounds, which may be leached from their litter and which inhibit the germination of seeds.

Defensive compounds and toxic chemicals are just one aspect of biochemical ecology; other compounds influence attraction to food plants, palatability and feeding responses (Harborne, 1982). Chemical cues, known as **pheromones**, which may either attract or repel, are important in the lives of many animals, especially insects and mammals. They play a vital role in the relationship between the western bark-beetle and the ponderosa pine which it frequently damages severely (section 6.6).

Similarly the size, number and composition of the seeds and fruits influence not only the efficiency with which the tree reproduces, but also the organisms which feed on many of these structures. Most trees are **polycarpic**, seeding many times, but some such as the talipot palm (*Corypha umbraculifera*) are **monocarpic**, producing a single large harvest and then dying.

The form of the **root system** differs both between and within species; beech (*Fagus sylvatica*) has extensive shallow roots when growing on chalk while Scots pine (*Pinus sylvestris*) develops long tap roots on deep soils. The ability of certain trees to form **root nodules** containing symbiotic nitrogen-fixing microorganisms [*Rhizobium* bacteria in leguminous trees, actinomycetes in alder (*Alnus*), she-oaks (*Casuarina* spp.) and some other non-leguminous trees] makes a very important contribution to the amount of fixed nitrogen available in some forest ecosystems, e.g. the succession at Glacier Bay, Alaska (section 5.4). Soil erosion is often reduced by tree roots and avalanches prevented by planting protection forests.

Because of their economic importance an increasing proportion of forests and woodlands is planted by man, often leading to a loss of species diversity, and sometimes a danger of rare species or varieties becoming extinct. Ecological studies of woodlands are increasingly concerned with plantations and semi-natural communities, and can make a worthwhile contribution to forestry, one of the few primary industries dealing with a **renewable resource**.

1.2   The spatial framework and the changing seasons

Trees form the most obvious feature of a wood; their aerial parts tower above the shrubs, herbs and bryophytes which form the other three layers or strata commonly present in natural temperate woodlands. In woods which have been maintained as **coppice-with-standards** all four of these strata are present, while in those which have been kept as **wood pasture**, used by cattle or sheep, there are fewer shrubs and the field layer is much less diverse. Relatively recent **plantations** have no shrubs and the field and bryophyte layers are species-poor. The three-dimensional structure of tropical rainforests is often so complex that attempts to identify strata may be misleading, though some writers discuss at least five. Tall plants modify very considerably the microclimate experienced by smaller organisms associated with them; trees are the dominant plants in woodlands, largely determining the conditions of life of smaller plants. Temperate trees are commonly **sun plants** (section 2.4) with a fairly high requirement for light energy. Young trees compete very vigorously, particularly for light and water, so the number of saplings present in an area that has experienced an exceptionally good mast year, in which seed production is very high, falls very rapidly and the survivors are often very drawn up unless the area is further thinned by a forester.

The **stratification** shown by the aerial systems of woodland plants (Figure 1.1) is paralleled by the **layering** of their underground systems,

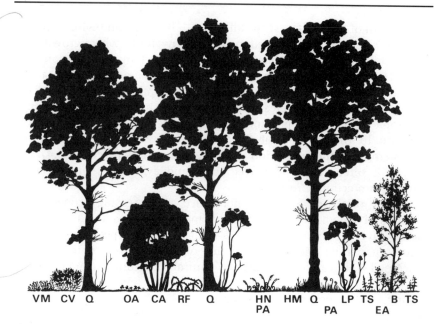

Figure 1.1 Stratification of shoot systems in the Wyre Forest, Shropshire. The oak trunks have the basal curve characteristic of individuals which have re-grown from coppice stools, while the mosses and liverworts of the bryophyte layer are too small to depict. The most acid area is on the left, while near the centre is an aspect society in which bluebell, creeping soft-grass and bracken follow each other in a seasonal sequence. (Drawn by P. R. Hobson.)

VM, bilberry; CV, heather; Q, oak (*Quercus petraea* or Q. *robur*); OA, wood sorrel; CA, hazel; RF, bramble; HN, bluebell; HM, creeping soft-grass; PA, bracken; LP, honeysuckle growing over weak oak coppice regeneration; TS, wood sage; EA, wood spurge; B, silver birch.

those of the largest plants usually penetrating to the greatest depths. The deeper roots of trees absorb water and anchor the trunks securely, while the shallower or feeding roots absorb most of the mineral nutrients taken up by the tree, often assisted by mycorrhizal associations with bas-idiomycetes. Adjacent trees of the same species are commonly joined by **root grafts**, which occasionally take over the root system of a neigh-bour whose trunk has been cut down; grafts are also a means of trans-mitting disease. Very occasionally the cut stump of a conifer whose roots are grafted to another will completely heal over and even slightly increase in girth; under such circumstances many hardwoods would recover by coppice growth. A notable healed and living Douglas fir stump, devoid of leaves, occurs in a thinned plantation south of Exeter.

Herbs live in microhabitats where trees shade out much of the light and deplete the soil of water and nutrients. Though root layering may diminish competition between herbs and trees it does not eliminate it; indeed it is most important to reduce competition from other plants when attempting to establish young trees (see Figure 10.10). The bare areas of many young beechwoods often result as much from the permeation of the ground by their roots as from the heavy shading caused by a virtually unbroken canopy. On the other hand, trees provide a windbreak and the microclimate of a forest is much better buffered than that of open country, temperatures are less extreme and relative humidity is usually high enough to encourage the growth of small herbs, liverworts and mosses. The development of the understorey is greatly influenced by the species, size and spacing of the trees present; that beneath English oak (*Quercus robur* and *Q. petraea*) is much more extensive than beneath beech. There is normally a **mosaic** on the forest floor; variations in humus and nutrient contents, pH, soil moisture (section 2.5), soil aeration and available light are reflected in the distribution of shrubs, herbs and bryophytes, which can often be used as **indicators** of environmental conditions.

One of the most fascinating features of deciduous woodlands with rich herb communities is the sequence of flowers, many brightly coloured, produced by different species in spring and early summer (Figure 3.9). These form **aspect societies** in which the appearance of an area varies as the flowering periods of different species follow each other. Some of the herbs involved, for example bluebell (*Hyacinthoides non-scripta*) and wood anemone (*Anemone nemorosa*), are sun plants which use the relatively large proportion of the available light reaching the forest floor before the tree canopy expands to grow, reproduce and accumulate the food reserves that enable them to survive another year.

In temperate deciduous woodlands early spring growth by many of the herbs is followed by the expansion (**flushing**) of the leaves of the trees and shrubs, thus diminishing the proportion and absolute amount of solar radiation reaching the forest floor. This renewed activity in the spring is initiated by rising temperatures which help cause the hormonal changes that break dormancy. Provided water availability is not limiting, the overall primary production of the woodland reaches its peak in midsummer when temperatures and solar radiation are high. With autumn comes the maturation and dispersal of seeds and fruits, followed by leaf fall and a period of relative quiescence. The aerial parts of many herbs now decay though some, such as wood sorrel (*Oxalis acetosella*) and yellow archangel (*Lamiastrum galeobdolon*), are wintergreen and may accumulate dry weight into autumn. Many other seasonal changes occur. The fruiting bodies of the larger fungi are a notable feature of the autumn (when a 'fungal foray' is held in many areas). Breakdown and ultimate mineralization of plant litter release nutrients into the soil from which they may be absorbed for spring growth.

Seasonal changes in climate can have a direct influence on animals (Figure 1.2); for example, rut in fallow deer is stimulated by the effects of short day length. There are many other examples of the direct effects of climate on animals, but their activity is largely dependent on food availability which may vary seasonally. The reproduction of host-specific herbivores, such as sycamore aphids and bruchid beetles, may be closely geared to such aspects of plant phenology as leaf flushing or fruit ripening. More catholic species, for example the great tit (*Parus major*), vary their diets according to availability but may still show seasonality in breeding, related to an abundance of caterpillars or other seasonal herbivores. Similarly, the breeding seasons of some litter-dwelling invertebrates may be correlated with temperature, day length or stage of decay of plant debris. Certain worms and millipedes avoid extremes of cold or drought by burrowing deeper into the soil; wider scale migrations by birds and mammals are more familiar avoiding strategies. Finally, unfavourable seasons may be endured in an inactive condition, as in hibernating mammals or in those arthropods which can enter a state of suspended development (**diapause**) at various stages in their life history.

Woodland structure and nature conservation

Woodland structure is of vital importance to the continued existence of a wide range of organisms. The nightingale (*Luscinia megarhynchos*) is favoured by a particular stage in the coppice cycle and it was the loss of this species from the Ironbridge Gorge, Shropshire, that led to the re-establishment of coppicing in Lloyds coppice on the north bank of the River Severn. Many of Britain's 54 species of butterfly are associated with particular forms of woodland; in this century the majority have suffered population declines because of subtle changes in climate and loss of suitable habitat. The most dramatic losses have occurred in areas which have changed from ancient landscape to large-scale modern agriculture.

Parts of the Wyre Forest, an early post-glacial woodland, still encompass many features typical of ancient landscape, with a dense canopy of sessile oak (*Quercus petraea*) occasionally broken by the presence of old, herb-rich meadows, most of which are strung out along Dowles Brook. These remaining semi-natural habitats provide a haven for 60% of the country's species of butterfly.

Three nymphalid species in particular, White Admiral (*Ladoga camilla*), Pearl-bordered Fritillary (*Bolonia euphrosyne*) and High Brown Fritillary (*Argynnis adippe*) draw attention to the impact of land management on this group of insects. Over the last 60 years the two species of fritillary

Figure 1.2 The main groups of animals from Western European oak-beech forest. (From Burrows 1990, modified from Duvigneaud 1967.)

Figure 1.2 (Cont.)

		Feeding habits				Feeding habits
I Vertebrates				**Birds**		
Mammals*						
1 Fox	*Vulpes vulpes*	C	14	Buzzard	*Buteo buteo*	C
2 Stoat	*Mustela erminea*	C		Sparrow hawk	*Accipiter nisus*	C
Polecat	*Putorius putorius*	C	15	Tawny owl	*Strix aluco*	C
3 Feral cat	*Felis domesticus*	C	16	Pheasant	*Phasianus colchicus*	H(O)
4 Wild pig	*Sus scrofa*	O	17	Wood pigeon	*Columba palumbus*	H(O)
5 Red deer	*Cervus elephas*	H	18	Green woodpecker	*Picus viridus*	C
Fallow deer	*Dama dama*	H		Cuckoo	*Cuculus canorus*	C
Roe deer	*Capreolus capreolus*	H		Jay	*Garrulus glandiarius*	O
6 Badger	*Meles meles*	O		Magpie	*Pica pica*	O
7 Hedgehog	*Erinaceus europaeus*	C		Blackbird	*Turdus merula*	O
8 Pipistrelle bat	*Pipistrellus* spp.	C		Songthrush	*T. ericetorum*	O
Noctule bat	*Nyctalus noctula*	C	19	Chaffinch	*Fringilla coelebs*	O
9 Squirrel	*Sciurus vulgaris*	H		Bullfinch	*Pyrrhula pyrrhula*	O
10 Hare	*Lepus capensis (europaeus)*	H		Robin	*Erithacus rubecula*	C
11 Mole	*Talpa europaea*	C		Great tit	*Parus major*	O
12 Shrews	*Sorex* spp.	C		Wood warbler	*Phylloscopus sibiliatrix*	C
13 Voles, mice	*Cleithrionomys* spp. *Arvicola* spp.	H(O)		Dunnock	*Prunella modularis*	C
	Sylvaemus spp.	H(O)		etc.		

		Feeding habits
Reptiles		
20 Adder	*Vipera berus*	C
Amphibia†		
Toad	*Bufo bufo*	C
Newt	*Triturus* spp.	C

II Invertebrates‡

Mollusca	Snails, slugs	H	**Insecta**		
			Hymenoptera	Bees, wasps, ants	H,C,X,Y
Arachnida	Spiders, mites	C,H,D,X,Y	Siphonaptera	Fleas	X
	Harvestmen	C	Myriapoda	Millipedes, centipedes	C,D,H
Insecta					
Collembola	Springtails	H,D	Crustacea	Slaters, hoppers	D,H
Orthoptera	Cockroaches, locusts	H	Annelida	Earthworms	D
Psocoptera	Book lice	D	Rotifera	Rotifers°	D
Anopleura	Sucking lice	C,X			
Thysanoptera	Thrips	H,D	Nematoda	Roundworms°	D,C,H,X,Y
Hemiptera	Cicadas, plant bugs, aphids, mealy bugs	H,Y	Platyhelminthes	Flatworms	C,D,X
Lepidoptera	Moths and butterflies	H			
Diptera	Flies	H,C,D,X,Y			
Coleoptera	Beetles	H,C,D	Protozoa	Amoeba etc.°	D,X

* The larger mammals are present only in extensive areas of forest, or in remote localities.

† Only in localities where there is standing water.

‡ Only the main groups are listed. Very large numbers of species and of individuals from certain groups are present. They are not shown on the diagram but, broadly, spiders and certain insect groups are arboreal, while most of the rest occur on the ground or in the soil. In particular, many species of insects are present, filling many different roles. Often their larvae feed in very different ways from the adults.

o, microscopic; C, mainly carnivorous; O, omnivorous; H, herbivorous; D, detritivorous; X, parasites on animals; Y, parasites on plants.

have shown a marked decline in both numbers and distribution throughout Wyre Forest. This trend coincides with changes in forestry practice, particularly the cessation of coppicing. The remaining strongholds of both species are confined to well-defined areas within existing semi-natural woodland.

The Pearl-bordered Fritillary, like so many butterflies, is a specialized feeder, and is dependent on the violet (*Viola* spp.) as a food plant during the larval stage. Adult females seek out the host plant along the sunny edges of moist glades or old herb-rich meadows. Bracken and sparsely vegetated ground are important components of sites selected by ovipositing females. Recent events in Wyre Forest have helped to secure the future of this attractive fritillary. The hot summers of 1989 and 1990 produced an upsurge in local population numbers at a time when the Nature Conservancy Council had commenced coppicing an area of oak woodland which included sites that had not been cut for 180 years. Butterflies from existing colonies quickly spread into recently coppiced sites with the effect of increasing their distribution two-fold, and the females found suitable new sites for egg laying.

The life history of the White Admiral, in contrast to the two fritillaries, favours old derelict woodland punctuated by glades or old meadows. Centuries of efficient woodland management restricted the distribution and numbers of this species which became limited to scattered locations in southern England. By the beginning of the twentieth century coppicing was being phased out and woodlands like those of Wyre Forest were tending to become derelict. The more mature stands of oak favoured rampant growth of honeysuckle, food plant of the White Admiral. This species of butterfly now tends to occur in the riparian woodlands whose alder and sallow are not managed very intensively.

1.3   Woodland ecosystems

The term **ecosystem** coined by Tansley (1935) to describe a level of organization which integrates the living and non-living components of communities and their environments into a functional whole, was voted the most important concept in ecology in a recent survey of ecologists (Cherrett, 1989). It provides an invaluable conceptual framework when attempting to visualize the structural and dynamic organization of a community of plants, animals and microbes, especially in relation to the major ecosystem processes of energy flow and nutrient cycling. It should be noted, however, that certain authors (e.g. Begon *et al.*, 1990) refuse to distinguish ecosystems as a level set apart from community studies.

This framework is normally based on feeding relationships, notably in

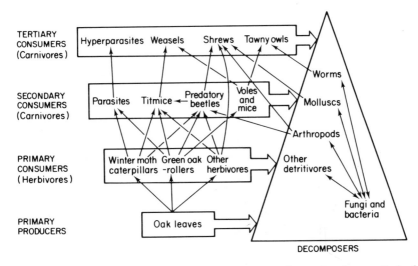

Figure 1.3 Simplified food web based on oak at Wytham Wood, near Oxford. (Data from Varley, 1970.) Note the two-way interactions among the decomposers.

the form of food chains and food webs (the latter being called food cycles by Elton, 1927). Although many published food webs are over-simplified, especially with regard to temporal and spatial variability, to the point that they are caricatures of nature (Pimm, 1982), they enable us to gain some idea of the position of a particular species *vis-à-vis* its food sources and its predators. So long as these relationships are recognized as flexible – a meshwork of movable feasts – they can be used to increase our understanding of community patterns, such as the proportional representation of predators and prey, and the lengths of food chains (Lawton, 1989).

The concept of trophic levels is fraught with the danger of rigid categorization of the status of individual species, especially among decomposers (Figure 1.3). Since the concept 'conceals more than it reveals' (Lawton, 1989) some authors avoid using this term.

Rather than becoming enmeshed in the minutiae of food webs, the fundamentals underlying the transfer of energy and matter through a woodland ecosystem can be outlined using a systems approach. Ecosystem processes can be visualized as occurring within and between three **subsystems** (Figure 1.4): the plant subsystem, the herbivore subsystem and the decomposition subsystem, the last two being virtually equivalent to the more familiar grazing and detritus food chains (Swift *et al.*, 1979).

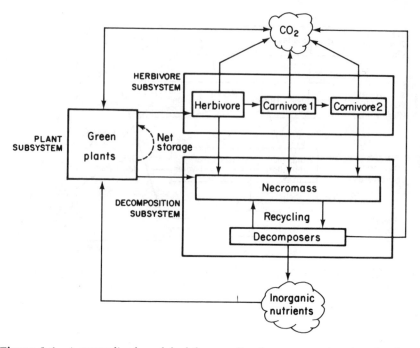

Figure 1.4 A generalized model of the woodland ecosystem showing the three subsystems. Arrows indicate major transfers of matter between organic matter pools (rectangles) and to and from inorganic pools ('clouds'). (Modified from Swift *et al.*, 1979.)

Plant subsystem

The **plant subsystem** involves the primary producers or autotrophs, the green plants of woodlands. Through photosynthesis these fix carbon, from atmospheric CO_2, and trap a small proportion of the incident solar radiation. The resultant **gross primary production** is in the form of simple sugars, from which other compounds can be synthesized for the construction of plant tissues, for storage, or as secondary compounds. These transformations may also involve additional elements, such as nitrogen; all require energy, as does the maintenance of living tissues. This energy is released by respiration, involving the catabolic breakdown of organic substrates, with the production of CO_2. Complying with the Second Law of Thermodynamics, a proportion of the substrate's energy (in the case of glucose, one-third) is lost as heat, which can play no further direct role in metabolic processes. The amount of carbon actually incorporated into the plant (**net primary production**) represents the difference between uptake in photosynthesis and losses during construction and maintenance respiration. (Photorespiration, which accompanies photosynthesis in C_3 plants, also yields CO_2 but no utilizable energy.)

Because a high proportion of the living cells of a tree is in non-photosynthetic tissues such as roots and sapwood, respiratory costs are high relative to plants dominated by foliage and seed production. The role of stands of trees as sources or sinks in the local carbon balance varies with time: diurnally, seasonally (including the disproportionately greater influence of high temperatures on maintenance metabolism than on photosynthesis) and with stand age.

Within the plant, net production may be represented by relatively short-lived tissues such as most leaves, reproductive structures and fine roots, or by long-term sequestering in larger roots, trunks and branches (as measured by incremental growth or net storage). All is potentially available to heterotrophs in the two remaining subsystems.

Herbivore subsystem

Herbivores subsist on living plant tissues, including leaves, flowers and fruits, and parts of roots and stems such as sapwood. (Biotrophic bacteria and fungi should be included with herbivorous animals here, although their influence on energy and nutrient transfer is rarely considered.) Conversion of tissues into faeces represents a loss from this subsystem, but assimilated (digested) compounds contribute to that portion of secondary production represented by the body tissues of herbivores and their offspring, after allowing for losses of energy and CO_2 associated with the animals' respiration. The amount of energy available to the remainder of the subsystem, the various categories of carnivores and parasites, in the form of prey tissues, is thus progressively depleted (by defaecation and respiration) on ascending a food chain. For terrestrial grazing chains the average efficiency for energy transfer between successive links is probably less than 10% (Heal and MacLean, 1975).

Decomposition subsystem

Decomposition can involve biotic or abiotic agencies, but usually both. **Decomposers** include microbes (**microflora**, especially fungi and bacteria) and various animals, such as certain arthropods, worms, molluscs and protozoans. These organisms depend on dead organic matter (**necromass**) predominantly in the form of plant remains, but also as corpses, cast skins and faeces of animals. This faunal input is partially a by-product of the herbivore subsystem but also derives from the decomposers themselves, i.e. from animals exploiting necromass and its associated microflora, and from their predators and parasites. Faeces and corpses of these decomposers, together with dead remains of fungi and bacteria, all contribute to necromass, which still contains resources of energy and nutrients.

The complete decomposition of a unit of plant necromass, such as a dead leaf, will normally require several passes through the subsystem, with numerous opportunities for the exploitation of different sizes and types of necromass, and a host of points where heat and CO_2 may be liberated. The most recalcitrant remains accumulate as humus, which may lock up resources, including carbon, for hundreds of years. Where decomposition is retarded, as under waterlogged, anaerobic conditions, the result is peat or, ultimately, coal.

The reprocessing of organic matter is a significant distinction between the decomposition and herbivore subsystems: herbivores get only one bite at the cherry. It is important not to confuse this reworking, which gradually dissipates energy, with recycling of energy, which is thermodynamically impossible: each joule can only be used once, whether by plant, herbivore or decomposer.

This situation contrasts fundamentally with that of **nutrients**, which are globally in finite supply, and which are continually being routed through biogeochemical cycles, including their incorporation into organisms and their subsequent release. Hence the CO_2 evolved by respiration from the three subsystems (the ecosystem respiration, R_E) can be assimilated once again by photosynthesis. In a climax forest, carbon sinks (uptake in photosynthesis) are balanced by sources of CO_2 (R_E), but the area can all too easily become a net source, with implications for the greenhouse effect, if such forests are destroyed and replaced by less productive crops.

Nutrients such as nitrogen, phosphorus and sulphur, incorporated into organic molecules by organisms, are said to be **immobilized**. Most are transferred in this form during feeding, but are eventually returned to an inorganic (**mineralized**) state by catabolic breakdown of organic matter (which also releases energy). Animals effect some of these transformations (e.g. the release of ammonia from proteins) but most are the responsibility of the microflora. During decomposition the balance between immobilization and mineralization is dictated by the relative availability to the decomposers of carbon (representing energy resources) and of other elements, which may be present in limiting amounts. The situation can be summed up in terms of energy to nutrient ratios, e.g. C:N, C:P. Normally a nutrient will only be released (so becoming potentially available for uptake by plant roots) when it is no longer limiting, i.e. when the ratio approaches that of the decomposer's own tissues.

This explains why the incorporation of materials of high C:N ratio, such as straw, into the soil tends to lock up any available inorganic nitrogen from the soil in the form of microbial biomass; the resultant nitrogen deficiency is eventually redressed, as carbon is lost through respiration, and the C:N ratio falls below that of microbial tissues (*ca.* 20:1).

Finally, abiotic components of the decomposition system, such as fire, percolating rainwater, and alternate drying and wetting, may supplement or even supersede the activities of organisms in decomposition and nutrient cycling.

Characteristically, an ecosystem functions as a largely self-contained entity. Energy is imported as sunlight and exported as longwave radiation, but, with the exception of losses in harvests or in drainage water, there is usually only minimal transport of fixed energy or nutrients across ecosystem boundaries.

1.4 Tropical forests: the primary complexity

Tropical rainforest is the archetype of all forest: indeed Corner (1964) considers it to be the cradle of angiosperm evolution. The trees of coal forests are likely to have evolved in conditions of high temperature and constant humidity, as are the first seed plants and the first flowering plants. It should not, however, be regarded as a stable archaic relic: extensive climatic changes have affected its location and species composition over millions of years (Ayensu, 1980). It may develop where rainfall is between 200 and 600 cm or more (with no month having less than 10 cm) and temperatures are between 21 and 34°C. In striking contrast to the boreal forest the diurnal temperature range is greater than the annual range. Soils may be nutrient-poor and acid (pH 4.5– 5.5); **laterization**, the leaching of basic ions and silicic acid, commonly causes accumulation of iron and aluminium sesquioxides. Much of the nutrient capital is in the above-ground biomass, derived from times when plant roots were in contact with fragmented parent rock, perhaps as far back as the Tertiary Period (Walter, 1973). Dead wood is destroyed by termites and fungi, especially decomposer basidiomycetes, while recycling of nutrients occurs by mineralization and root absorption often with the aid of mycorrhiza. Forest clearance results in leaching, so secondary forest is less luxuriant than primary. There is a great diversity of tree species, for example 1 ha of Sarawak rainforest has been found to contain 214 species, at a density of 700 trees per hectare (Proctor *et al.*, 1983).

The trees are typically 50–60 m tall, with slender trunks and thin bark and a canopy small in relation to their height, closed and intermingled with neighbours, or emergent above them. When the soil is waterlogged roots spread over the surface or are close to it, extending for great distances – those of a strangler fig in Daintree, Queensland, had a radial spread of 30 m. In place of a tap root there are likely to be aerial prop-roots or buttress roots running from as far as 9 m up the trunk. Flange and plank 'buttresses' are more likely to provide support under tension than compression; it is even possible their function is primarily gaseous exchange.

Evergreen leaves emerge limp and hanging, sometimes with striking red, white or purple colouring at first. They are frequently large and shed rain with drip tips, and clatter down to the forest floor when shed after a life of several months, 4.5–12.5 g dry weight falling per m^2 per day. Phenology is not tied to an annual cycle: branches of a single tree may be in different phases of leafing, flowering and fruiting. Leaf flushes ensure that at least some of the new leaves will escape consumption by herbivores. Development of flowers directly on the trunk, **cauliflory**, or on branches, **ramiflory**, is common and may be related to the weight of large fruits (e.g. jackfruit) or the need to be near the ground where beetles may pollinate them. Pollination by bats is also common. Vertebrates are wide-ranging but destructive pollinators and seed dispersal agents: they contribute to the diversity and sparse distribution of tropical tree species. Fruits are frequently large and often toxic to higher mammals, which suggests dispersal by mammals that are now extinct or by agents which can tolerate the toxic substances. Some trees have winged seeds which spiral outward away from their parent as they fall, e.g. the shuttlecocks of dipterocarps.

Shrub and herb layers grade into each other and are often sparse because of a lack of light and nutrients. They include miniature trees and saplings which are amazingly slender and strong for their height, rather than multi-stemmed shrubs, and persistent tuberous plants of the Araceae and similar families. When cultivated, these often survive the dark difficult conditions imposed upon indoor house plants and may have attractive variegated leaves. Conditions on the forest floor tend to be constant: there are few temporary habitats for annuals and non-woody herbs, other than by streams.

Symbioses with ants occur, the plant providing galleries for ants within its tissues and receiving protection from predators and epiphytic growths including fungi. Such **myrmecophilous** plants may be epiphytes, where the swollen gallery bases are often suspended from branches, or ant trees, where weakened patches of cork provide access to galleries in the trunk.

Epiphytes are abundant, especially in misty, high altitude forests. Various orchids absorb water through the root velamen or store water in leaf base tubers. Bromeliads, tank plants, of the New World use their roots only for anchorage, absorbing water from the leaf base tanks through scales. Basket ferns collect detritus and water in overlapping basal fronds.

Lianes, or vines, are draped around trunks, their knotted stems providing handholds for human climbers. Their stems remain pliable because of anomalous phloem bands and broad parenchymatous rays within the xylem: they may remain aloft when their original tree support falls, or slip and re-ascend from their growing tip, sometimes achieving total lengths of over 100 m. Prime amongst these abundant vines are the climbing palms, especially rattans of the genus *Calamus*. They possess divaricating branches and sprawl across growing saplings in regeneration gaps, clinging with viciously hooked leafsheaths, petioles or extended main veins and can reach 240 m length after repeated looping. Hemi-epiphytes such as *Monstera* germinate on the ground and ascend as root climbers, with aerial roots growing down to the ground. On the other hand, stranglers originate as epiphytes and send down aerial roots: when these reach the ground, the shoot system grows rapidly and the roots enmesh the supporting tree, often preventing further secondary thickening. The strangler remains erect but hollow when the support finally dies and rots away.

Except where felling or a regeneration gap has allowed light to penetrate and produce dense secondary growth, the forest floor is generally open, dark, and covered with a thin layer of shed leaves which rot quickly without a noticeable humus or leaf mould layer. Human progress is impeded by vines, including the hooked rattans, justly named wait-a-while or lawyer canes, or by occasional groups of thin shade-suppressed saplings, rather than by dense vegetation.

These ecosystems have the richest invertebrate faunas of all forests and there are many niches for larger animals. Many mammals roam the forest floor, including species adapted to digging for termites and insect larvae (e.g. pangolin, armadillo and anteater), those specializing in rooting out tubers and fungi (e.g. pigs, peccaries and tapirs), small browsing herbivores (e.g. agoutis and chevrotains) and large browsers (e.g. okapi and banteng). Their predators are likely to be cats great and small (e.g. jaguars, tigers and civets) which as a tribe have the advantage of being able to move stealthily across the forest floor and to climb rapidly into the canopy, where there are many animals adapted to arboreal life. The high dense canopy favours flying species; bats, as well as birds and insects, exploit nectar and pollen as food sources. A number of amphibians, reptiles and mammals have adopted the habit of gliding from tree to tree. An alternative is to leap and swing, as do many primates, who can reach out to gather fruits, nuts and shoots. The New World platyrrhine monkeys may be assisted by a prehensile tail, an adaptation found in some other mammals such as Australian ringtails. Ground-dwelling birds, such as junglefowl and pheasants, and the larger flightless cassowary, feed on the seeds and fruits dropping down from the high canopy. The rich and colourful insect life is dominated by the termites, ants and beetles, which play a vital part in the comminution of dead organic matter.

1.5   Forest biome-types

From the luxuriant regions of tropical rainforests, trees spread into and ameliorate less favourable environments. Their adaptive radiation has led to a spectrum of communities whose nature is primarily determined by climate, but frequently interrupted by geographical barriers. This range of vegetation, with its associated animals, is commonly divided into **biome-types**. The exact species found within the same biome-type vary from continent to continent, giving rise to different biomes. Species of related biomes sometimes belong to the same genera (e.g. ***Nothofagus*** in Chilean and New Zealand rainforest) which suggests a former proximity of the continents.

In principle, communities are assigned to particular biomes on the basis of physiognomy – their form and structure in nature – but in practice environmental factors are also employed. Of the six major terrestrial physiognomic types listed by Whittaker (1975) – forest, wood-land (of small trees generally in open spacing and with well-developed undergrowth), shrubland, semi-desert scrub, desert and grassland – trees dominate the first two. Each of these six physiognomic types occurs so widely that more than one biome-type is defined within it on the basis of climate. There is an almost infinite variety of forests, woodlands and other communities in which trees play a role; our purpose here is to contrast the distinctive characteristics of their major groupings (Figure 1.5).

The other major biome-types dominated by trees can be regarded as being derived from **tropical rainforest** (sections 1.4 and 2.1) as mean temperature and rainfall decrease. In most regions of the world this decrease occurs seasonally causing interruption of plant growth. Leafing out, stem extension, flowering, fruiting and leaf fall follow each other in the inevitable sequences, phased with the seasons, which are characteristic of biomes of drier and cooler regions. Temperature decrease may be due to higher latitude or higher altitude.

Tropical montane forests are subject to both lower temperatures and greater exposure to wind. In the cloud belts associated with the mountains, dry periods are unusual and cryptogams, sometimes including tree ferns, are more common. **Elfin forests**, whose gnarled and stunted trees are draped in epiphytes, are found on many tropical mountains.

Where temperature decrease is due to higher latitude **subtropical rainforest** develops. It is sustained by less rain, decomposition processes are slower and the pool of nutrients is divided more evenly between the biomass of vegetation and the decomposing litter. The diversity of species is less and shifts from the tree canopy to the lower shrub and herb strata, and the forest is difficult to traverse. Epiphytes and climbers become less common, mean leaf size is smaller and buttress roots and cauliflory occur less often.

In the **temperate rainforests** of western Tasmania, New Zealand south of latitude 38° and southern South America, tree diversity is further reduced, with fewer lianes and epiphytic angiosperms. However, epiphytic cryptogams may be abundant and trunks and branches support a rich diversity of lichens and mosses. Such forest is characterized by podocarp conifers, abundant tree ferns and evergreen species of southern beech (**Nothofagus**) which become dominant in the highlands. While the diversity may be reduced, there are still over 180 native species of trees and shrubs in New Zealand (Salmon, 1986) in contrast to some 35 in Britain. The temperate rainforests of North America are dominated by **Tsuga**, **Thuja** and **Pseudotsuga**. Where there is no pronounced cold season and no summer dry period, exceptionally tall forests of trees over 100 m may develop, e.g. **Eucalyptus regnans** in Tasmania and the coastal redwood (**Sequoia sempervirens**) in California, which benefits from summer fog rather than rain.

Further reduction in temperature with latitude leads to **temperate deciduous forest**. Chilling of the ground largely prevents tree roots from absorbing water and mineral nutrients in winter. Water loss in trees is restricted by a deciduous strategy or by possession of evergreen needles or scale leaves as in the conifers (section 2.6). Where deciduous trees predominate there is sufficient light and warmth in early spring for prominent herb and shrub strata to flourish (section 1.2). This gives rise to woodland, a more open community than forest, with abundant shrubs and flowering herbs and an aspect society (section 1.2) varying with the seasons. Seasonal low temperatures bring the danger of frost, the effects of which on vegetation, especially introduced species, may be severe and not necessarily apparent from an examination of mean temperature records. The exceptionally low temperatures of -25°C in Shropshire, England, in December 1981 killed many shoots of the introduced **Nothofagus**, which had shown promise as a species for forestry plantations. The timing of frosts is even more important than their severity: for example, leguminous trees such as **Albizzia** thrive in the continental climate of New York State, but fail when introduced to Britain with its unpredictable late spring frosts.

At yet higher latitudes in the Northern hemisphere is the boreal forest or **taiga**. This vast belt of subarctic–subalpine coniferous needle-leaved forest, which sweeps across North America, Scandinavia and on into Siberia, is dominated by relatively few species of spruce, fir and pine, with deciduous larch (including **Larix sibirica** and tamarack, **L. laricina**)

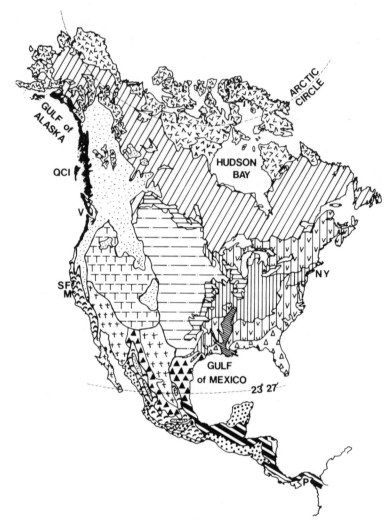

Figure 1.5 Zonation of the vegetation of North America showing the main forest and woodland types. Mangroves fringe the coast southwards from the Gulf of Mexico. The key shows a few of the important trees growing within the major temperate forest zones. *indicates North American species now widely planted in other parts of the world. M, Monterey; NY, New York; P, Panama; QCI, Queen Charlotte Islands; SF, San Francisco; V, Vancouver. *Pinus radiata** and *Cupressus macrocarpa* grow naturally on the Monterey peninsula. Zonal boundaries redrawn and much simplified after Schmithüsen (1976).

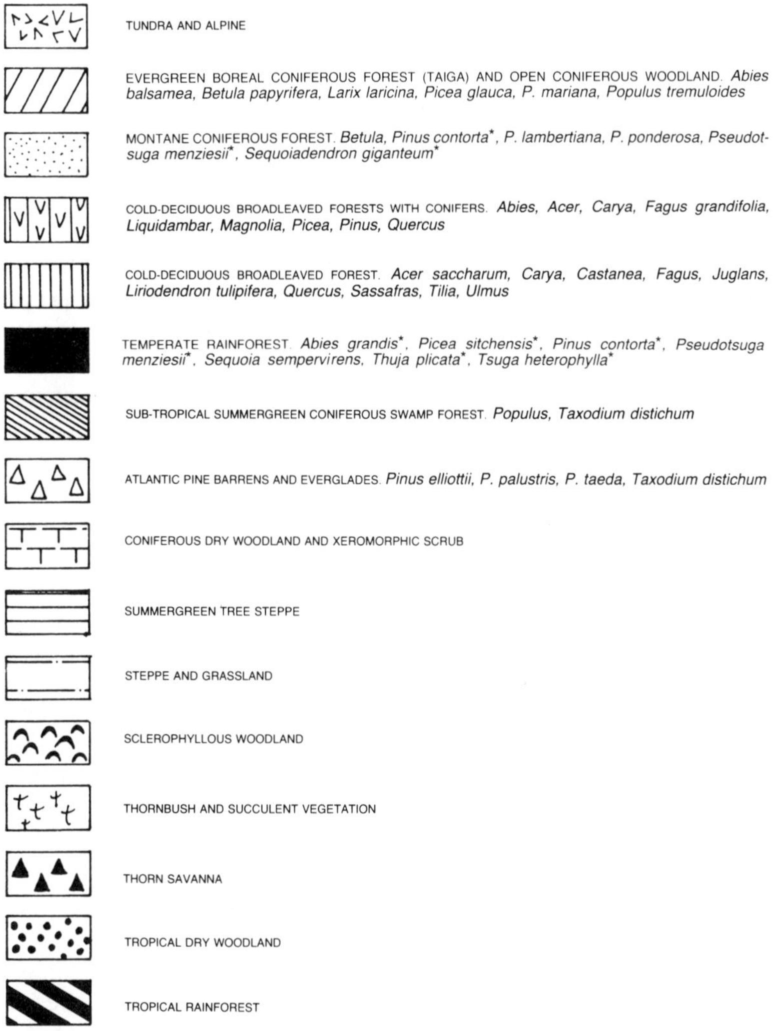

TUNDRA AND ALPINE

EVERGREEN BOREAL CONIFEROUS FOREST (TAIGA) AND OPEN CONIFEROUS WOODLAND. *Abies balsamea, Betula papyrifera, Larix laricina, Picea glauca, P. mariana, Populus tremuloides*

MONTANE CONIFEROUS FOREST. *Betula, Pinus contorta**, *P. lambertiana, P. ponderosa, Pseudotsuga menziesii**, *Sequoiadendron giganteum**

COLD-DECIDUOUS BROADLEAVED FORESTS WITH CONIFERS. *Abies, Acer, Carya, Fagus grandifolia, Liquidambar, Magnolia, Picea, Pinus, Quercus*

COLD-DECIDUOUS BROADLEAVED FOREST. *Acer saccharum, Carya, Castanea, Fagus, Juglans, Liriodendron tulipifera, Quercus, Sassafras, Tilia, Ulmus*

TEMPERATE RAINFOREST. *Abies grandis**, *Picea sitchensis**, *Pinus contorta**, *Pseudotsuga menziesii**, *Sequoia sempervirens, Thuja plicata**, *Tsuga heterophylla**

SUB-TROPICAL SUMMERGREEN CONIFEROUS SWAMP FOREST. *Populus, Taxodium distichum*

ATLANTIC PINE BARRENS AND EVERGLADES. *Pinus elliottii, P. palustris, P. taeda, Taxodium distichum*

CONIFEROUS DRY WOODLAND AND XEROMORPHIC SCRUB

SUMMERGREEN TREE STEPPE

STEPPE AND GRASSLAND

SCLEROPHYLLOUS WOODLAND

THORNBUSH AND SUCCULENT VEGETATION

THORN SAVANNA

TROPICAL DRY WOODLAND

TROPICAL RAINFOREST

in places. Hardwood species, such as alder, aspen, birch and maple are of relatively minor importance. The forest, however, is not so monotonous as it first appears, the shrub and field layers varying considerably with soil type and climate. Moreover, a species may change clinally, for example the geometric form of **Picea abies** becomes a very slender cone, better able to shed snow, in its more northerly stations. There are striking functional and taxonomic parallels amongst the plants and animals of the North American and Eurasian taiga. Since the land masses were severed by continental drift, some plants have evolved sufficiently to be regarded as separate but closely related species, for example the may lily

(*Maianthemum bifolium*) in Europe and *M. canadense* in North America. The mammals of the taiga have many species in common with the deciduous forests to the south, and in North America particularly there are huge populations of migratory deer. The elk, or moose as it is called in North America, (*Alces alces*) is common to both North America and Eurasia but the American form is heavier, reaching 550 kg. Great carnivores, including wolves, bears, lynx and eagles, formerly hunted throughout the taiga; the excessive numbers of elk now present in Sweden result from extreme reduction of predator numbers by shooting and trapping. Lesser carnivores such as stoats, weasels and foxes prey on shrews, voles and woodmice. Amongst the birds, the woodpeckers, sapsuckers and crossbills are notable for their dietary dependence upon trees, either directly or because their trees form a home for their insect prey. Where it is adjacent to mountain chains, such as the Rocky Mountains and the Scandes, the taiga passes into **montane coniferous forest**, in which pines are prominent until the treeline is reached. Along its northern margin, trees become more and more scattered and the undergrowth is dominated by a mosaic of dwarf shrubs, sedges, grasses, mosses and lichens, typical of the tundra lying to the north.

Tropical rainforest changes to **seasonal tropical rainforest**, with tall deciduous species and an evergreen understorey, as a result of a decrease in rainfall, rather than of temperature. A dry period extending beyond 4 to 5 months may lead to fully **deciduous woodland**. If leaves are retained by trees during the dry season, they are likely to be small, leathery and toxic to the herbivores that shelter under them. Since adjacent trees may be leafless, light penetrates better and the crowns are denser and more globular. The additional light admitted to the forest floor before a new season's growth is fully leafed out leads to the denser growth of shrubs and herbs already encountered in the deciduous woodlands of the ecocline established by temperature.

Further decrease in rainfall restricts the biomass that can be supported and leads to the development of **sclerophyllous woodland**. The forest becomes more open, trees are more dispersed and shrubs may become a more significant component of the vegetation, leading once again to the woodland type of physiognomy. Prolonged drought leads to the danger of fire caused by lightning strikes which are irregular but inevitable. Such woodland is strongly developed in Western Australia. The dominant trees are usually species of **Eucalyptus** in which hard sclerotic tissue prevents leaves from wilting, palisade tissue is relatively more prominent than spongy mesophyll and there is a thick cuticle with a layer of whitish wax. Their leaves hang down and admit light which enables a rich understorey to develop. This may include the she-oak **Allocasuarina fraseriana** where the young stems are photosynthetic, leaves are reduced to scales, and scarce fixed nitrogen is supplemented through a symbiosis with blue-green algae. Abundant epicormic buds below the resistant cork layer of the branches allow the she-oak to recover well after fire. Other sclerophyllous plants include many **Banksia** species, which are similarly fire-resistant, grass trees, **Xanthorrhoea preissei**, and the cycad **Macrozamia**. In more extreme dry conditions the eucalypts will be of mallee species, regenerating after fire or drought from underground lignotubers.

1.6   Threats to the forest

Major threats to the forest and woodlands of the world result almost entirely from a human population that is increasing – in its numbers and in its expectations – at an alarming rate. In recent centuries a far greater area of forest has been destroyed than has been planted or allowed to regenerate. Moreover, many of the ancient forests were of great complexity; the new plantations may provide timber and forest products but are much less diverse. Demand for land is now so great that few tree communities will remain untouched, almost all will be used as watersheds, for recreation, to provide timber and other forest products, or for some other purpose. In addition to their traditional uses as firewood and timber, trees are now the basis of the paper, cellulose, fibreboard, chipboard, plywood and veneer industries. Modern harvesting systems use ever more of the tree; bark is made into compost, and stumps are frequently extracted as raw material for paper pulp.

Though the motives and mechanisms may vary, man has everywhere caused a reduction in forest area or quality. Changes which occurred in Britain are described in section 5.6; although we now have a lower proportion of wooded land (10%) than any other country in Europe, new plantings have ensured a steady increase since 1900. Losses of ecologically valuable, ancient semi-natural woodland, however, have greatly increased since 1945 as broadleaves have been replaced by conifers. By the mid-1980s only 300 000 ha of ancient semi-natural broadleaved woodland, some 14% of total woodland, remained in Britain. In England forest clearances were mainly designed to provide land for agriculture; as lower grade land is removed from this use fresh initiatives are likely to result in new woodlands designed for diversity, so there is hope for the future.

In many parts of the world timber is being systematically mined rather than grown on a sustainable basis. It is, for example, very unlikely that fellings of tropical rainforest in Queensland will be sustainable as claimed. Indeed extensive areas of such vegetation, once devastated, are likely to take hundreds of years to recover fully. New regulations designed to increase wood production in Sweden will lead to the destruction of ecologically valuable old growth forests, particularly at high altitudes and latitudes where regeneration will be both slow and uncertain. In

North America there is a continual battle between the timber industry and the conservation interest, which attempts to retain substantial blocks of virgin forest. Such forest blocks represent gene banks: in Scotland timber extraction has left all too little of the original Caledonian pine-woods.

In many of the poorest countries wood is the major fuel. Use for firewood has largely destroyed numerous montane woodlands in Asia, leading to greater erosion, loss of soil fertility, and increased flooding in the lowlands. In Africa loss of woodlands and forest, often associated with excessive numbers of livestock, is a major cause of desertification and likely to promote still further that tendency to warmer world weather known as the greenhouse effect (section 11.1). On the positive side, wide-spread attempts to promote agro-forestry (section 10.7), particularly in the tropics, are achieving considerable success, as are the new community forests in India where whole populations are involved with the work and benefit from its success.

Several threats to the forest, such as acid precipitation (section 11.2), are discussed in detail later. Herbivores, whether rabbits, goats, deer or elephants, largely prevent regeneration and damage existing trees if their numbers rise too high. The regulation of such populations, which is of great importance to the forester, is discussed in Chapter 7. Fire is another major potential threat to forests, though it may be advantageous under certain circumstances (section 10.6). The replacement of native trees by exotic species is often criticized because it inevitably causes major changes in existing communities and often alters the soil as well. The large-scale planting of the 500 or so species of *Eucalyptus* in many parts of the world, from California to India, is a case in point. Nevertheless, *Eucalyptus* spp., though nutrient- and water-demanding as well as prone to fire, grow very rapidly, producing wood of high calorific value as well as other products. Many of the difficulties experienced with plantations result from planting programmes based on an insufficient knowledge of the trees.

The greatest losses of all are being experienced in the tropical forests of South America, Africa and Asia. These forests are frequently exploit-ed for their hardwoods or cleared to grow crops such as oil palms or rubber. Though shifting cultivation has its uses on a small scale, attempts to replace large areas of complex tropical forest by grazing for cattle or for cropping are usually disastrous. The annual burning of the ever-shrinking Amazonian Jungle is an appalling waste likely to modify the very climate of the planet (section 11.1). In such forests much of the mineral nutrients are usually incorporated in the vegetation and when this is cleared a rapid loss of nutrients by leaching, as well as extreme erosion, may in the words of Goodland and Irwin (1975) cause a change from 'green hell to red desert'. Unfortunately politicians, impressed by the immense standing crop of tropical forests, frequently embark upon schemes which ultimately change forests of great value and diversity into poor scrubland. In fact sustained tropical silviculture is perfectly possible; Matthews (1989) gives well-documented examples of its successful implementation.

Tropical rainforests are particularly threatened by mineral extraction, roads, railways and pipelines. Notable illustrations of damage caused by roads are provided by the Trans-Amazonian Highway and the road from Cape Tribulation to Bloomfield River. The latter is suitable for 4-wheel drive vehicles only, and to provide it a swathe 60–100 m wide has been bulldozed through the last stretch of unlogged coastal rainforest in Queensland. Prior to this, there was a unique ecocline from high montane tropical rainforest via lowland forest to fringing coral reef. It is feared that the road will act as an 'avenue of infection' (Borschmann, 1984) bringing in alien weed species, *Phytophthora* dieback, cane toads, feral pigs, fern and orchid collectors, rubbish and fires to both forest and beach while releasing choking sediments on to the coral reefs. The construction of the road was actively opposed by many conservation and scientific agencies, but the Queensland Government wished to open up the area in spite of its World Heritage quality.

In contrast is the response of the New Zealand authorities to the recent dumping of spoil from a playing field site into a mangrove swamp in Auckland. The mangroves were growing at virtually their southern limit and the contractor was required to remove the spoil and make good the damage. A larger decision resulting from public pressure on an international scale was the preservation of the South-West Tasmanian wilderness when the plan to dam the Franklin River was abandoned. Forests and woodlands are threatened almost everywhere; they will only survive in good order under effective management which is itself dependent upon political goodwill.

1.7   Silviculture and the replacement of trees

Silviculture treats forest trees as crops which are established, tended, harvested and then replaced by others. A **silvicultural system** encompasses (a) the regeneration of the trees, (b) the form of the crop produced, and (c) the orderly arrangement of the crops over the forest as a whole. The many different variants (Matthews, 1989) of the four main systems described below produce woodlands of distinctive character; the choice of which system to use is an important aspect of woodland management, particularly where soil and avalanche protection are involved. European foresters usually aim for yields which are regular and sustained rather than intermittent or spasmodic. From this concept of

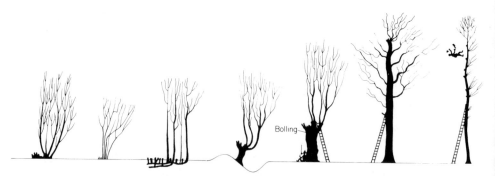

Figure 1.6   Methods of producing wood from trees. Left to right: coppice stool above ground (e.g. ash); coppice stool below ground (e.g. hazel); clone of suckers (e.g. elm); stub on boundary-bank; pollard, high pollard; shredded tree. The left-hand of each has just been cut; the right-hand half is fully regenerated and is about to be cut again. (From Rackham 1980; courtesy of the author.)

regularity arose the ideal of the **normal forest**, from which the same quantities of timber and other forest products can be taken each year or period of years. In such a forest there is a complete succession of age classes; these classes are so balanced and distributed that as each class becomes mature it can be harvested (and then regenerated) to provide a similar yield to each of the classes that preceded it. In practice there are pressures to fell more than usual when the value of timber is unusually high, or to refrain from felling in periods when the market is saturated with timber salvaged after great storms. In many parts of the world, however, forests are not being managed in this way and in others they are being destroyed rather than managed.

In natural forests more trees are maidens which have grown directly from seed, but in the past many managed woods were **coppiced**, the trunks being cut off near the ground so that fresh shoots grew from the stools every time a crop of poles was taken. Many species respond favourably to this treatment (Figure 1.6) and are effectively rejuvenated by it; a hollow stool of *Fraxinus excelsior* on a rather damp site in west Suffolk is at least 1000 years old and has a diameter of 5.6 m. Maple (*Acer*), wych elm (*Ulmus glabra*), oak, sweet chestnut, hazel, ash, alder, bass and lime (*Tilia* spp.), and the tulip tree (*Liriodendron tulipifera*) are amongst the species coppiced in various parts of the world, and some tropical trees such as teak (*Tectona grandis*) and *Eucalyptus* also shoot from the base after cutting. The coastal redwood (*Sequoia sempervirens*) is one of the few conifers which do this.

Many ancient woodlands were maintained for centuries as coppice-with-standards, often with hazel coppice and oak standards. The portion of a coppice system cut in any one year is known as a **cant**, **hagg** or **fall**. Standard trees are sometimes known as **storers** and the practice of singling, removing all but one of the poles from a coppice stool so that the coppice can return to high forest, is storing.

When aspen (*Populus tremula*) and most elm species are cut down the stumps normally die, but fresh trees arise from genetically identical **suckers**. Young shoots from freshly coppiced stools are easily browsed by cattle so the sites of old trackways in ancient woods may be marked by lines of **pollards**, whose trunks were cut between 1.8 and 4.5 m above the ground, leaving a permanent trunk or **bolling** beneath. Very short pollards or **stubs**, 1.2–1.8 m high, often mark land boundaries and are frequently associated with the wood-banks used to mark the boundaries of medieval woods. Such banks can sometimes be traced today long after the woods have gone. In **shredding**, the side branches are repeatedly cut off leaving a tuft at the top. Trees managed in this way are no longer grown in Britain but some still exist in France and Italy.

Though seldom seen in modern silviculture, vegetative reproduction of trees is quite common in unexploited forests, being particularly important near the limits of tree growth and under heavy shade (Koop, 1987). Many species produce suckers or form reiterative shoots from fallen trunks, while living branches or trunks often form adventitious roots if they come into contact with the soil. Clonal spreading is common in aspen and many elms, and may even occur in beech (*Fagus sylvatica*).

Using the methods illustrated in Figure 1.6, woodsmen could maintain stands indefinitely, even without seedlings. Modern methods of forest management are usually variants of **clear cutting, shelterwood** or **selection systems** which involve the growth of large numbers of seedlings, now often produced in enormous numbers in forest nurseries. If commercial seedlings are employed the trees are far more standardized, due to selection, than in older forests where individuals of such trees as *Quercus robur* show marked variation in size, leaf form and canopy shape.

Clear cutting involves harvesting trees of all sizes over a considerable area, a method frequently used with plantation monocultures. Sometimes such forests are allowed to regenerate by shooting from stumps or from seed; in the latter case trees are often cut down in strips or wedges so that seed can blow in from neighbouring stands. Clear-cut areas are, however, more usually planted with seedlings, particularly in the case of conifers which are now often set out in Japanese paper pots, or other containers which provide the roots with an appropriate substrate and greatly lessen losses in the season after planting out.

Where scenic beauty is important, or where the forest is on steep slopes and prevents soil erosion or avalanches, **selection systems** come into their own. Here the aim is to maintain a balance between trees of all sizes, and often between different species; ideally the volume of timber felled is equal to the net increment each year. Unthrifty trees will be removed along with a number of others of various sizes, and the forest structure remains broadly the same for long periods of time. This system demands a high degree of skill on behalf of both management and workers, while the scattered logs are difficult to collect even with skidders. On the positive side, gaps in the canopy of protection forests using this system close swiftly, and the soil is never liable to severe erosion. Trees in such forests are wind-firm and are said to be less liable to fire and pest damage (Ovington, 1965), but if a species is to arise from seed under such conditions it needs to be shade-tolerant.

In **shelterwood systems** existing trees are harvested on two or more occasions, allowing the seedlings which will form the basis of the future woodland to become well-established before the canopy is completely removed. In a previously poorly tended forest the first stage is to remove inferior trees so that seeds falling on the forest floor come from good quality parents. After some good seed years many of the parent trees are felled, thus providing an economic return and affording satisfactory light and moisture conditions for the developing seedlings. Finally, the last of the mature trees are removed when the seedlings are well-established. With the shelterwood system commercial seedlings have to be used if natural regeneration is patchy, especially in the rather extreme variation known as the **seed tree system** where very few parent trees are left. These trees are vulnerable to windblow, and offer little protection to the young seedlings which may be rapidly overgrown by weeds.

In the temperate zone there are now very few examples of truly wild or primaeval forests – **wildwood** in the sense of Rackham (1976) – and soon there may be none. In the near future all tree communities are likely to become subject to one or other of the four main silvicultural systems described. Of these, selection systems give rise to forest most similar to wildwood, though the latter frequently contained more tree and shrub species. The chief differences lie in the possession by the wildwood of senescent trees, standing dead timber and decomposing logs, which encourage a diversity of birds, insects and other organisms. However, many semi-natural woodlands will continue to exist, and in these **gap regeneration** (section 5.1) is often similar to that in wildwood.

The ecological effects of forestry practices in long-established woodland and their implications for nature conservation are emphasized by Mitchell and Kirby (1989); the rapidly developing practice of **agroforestry** (section 10.7) will doubtless have ecological implications of its own.

Figure 1.7   Even-aged crown classification in a stand of Douglas fir (*Pseudotsuga menziesii*) as it grows with naturally regenerating saplings (unlabelled) in the Wyre Forest, England. The tree on the extreme right, adjacent to a glade, is as close to a wolf tree as occurs in this area of Worcestershire. The tree fourth from the right is dead. D, dominant; C, co-dominant; I, intermediate; S, suppressed.

Classification of trees

Effective silviculture frequently involves thinning and unless this is done on a systematic basis, such as the removal of alternate rows (section 10.4), judgements have to be made about individual trees which in even-aged plantations are often placed in five classes (Daniel *et al.*, 1979). The crowns of **dominant trees** receive full light from above and, to some extent, from the side. **Co-dominant trees** are slightly shorter: their crowns receive full light from overhead but tend to be hemmed in laterally by the dominants. Trees of these two classes are the most thrifty and form the main canopy (Figure 1.7).

The crowns of **intermediate trees** receive some direct light through holes in the canopy, but are lower and subject to severe lateral competition from larger trees. **Suppressed trees**, being markedly overtopped by those of the previous three classes, are dependent on light filtering through the canopy or upon sun flecks. They are usually weak, slow growing and frequently destined to fall into the last category of **dead trees**, of which considerable numbers can be observed in unthinned plantations. The largest dominant trees may lack effective lateral competition and grow to be coarse, heavy-limbed and with crowns so broad that they inhibit the growth of more thrifty individuals. Such **wolf trees** are more common in stands that are not completely even-aged; though they may be of conservation interest they should be removed from timber crops.

Various other classifications have been devised for stands whose trees are of an uneven age. Here the main criterion is not size but degree of vigour, which will determine the effectiveness with which a tree will respond to release if the older individuals surrounding it are felled. Vigorous trees are less susceptible to both disease and insect attack.

2
Primary production: the autotrophs

2.1 Plant life forms

Botanists have for centuries attempted to classify the life forms of green plants responsible for primary production in forest ecosystems, and the categories of woody and herbaceous plants have long been recognized. The most widely known scientific description and classification of life forms is that developed by Raunkiaer (1934), who was the first to use life form to construct **biological spectra**. The main feature of his simple but ecologically valuable system is the position of the vegetative perennating buds or persistent stem apices in relation to ground level during the cold winter or dry summer which forms the unfavourable season of the year. Figure 2.1 illustrates the main life forms in a sequence which shows successively greater protection from desiccation, indicating the position of the vegetative buds when the plant is dormant.

Raunkiaer assumed that when the flowering plants evolved the climate was more uniformly hot and moist than it is now, and that the most primitive life form is that still dominating tropical vegetation. In such a climate large terrestrial plants, **phanerophytes**, can grow continually forming stems, often with naked buds, projecting high into the air. Other forms whose buds are protected from cold or desiccation by bud scales can be considered to be more highly evolved.

Tropical evergreens such as *Eucalyptus orientalis* are without the protective bud scales of evergreen phanerophytes of the temperate zone, such as holly (*Ilex aquifolium*) and Scots pine (*Pinus sylvestris*). Ash and larch belong to a third group formed by deciduous phanerophytes with bud scales. These large plants may be divided into four height classes: **nanophanerophytes**, woody plants with perennating buds between 0.25 and 2 m above the ground; **microphanerophytes**, between 2 and 8 m; **mesophanerophytes**, between 8 and 30 m; and **megaphanerophytes** of over 30 m. The two criteria of height and bud protection enabled Raunkiaer to divide the majority of phanerophytes into 12 groups, but he recognized others, such as the epiphytic forms (including many aroids and orchids) which often grow on the trees of tropical and subtropical forests.

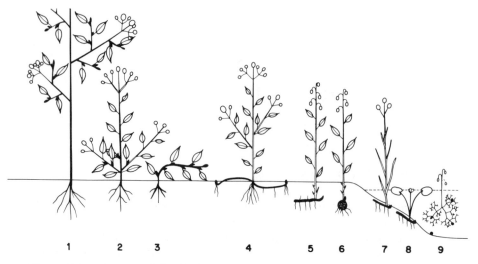

Figure 2.1   Diagram of the chief types of life form, apart from the therophytes, based on Raunkiaer's classification: 1, phanerophytes; 2–3, chamaephytes; 4, hemicryptophytes; 5–6, geophytes; 7, helophytes; 8–9, hydrophytes. The parts of the plant which die in the unfavourable season are unshaded; the persistent axes and perennating buds are in black. The sequence shown represents increased protection of the surviving buds, which are most exposed in the phanerophytes. (From Rankiaer, 1934; courtesy of Clarendon Press, Oxford.)

Temperature is the most important factor controlling leaf type and duration in the major type of tree present in any area with adequate rainfall (Woodward, 1989). Broadleaved evergreens tend to dominate from the equator to the Mediterranean region. Of these holm oak (*Quercus ilex*) can tolerate the lowest temperatures, surviving brief periods at −15°C, but none are present in the northern forests. Broadleaved trees, such as *Q. robur*, which can survive minimum temperatures below −15°C are almost all deciduous; many can survive a minimum temperature of −40 to −50°C. Where minimum temperatures fall below −50°C, the forests are dominated by needle-leaved conifers – typically pines, firs and spruces – of which a few (*Larix* spp.) are deciduous. Even here a few broadleaved deciduous species of birch and poplar, for example aspen (*Populus tremula*), manage to survive.

Woody climbers such as ivy (*Hedera helix*), honeysuckle (*Lonicera periclymenum*), Old Man's beard (*Clematis vitalba*) and the tropical lianas are specialized phanerophytes which profit from the stature of their neighbours. The more complex life form system given in Mueller-Dombois and Ellenberg (1974 p. 449) uses five main stem or trunk forms (normal woody, tuft trees, bottle trees, succulent and herbaceous stem trees) in the subdivision of the phanerophytes.

Chamaephytes are woody or herbaceous low-growing plants whose perennating buds are on aerial branches not more than 25 cm above the soil, and frequently much lower, where the wind is not so strong and the air is damper. The perennating buds of **hemicryptophytes** are at the surface of the soil where they are even better protected, while those of **geophytes** are buried beneath the soil on rootstocks, rhizomes, corms, bulbs (Figure 2.2) or tubers. Geophytes are particularly abundant in loose and soft soils. **Therophytes** survive the unfavourable season as seeds, and are abundant in deserts and open habitats, in waste places and as weeds of cultivated land. Common in the early stages of the reversion of waste land to scrub, they become rarer as it progresses to mature woodland.

Life form is primarily determined by heredity and selection; it may be regarded as an adjustment of the vegetative plant body and life history to the habitat. Under some circumstances, however, the environment directly influences life form, for example stinging nettle (*Urtica dioica*) may overwinter as a herbaceous chamaephyte under favourable conditions, but is normally a hemicryptophyte. Conversely, severe winter conditions may kill the upper buds so that individual plants fall into the life form below that normal to the species, as in the dwarfing of trees growing at high altitudes or subject to almost constant wind.

Raunkiaer employed six sizes of leaf in the finer details of his classification. **Leptophylls** each have an area of up to 25 mm², while the upper area limits of the next four members of the series (**nanophyll, microphyll, mesophyll** and **macrophyll**) increase by a factor of nine in each instance, the largest leaves (**megaphylls**) exceeding a nominal value of 164 025 mm² in area. Leaves tend to be large in hot wet regions (tropical rainforest), medium sized in temperate woodlands, and small in the cold or dry conditions of tundra and heaths. There is thus some justification for the use of the **leaf size spectrum**, based on the percentages of the different leaf sizes present, to characterize different vegetation types. However, light intensity and edaphic conditions, particularly soil nitrogen and phosphorus, are also important determinants of leaf size even within the same genotype.

Leaves may be needle-like, simple or compound. Those of the tropical aroid genus *Monstera*, climbing shrubs which often mature as epiphytes with aerial roots reaching the soil, have gaps or rounded holes when mature. In tropical rainforests at least 80% of the species of trees present have leaves of the mesophyll class (2025–18 225 mm²): most are unlobed sclerophylls ('hard leaves') with a pronounced point known as a 'drip tip'. Microphylls predominate in montane rainforests where the climate is cooler.

Trees grade from **pachycaul** forms, such as tree ferns and palms, with thick, unbranched or little branched main stems bearing a terminal crown of large compound leaves, to 'twiggy', much branched **leptocaul**

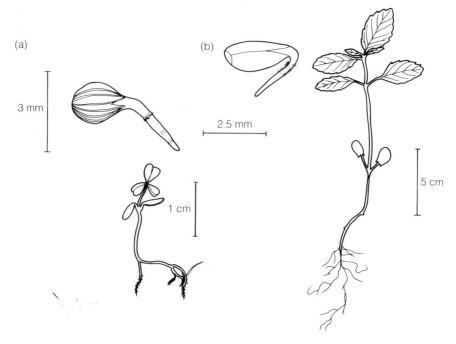

Figure 2.2 Seedlings of two woodland chamaephytes (a) wood sorrel *Oxalis acetosella* and (b) yellow archangel *Lamiastrum galeobdolon*. The seeds of these species appear to require a period of chilling to break dormancy and are difficult to germinate experimentally. In nature considerable numbers of seedlings often appear together in early spring. Very occasionally *Lamiastrum* seedlings have three cotyledons. Note that *Oxalis acetosella* can also exist as a rhizome geophyte or a rosette hemicryptophyte. An example of a bulbous geophyte (bluebell) is shown in Figure 5.20, p. 194.

forms bearing smaller undivided leaves such as the elms. Common ash (*Fraxinus excelsior*) with its pinnate leaves and stubby twigs, has a tendency to the pachycaul habit, but no British tree has the massive frost-sensitive apical meristem of a true pachycaul. Another method of biological classification is based on the principal agent of **seed dispersal**, usually wind, animals or water. Plants adapted to dispersal over long distances by wind are frequent in the taller strata. Herbs and trees, such as fireweed (*Chamaenerion angustifolium*) and birch, of pioneer communities, usually spread more efficiently than species typical of climax vegetation.

2.2   Biological spectra

A **biological** (or **life form**) **spectrum** for a particular area is constructed by expressing the numbers of the species in each life form class as percentages of all the species present. As a standard of comparison Raunkiaer used the **normal spectrum** derived from 1000 plants taken as a representative sample of the world flora (Figure 2.3). There is a strong correlation between the climate of an area and the life forms of the plants present: a **phytoclimate** is characterized by the life form which most greatly exceeds the percentage for its class in the normal spectrum. The phytoclimate of Raunkiaer's native Denmark is hemi-cryptophytic, like that of most of the cool temperate zone including Britain. In this system every species carries equal weight regardless of abundance or importance. Consequently, although the relatively few tree species native in Britain since the last glaciation were dominant almost everywhere until the advent of large-scale agriculture, the phytoclimate is not classed as phanerophytic. The regions in which the four major world plant climates (**phanerophytic, therophytic, hemicryptophytic** and **chamaephytic**) occur, with their subdivisions, may be delimited by lines along which the biological spectra are similar.

The tropics are **phanerophytic** where rainfall is not deficient. Within this zone a greater proportion of the larger forms are found in the wetter areas. Subtropical desert areas are **therophytic** and spring to life after the very occasional periods of heavy rain. Geophytes are best repre-sented in regions with a Mediterranean climate where the unfavourable season is the hot dry summer. Plants with this life form are at a dis-advantage where the soil warms slowly in spring and the growing season is short. The resting buds of **hemicryptophytes** in the cool temperate zone are protected by snow in hard winters, but are warmed by the sun as soon as it melts. **Chamaephytes** characterize the cold zones near the poles where cushion forms in particular derive protection from snow, but grow as soon as the spring melt commences.

Shimwell (1971) presents data for eight temperate broadleaved wood-lands in Britain and North America. Hemicryptophytes are abundant in all, but especially in the northern sites, while cryptophytes are more important in the southern and drier sites. The type of life form spec-trum in which all species count equally can also be employed to contrast woodland floras subject to different environmental influences (section 10.6).

Variation within a Swedish primeval forest

Figure 2.3 gives the biological spectra and numbers of vascular species from five relevés whose positions in the primeval forest of Fiby urskog are shown in Figure 4.9 (see also sections 5.1 and 5.2). The species-rich

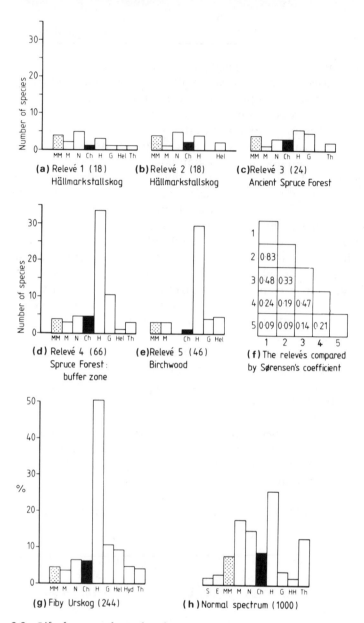

Figure 2.3 Life form analyses for the primeval forest of Fiby urskog, central Sweden. (a–e) Numbers of species in the various life forms in relevés 1–5 (relevés 1, 4 and 5 each had an area of 0.3 ha, relevés 2 and 3 each measured 0.4 ha). (f) The relevés compared by means of the Sørensen coefficient of community (see 4.3). (g) Biological spectrum for the vascular plants of the entire Fiby urskog forest reserve compared with (h) the normal spectrum taken by Raunkiaer from the world list. Figures in parentheses are numbers of species. S, stem succulent phanerophytes; E, epiphytic phanerophytes; MM, meso- and megaphanerophytes; M, microphanerophytes; N, nanophanerophytes; Ch, chamaephytes; H, hemicryptophytes, G, geophytes; Hel, helophytes; Hyd, hydrophytes; HH, helophytes and hydrophytes; Th, therophytes. (Unpublished data of Packham, Hytteborn, Claessen and Leemans.)

spruce forest of the buffer zone (relevé 4), which has been subject to considerable interference in the past, has far more herb species than the other conifer-dominated relevés (though parts of the primeval forest are equally species-rich) and the greatest number of the extra species are hemicryptophytes. The next most important group is that of the geophytes. Many of these are rhizome geophytes such as *Gymnocarpium dryopteris*, *Actaea spicata*, *Anemone nemorosa*, *Hepatica nobilis*, *Maianthemum bifolium* and *Trientalis europaea*. *Oxalis acetosella*, which is abundant in northern spruce forests, can be a herbaceous chamaephyte, a rosette-hemicryptophyte or a rhizome geophyte according to the circumstances of its growth. Relevé 5 is a damp *Betula pubescens* wood now being invaded by spruce. **Helophytes** (marsh plants permanently rooted in mud) such as *Carex elata*, *Equisetum fluviatile*, *Galium palustre*, *Lysimachia vulgaris* and *Potentilla palustris* are a conspicuous element, while therophytes are absent.

When compared by Sørensen's coefficient of similarity (Figure 2.3 and section 4.3) the floras of the two areas of **hällmarkstallskog** (ridge-ground pine forest) are seen to be very similar; both are progressively less similar to the series ancient spruce forest, spruce forest of the buffer zone and the seral birchwood. Although relevés 1 and 2 both had only 18 species of vascular plants, they were in an area with many bryophyte and lichen species, particularly where the cryptogamic mat was undamaged. The biological spectrum for the Fiby urskog reserve has more than twice as large a representation of hemicryptophytes as the normal spectrum and the same is true of the geophytes. The comparatively large representation of helophytes and hydrophytes is accounted for by the vegetation of the lake and of the mires which are a characteristic feature of Scandinavian forests.

Other life form classifications

In general, central European phytosociologists now use life form systems more complex than those of Raunkiaer and also make full records of the bryophytes, lichens and fungi. Advances have also been made in incorporating **relative values for species** into biological spectra, rather than weighting each species equally. Such values can conveniently be based on cover-abundance ratings or on some quantitative measure such as cover, density or frequency.

In their treatment of three forest areas near Hamburg, including two (oak–birch and spruce) on dry soils, Mueller-Dumbois and Ellenberg (1974) give separate biological spectra for vascular and non-vascular plants. Further ecological differences between the three forests are brought out by indicating the proportions of the plants which are either winter- or evergreen and those which are spring- or summergreen. Spring- and summergreen rhizome geophytes are very common in beech forest on moraine. Evergreen vascular plants and thallophytes are far less common here than in the less shaded oak–birch forest. Given the same climate and tree cover, these plants usually become more abundant as the soil pH and nitrogen content decrease. They form the greater part of the ground vegetation in planted spruce forest whose evergreen foliage provides much the same shade throughout the year.

There is considerable scope for the use of life form analysis in further work on British **woodland bryophytes**. Five main types have been described: **cushions** (e.g. *Leucobryum glaucum*), **turfs** (e.g. *Dicranum majus* and *Polytrichum formosum*), **canopy formers** (**dendroid forms** such as *Climacium dendroides* and *Thamnium alopecurum*) **mats** and **wefts** (Gimingham and Birse, 1957). Rough mats, smooth mats and dendroid forms are the commonest types on rocks but a wide variety grow on soil (section 5.7).

2.3   Processes of primary production and the light climate

The life of all components of the food web is maintained by the latent chemical energy bound in organic compounds during photosynthesis. The overall course of the production of a hexose sugar by photosynthesis is represented by the equation

$$6CO_2 + 12H_2O \xrightarrow[\text{of chlorophyll}]{\text{light in presence}} C_6H_{12}O_6 + 6H_2O + 6O_2$$

This process utilizes only a small proportion of the solar radiation impinging on the woodland as most of it is immediately transformed into heat, some of which drives the hydrologic cycle (Figure 2.13). Some 40–45% of the energy of this radiation consists of wavelengths in the visible light range (380–740 nm), which corresponds to the bands absorbed by the photosynthetic pigments, though photosynthetically active radiation (PAR) is generally considered as extending from 400 to 700 nm (0.4–0.7 μm). PAR is bounded by ultraviolet (UV) radiation on the short wavelength side and by infrared (IR) radiation on the long wavelength side. Almost all the solar radiation received at the Earth's surface is included in the waveband 300–3000 nm.

The proportions of the radiation striking a leaf which are reflected, absorbed or transmitted through it vary according to wavelength, the angle at which light strikes the leaf, and the nature of the leaf itself.

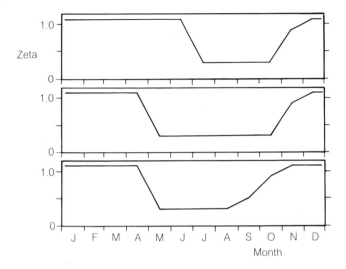

Figure 2.4 Estimated annual change in the R/FR ratio as zeta under three deciduous tree canopies. (a) Late leafer; (b) early leafer; (c) early leafer with early leaf fall. (From Mitchell 1992.) **Zeta** is the ratio of quantum irradiance in the 655–665 nm and 725–735 nm wavebands, which has a value around 1.15 in the open and down to 0.1 in the deep shade of a leaf canopy.

Transmission of visible light is greatest at wavelength bands where reflection is also high. Tree foliage is a selective filter; radiation which has passed through it on its way to the forest floor is rich in green light and in the near IR. It is deficient in the UV and also in the red and blue zones in which the chlorophylls absorb strongly, but the action spectrum for photosynthesis is much less sensitive to wavelength in the 400–700 nm range than the absorption spectrum of chlorophyll. Leaves absorb the vast majority of the far IR; the resulting heat load is dissipated by increased transpiration and by convection currents.

Tree foliage also changes the ratio of red (R, 600–700 nm) to far red (FR, 700–800 nm) light; this is of great importance in plant responses controlled by **phytochrome**. Physiologically active FR is proportionately far more abundant in light which has passed through tree leaves than in normal daylight, whereas the reverse is true of R. Seasonal variation in the R/FR ratio beneath deciduous canopies is markedly affected by the phenology of the trees involved (Figure 2.4). Light-requiring seeds are affected by the quality of light to which they are exposed. In the open the effects of R normally predominate and such seeds, when imbibed, germinate if exposed to light, while in woodland FR tends to inhibit germination. If a gap arises in the leaf canopy, seeds of certain herbs often germinate in large numbers; the seeds have detected changes in **light quality** by means of phytochrome. Grime (1979)

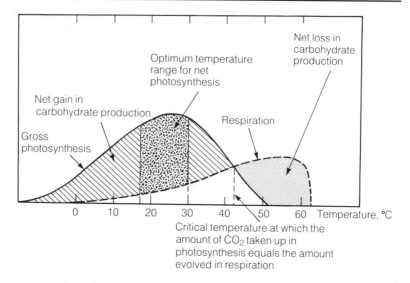

Figure 2.5 Generalized relationship between photosynthesis and respiration with increasing temperature, under conditions in which photosynthesis is not limited by the amount of CO_2 or light available to the plant. Net photosynthesis (the difference between gross photosynthesis and respiration) is indicated by the hatched area. (Redrawn from Waring and Schlesinger (1985) who rescaled the diagram of Daniel *et al.* (1979) to emphasize somewhat greater importance of photosynthesis over respiration at higher temperatures.

states that high germination rates occurred in seeds of both *Juncus effusus* and *J. articulatus* exposed to high intensity R alone; no germination occurred with FR. This fits with experience of freshly coppiced areas where rushes sometimes grow so vigorously from disturbed persistent seed banks as to reduce growth of the coppice poles.

The **photosynthetic pigments** of the primary producers are frequently well-suited to the spectral composition of the irradiation available to them. For example, the higher proportion of chlorophyll *b* relative to chlorophyll *a* found in some shade plants enhances light-absorbing capacity in the blue-green wavebands (between the main blue and red absorption bands of chlorophyll *a*) which are relatively abundant on the forest floor. The **light reactions** of photosynthesis involve the absorption of light by chlorophylls *a* and *b*, leading to the formation of $NADPH_2$ and of the energy-rich compound ATP, both of which are essential in carbon fixation during the ensuing thermochemical **dark reactions**. These, unlike the light reactions, are often limited by temperature under natural conditions. Respiratory rates are also influenced by temperature; Figure 2.5 shows how greatly **net photosynthesis** (gross photosynthesis – respiration) is dependent upon it. Forest trees, and most of the smaller plants of temperate lands, fix carbon by the C_3

pathway; the C_4 pathway and crassulacean acid metabolism are of negligible importance in the world's woodlands. The carotenoids, which are largely responsible for the yellowish colour common in the leaves of sun plants, seem relatively inefficient at light gathering for photosynthesis; their main function may be to afford protection to chlorophylls from photo-oxidation.

Rates of photosynthesis are frequently limited by temperature, the amount of CO_2 available to the plant, or the available light. CO_2 concentrations are frequently lower in actively photosynthesizing tree canopies than near the forest floor, where decomposer organisms respire actively as they break down the litter. The proportion of the available light which a plant can use in photosynthesis is influenced by many factors, including the nature and concentration of the pigments in its photosynthetic cells, the arrangement of the cells in the leaf, the surface texture of the leaf, the arrangement of leaves in space and the angles which they present to the direction of the light. Shading by the leaves of the same or of different plants is common, while variations in leaf size and shape, as well as the distances between any layers of leaves that may be present (section 5.4) can be crucial. Figure 2.6 shows the fate of visible radiation received by a temperate forest on a clear day in summer. On cloudy days most of the light incident on the forest canopy is **diffuse** (skylight), while even on clear days the only **direct light** (sunlight) reaching the forest floor away from glades and trackways is in the form of sunflecks produced when the sun's rays pass through gaps in the leafy canopy. In some instances the radiation received is so intense as to cause damage through overheating; Rackham (1975) quotes an example in which an area of *Mercurialis perennis* became scorched by the passage of a large sunfleck after rain.

Sunflecks vary in size, duration and periodicity. 'Windflecks' can flutter between sun and shade at around 20 cpm (cycles per second) as small apertures occur between moving leaves. Larger sunflecks caused by movement of the sun relative to more permanent gaps have been called 'timeflecks'. Plants strongly adapted to shade are often unable to make effective use of sudden brief increases of irradiance, but phytochrome-mediated responses have been detected within 10 min. Wood sorrel (*Oxalis acetosella*) closes its leaves in a photonastic response to natural sunflecks that protects the plant from potentially damaging light intensities. This takes place within 3 min and is mediated by blue light (Hart, 1988). Most photosynthetically active radiation, PAR, is absorbed in the crowns of trees and the amount reaching the ground can be very small indeed. This is important because at the low irradiance of most forest floors when the trees are in leaf the rate of photosynthesis is often limited by PAR, whereas at higher irradiance the rate may be governed by the thermochemical or dark reactions.

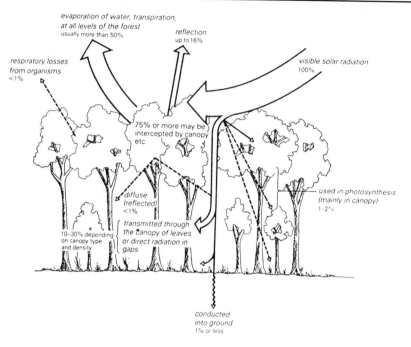

Figure 2.6   Modulating effect of the vegetation cover. The fate of visible solar radiation (wavelengths 0.4–0.7 μm) received by a temperate forest on a clear day in summer. Most of the radiation absorbed at and beneath the canopy is used in the evaporation of water or in transpiration. About 6–16% of visible radiation (mainly green (0.5 μm), some orange and red (0.6–0.75 μm)) as well as about 70% of the infrared (0.75–1 μm) is reflected. At ground level the distribution of radiation is spatially very uneven. Large canopy gaps may be well lit for part of the day. Beneath small gaps sunflecks may reach particular spots on the ground briefly (changing continuously during windy conditions). Other places under the densest part of the canopy will receive only weak transmitted and diffuse (reflected) light, possibly reduced to 2% of full sunlight or less, and with spectral composition very low in the portion (0.4 and 0.7 μm) useful for photosynthesis, somewhat higher in the green (0.55 μm) and high in far-red (0.75 μm). (From Burrows, 1990; after Larcher, 1975; and various other sources.)

Seasonal variations in the light climate

Comparative studies of sun and shade plants, and of the phenology of the communities to which they belong, have gone hand in hand with investigations of the quantity and spectral composition of the radiation which they experience. Light conditions within a woodland vary in respect of intensity, and of directional and spectral composition, from hour to hour and year to year as well as from month to month. The problems of making meaningful light measurements (Table 2.1) and of

Table 2.1 Units involved in the measurement of light

Unit of energy: $1 J = joule = 1 kg m^2 s^{-2}$
$kJ = 10^3 J$ $MJ = 10^6 J$
$1 cal (calorie) = 4.18 J$

Unit of power: $1 W = watt = 1 J s^{-1}$
$mW = 10^{-3} W$

Unit of illuminance: $lx (lux) = lumens m^{-2}$
$klx = 10^3 lx$

Quantum energy: 1 mole of light $= 6.023 \times 10^{23}$ photons or quanta
$1 \mu mol = 1 \mu E (\mu Einstein) = 6.02 \times 10^{17}$ quanta
$4.6 \mu mol m^{-2} s^{-1}$ at 550 nm $= 1 W m^{-2}$

Light is measured in units which differ according to the purpose for which it is measured. In this book the units given are those employed by the authors in their original measurements.

Radiometry is the measurement of radiant energy, including light, and is suitable for ecological work. Its instruments can measure total radiant energy falling on a surface and express it as irradiance, in $W m^{-2}$. Alternatively, filters can correct the measurement to allow for the varying energies of light quanta at different wavelengths within the spectral range of photosynthetic pigments (400–700 nm) and express it as the number of photons flowing on to unit surface, in $\mu mol m^{-2} s^{-1}$.

Photometry is the measurement of light for engineers and photographers. Its instruments contain filters to match the light to the spectral sensitivity of the young human eye and express the light falling on a surface as illuminance, in lux. Since humans are more sensitive to green light than to red or blue, the lux varies from less than $2 mW m^{-2}$ for green light to more than $120 mW m^{-2}$ for blue light.

---

creating suitable light regimes in shading experiments are complex (Evans *et al.*, 1975; Hart, 1988). Total radiant energy may be several hundred $W m^{-2}$ in the open but is usually less than a hundred in the shade. When energy is integrated over time, the megajoule per square metre per day is a convenient unit, ranging from the order of 2–5 in the open during winter to the order of 20–25 in summer at the latitude of Britain. The light of major significance in plant growth is **photosynthetically active radiation**, PAR, which is usually taken as being bounded by the wavelengths 400 and 700 nm. This is measured as quantum flux of PAR in micromoles (μmol) $cm^{-2} s^{-1}$, at particular times or integrated over a period. A direct reading of PAR below, or at a defined position within, a leaf canopy as a percentage of that above it can be valuable. The spectral quality of light, and of the instrument measuring it, must always be borne in mind when designing experiments or assessing their results. The light climate of woodlands is discussed in detail by Mitchell (1992), who provides comprehensive information concerning seasonal and

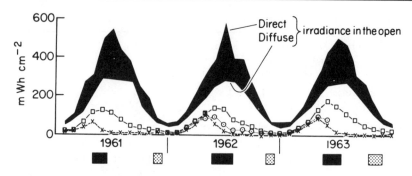

Figure 2.7 Monthly averages of daily irradiance in the open at three sites in Madingley Wood, with the contribution of direct irradiance in the open shown in black. □, large clearing (some 20 m in diameter); ⊙, small clearing; ×, photosite. The 'photosite' was in the middle of a stand dominated by *Ulmus carpinifolia* with some *Quercus robur* and an understorey largely of hazel. The small clearing was only 5 m from the photosite and canopy conditions there were similar, apart from a small patch of open sky near the zenith. The horizontal bars at the bottom of the figure represent tree leaf expansion (black) and leaf fall (stippled). Measurements of the light conditions were made with integrating electrolytic photometers and bimetallic actinographs; gaps in the record were filled by photographic estimates of direct and diffuse light. The more strongly shaded the site the more is its annual peak of irradiance shifted away from mid-summer towards the spring. (From Anderson 1964b, by courtesy of the *Journal of Ecology*.)

annual changes in PAR and R/FR beneath coppice and high forest canopies.

As early as 1916 Salisbury laid the foundations of the study of shade as an ecological factor, using the darkening of photosensitive paper when exposed to light to compare the amount of light in the open with that reaching the woodland floor. To present the results of shading as a percentage reduction of the light in the open, however, fails to take account of the great variation between sites in different climates and latitudes, and also between different occasions at the same site. This method of light measurement also ignores the quality of light penetrating the canopy. Young (1975) illustrated the importance of this aspect of the light climate by growing plants of *Impatiens parviflora* at various **R/RF ratios**. Though all the plants received the same intensity of radiant energy, lowering the R/RF ratio caused an increase in specific leaf area (Table 2.2) and in internode elongation. With modern techniques, radiation can be measured in absolute terms instrumentally as Anderson (1964a,b) did in her long-term studies (Figure 2.7) in Madingley Wood, Cambridgeshire.

Hemispherical camera photographs such as that shown in Figure 2.8 make it possible to estimate, with reasonable accuracy, the amounts of light reaching particular woodland sites without the continuous use of recording instruments. Estimations of direct light are made using solar track diagrams, whereas diffuse light is calculated using a grid which divides the photographs into a number of areas each of which, in the absence of interference by the tree canopy, would receive the same amount of skylight. Photographs taken in the summer are used to estimate the proportion of skylight able to pass through the foliage.

In evergreen forests the shade offered by the tree canopy is usually much the same throughout the year, so the monthly light totals can be expected to show a symmetrical seasonal trend similar to that in the open. The situation is different in deciduous woodlands where Salisbury recognized a **light phase** in which the trees are bare of leaves and a **shade phase** in which the tree canopy is present. In the English Midlands the light phase, which Rackham (1975) refers to as the **bare half-year**, usually lasts at least five months. In open sites in Britain approximately half the total light energy is **sunlight** direct from the sun and the other half is **skylight** diffused from clouds or blue sky. During the bare half-year the spectral composition of light inside and outside the wood is very similar, although shading by trunks and branches reduces the amount of skylight received by most of the forest floor to between one-quarter and one-third of that in the open. In December sunlight rarely penetrates to the ground, except in large glades, but by April its amount is almost as much as that of the skylight. The proportional increase in total light from the winter minimum to April is greater inside the wood than in the open.

In the shade phase ('**leafy half-year**') the light climate in different parts of the wood is much more variable. Skylight can be reduced to a fraction of 1% (of that present in the open) under the densest shade, while the light is largely green after transmission through tree foliage. The more heavily shaded a woodland site is during summer, the greater is the proportion of the year's total light received in spring (Figure 2.7). Variation in the time at which the tree canopy expands, which differs considerably from year to year, greatly affects the total light received by the ground vegetation in spring, while daily light totals have been shown to vary at least five-fold within a week at Madingley Wood.

Regional variations in **day length** are considerable. Annual variation for forests at the latitude of Bombay (19°N) is less than 3 hours, whereas in Oslo (60°N) day length changes from less than 6 hours in mid-winter to over 18 hours in mid-summer.

Most trees flourish when light is abundant, being **sun plants** (heliophytes) while many of the small plants growing beneath them are **shade plants** (sciophytes). The aerial environments of these two groups differ, and the adaptations to these conditions involve major morphological and physiological differences considered in the next section.

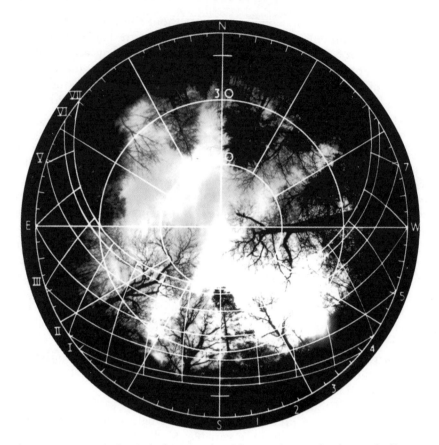

Figure 2.8 Hemispherical photograph with superimposed solar track diagram for 52° 13′N, of the large clearing (Figure 2.7) in Madingley Wood, Cambridge-shire. The fisheye lens has recorded a hemisphere on a single photograph, covering a vertical arc of 180°.

This photograph was taken at 1450 hours GMT on 2 March 1957 with a bright but hazy sky; the sun was mid-way between the solar tracks for 21 February (III) and 21 March (the equinox), and a little to the left of the 3 pm hour line. The spruce (*Picea*) a few degrees west of south, an ash (*Fraxinus*) 30° west of south, and an oak (*Quercus*) almost due west, were felled before the records shown in Figure 2.7 were made, but the basic pattern of the canopy was similar (cf. Anderson, 1964a, photograph 2, taken 10 January 1962).

The superimposed grid consists of concentric circles for 0°, 30° and 60° of altitude, together with radiating lines for each 30° of azimuth. 1–7 indicate hours from apparent noon. The seven solar tracks were computed by the method of Evans and Coombe (1959) and apply to the following 12 days of the year, upon each of which the sun enters a sign of the zodiac:

I 22 Dec II $\frac{22\ \text{Jan}}{21\ \text{Nov}}$ III $\frac{21\ \text{Feb}}{22\ \text{Oct}}$ IV $\frac{21\ \text{Mar}}{23\ \text{Sept}}$ V $\frac{22\ \text{Apr}}{22\ \text{Aug}}$ VI $\frac{22\ \text{May}}{22\ \text{July}}$ VII 22 June

2.4   Sun and shade plants

The most generally limiting factor for herbs, shrubs and young trees in woodland situations is light and many of them show remarkable adaptations to different levels of PAR. In deciduous woodlands, herbs such as *Anemone nemorosa*, *Hyacinthoides non-scripta* and *Ranunculus ficaria* largely evade severe shading by having their main period of growth before the tree canopy expands, indeed *Hyacinthoides* is well known as a sun plant (Blackman and Rutter, 1946, 1954). Others are truly **shade tolerant** (e.g. enchanter's nightshade, *Circaea lutetiana*), and in some the shade leaves are modified to such an extent, by increasing the light-catching area and chlorophyll content, that for the same expenditure of dry weight they achieve much the same rates of photosynthesis as sun leaves when under dense canopy; the **sun leaves** which develop under the yellow archangel (*Lamiastrum galeobdolon*), which possesses great **phenotypic plasticity** and grows along hedgerows as well as in woodlands, show differences between their sun and shade leaves (Figure 2.9) considerably greater than those in *Oxalis acetosella*, a plant well adapted to shade but not to sun (Packham and Willis, 1977, 1982). These and many other plants produce large, deep green, relatively thin, fragile leaves when under dense canopy; the sun **leaves** which develop under high light intensities are smaller, thicker and often yellowish-green. **Shade leaves** in many species commonly have a spongy mesophyll with very large air spaces and palisade cells, known as **funnel cells**, which taper downwards from the margin adjoining the upper epidermis. They also have a thinner cuticle, large epidermal cells and a lower vein density than do sun leaves.

Lamiastrum receiving PAR filtered to give a low R/FR ratio, as under a deciduous tree canopy in summer, flowers less vigorously. Its internodes and petioles are also more elongated when compared with plants receiving

Given the mean hourly irradiance data for the appropriate latitude and climate (see Anderson 1964a) such solar track diagrams can be used to estimate the percentage of direct light reaching the site where the photograph was taken at any time of the day or period in the year. The longer the period taken the more accurate will be the estimate. Pope and Lloyd (1975) deal with the special considerations that apply to sloping sites.

This photograph has been prepared by Dr D. E. Coombe from the original negative but with a more accurate orientation of the solar track than that shown in photograph 6, Evans and Coombe (1959). (Coombe's 1957 photograph was taken at a point about 2 m NW of Anderson's 1962 photograph so that the orientations of the trees vary somewhat.)

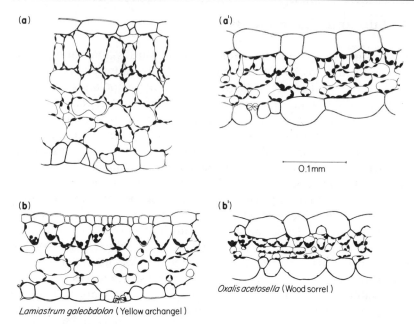

Figure 2.9   T. S. leaves of *Lamiastrum galeobdolon* subsp. *montanum* (left) and of *Oxalis acetosella* (right). Both species are hypostomatous (stomata on the underside of the leaves only). (a), (a') Sun leaves; (b), (b') shade leaves, whose palisade mesophyll consists of funnel cells. (From Packham and Willis, 1977, 1982; courtesy of *Journal of Ecology*.)

the same amount of unfiltered light (Mitchell and Woodward, 1988); such responses assist foraging for PAR amongst the ground vegetation.

Adaptations to shading

Growth analysis techniques are valuable when studying the ways in which plants react to changes in light level, humidity regime, temperature and other features of the environment which are altered by shading. Chlorophyll content per unit dry weight is usually considerably higher in shade leaves than in sun leaves, except where shading is so extreme as to result in the production of relatively small etiolated leaves. Differences in chlorophyll contents between sun and shade leaves appear much greater when expressed on a dry weight than a fresh weight basis, while chlorophyll values per unit area may actually be lower for shade leaves than sun leaves. Thin leaves are not always characteristic of shade plants; many rainforest species such as *Cordyline rubra* and *Lomandra*

Table 2.2 Definitions used in growth analysis

Relative growth rate (RGR) is the rate at which a plant increases its dry weight per unit dry weight. RGR can, at a particular instant, be resolved into three components:

1. Unit leaf rate (ULR) = $dW/dt \times 1/L_A$
2. Leaf weight ratio (LWR) = L_W/W
3. Specific leaf area (SLA) = L_A/L_W

   where L_A = total leaf area
   L_W = total leaf dry weight
   W = total plant dry weight
   dW/dt = rate of dry weight increase of the whole plant

RGR = ULR × LWR × SLA

Leaf area ratio (LAR) is a morphological index of plant form, the leaf area per unit dry weight of the whole plant.

4. Leaf area ratio (LAR) = L_A/W = LWR × SLA

In contrast, **unit leaf rate** (ULR), the rate of increase in dry weight of the whole plant per unit leaf area, is a physiological index closely connected with photosynthetic activity. A high LAR together with low ULR is characteristic of heavily shaded woodland herbs in termperate forests.

Relative leaf growth rate (RLGR) is analogous to RGR and is the rate of increase in leaf area per unit leaf area.

5. Leaf area index (LAI) = L_A/Ground area occupied by plant

$$\text{Stomatal index} = \frac{\text{Number of stomata per unit area}}{\text{(Number of stomata per unit area + Number of epidermal cells per unit area)}}$$

longifolia have thick leaves with unusually large chloroplasts concentrated in the upper palisade. This gives them a high chlorophyll content and a low specific leaf area (SLA). (This term is defined in Table 2.2, together with others used in growth analysis.) Stomatal frequency and pore size vary amongst rainforest plants but differences between sun and shade species are not significant.

Besides the major morphological and physiological differences between woodland species and plants of other habitats, environmental and genetic factors result in variations within individual species, a point illustrated by Figure 2.10 which shows the results of a shading experiment. Using artificial shading under controlled conditions, Hughes (1959) showed that the meristematic activities of the sun and shade leaves of small balsam (*Impatiens parviflora*) are fundamentally similar, and that the difference in structure at maturity is caused by the greater expansion of the shade leaves which consequently have lower stomatal frequencies. This situation is common amongst temperate plants and explains

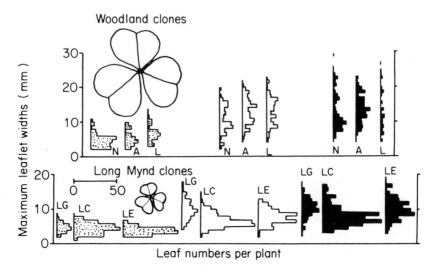

Figure 2.10 Variation in leaflet widths of clones of *Oxalis acetosella* from two contrasting habitats in Shropshire. Clones N, A and L were from a woodland streamside near Telford; clones LG, LC and LE from beneath *Pteridium aquilinum* in a montane pasture on the Long Mynd. Clones were gathered from the wild in late March 1971 and grown in cold frames under three light regimes: under clear polythene only (stippled); in light further reduced by a muslin shade (white); and in light greatly reduced by a muslin shade (black). The plants were harvested in August 1972. While phenotypic plasticity is very important, genetic influences are also involved. Clones from both sites show greater leaflet widths as the light levels are reduced, while the leaflet width/leaf number relationships of the two populations differ, plants from the Long Mynd tending to have smaller leaflets but a greater number of leaves than those from the woodland. Individual clones from the site also differ in their behaviour, as seen in the contrast between clones LG and LC. (Redrawn from Packham and Willis, 1977; courtesy of *Journal of Ecology*).

why the stomatal frequency can vary according to shading regime, while the **stomatal index** (Table 2.2) remains similar.

Young (1975) found that though the **leaf weight ratio** (LWR) of *I. parviflora* is markedly affected by the rooting medium, it is little altered by changes in total daily light. SLA varies with temperature, rooting medium, day length, total daily light and the 'physiological age' of the leaf or plant. As SLA is, over a substantial range of daily light level, inversely proportional to total daily radiation received, while **unit leaf rate** (ULR) is directly proportional to it, the net effect of these two relationships is that the **relative growth rate** (RGR) of *I. parviflora* remains approximately constant over the range concerned.

The proportion of total plant weight devoted to roots is commonly low in shade plants, although the **root/shoot ratio** appears to be related to soil moisture content as well as the intensity of radiation. Temperate shade species grown under lower PAR tend to possess high fresh weight/dry weight ratios and SLAs, while RGR is low. The very low levels of radiation at which some shade plants occur in nature are often well below the optimum for the species. Heavy shading, however, has the effect of eliminating competition from sun species of much higher potential RGR.

Above quite low radiation levels shade plants are unable to make use of additional light, while in sun plants net photosynthesis increases until very much higher levels are reached. The higher capacity for CO_2 fixation per unit leaf area in sun plants appears to be directly related to the greater amounts of the carboxylating enzymes and to the greater volume of the leaf per unit leaf area. The proportion of chlorophyll a to chlorophyll b increases in plants grown at high light intensities; even fully grown leaves may partially adapt to a change in light level by a modification of their enzyme systems.

The phenotypic plasticity which enables a given genotype to adapt to various light levels is effected by a wide variety of morphological, physiological and biochemical mechanisms; these are reviewed by Boardman (1977). He concludes that the light levels to which a genotype can adjust reflect a genetic adaptation to the conditions of its native habitat. Moreover, adaptation for great photosynthetic efficiency in strong sunlight precludes high efficiency in dense shade.

Radiation profiles and photosynthetic activity of trees

The density of the foliage and the shape of the crown cause great variations in the amount of light reaching individual leaves; every tree has a characteristic radiation profile. In cypress (*Cupressus*) the amount of light reaching the interior is about 0.5% of that present at the surface of the crown and no leaves are present in this **dysphotic zone**. Leafy branches are present even in the innermost region of the more open crown of the olive (*Olea europea*). The relative irradiance (mean value of radiation compared with that for an entirely open site) corresponding to the minimal requirements for shade leaves in trees with open crowns (e.g. *Betula, Larix, Pinus sylvestris*) is, according to Larcher (1975), between 10 and 20%, while in *Picea, Abies* and *Fagus* which have dense crowns it is 1–3% of the incident radiation. In parts of the crown where the average amount of PAR is insufficient to meet the minimal needs of the shade leaves, existing side shoots frequently wither and no new leaves are formed. It is probable that the critical radiation level which prevents this **self-pruning** process is that which just enables a branch to make a net gain in photosynthesis. Conversely, epicormic shoots become prominent in certain trees, such as larch, if strong light is again allowed to reach trunks which have become bare as a result of severe shading.

In trees differences occur in the intensity and seasonal duration of photosynthetic activity of sun and shade leaves borne on the same plant, as is illustrated by the investigations of Schulze (1970) on a single tree of *Fagus sylvatica* from the Solling Mountains of Germany. The rate of CO_2 uptake in sun and shade leaves was about the same in relation to leaf dry weight, but on a leaf area basis the assimilation in the shade leaves was approximately half that in the sun leaves. In relation to chlorophyll content the apparent assimilation of the shade leaves was only one-third of that of the sun leaves. By the time the leaves of *Fagus* reached maximal assimilation in August the photosynthetic activity of the sun leaves had already declined. Senescence of the sun leaves began much earlier than in shade leaves, as in cold frame experiments with the herb *Oxalis acetosella* (Packham and Willis, 1977). In both instances differential senescence was probably caused by the higher temperatures and by accelerated photo-oxidation of photosynthetic pigments at greater light intensities.

Variation in light compensation point

Investigation of a number of herbs growing beneath *Fagus* on the Solling Mountains (Schulze, 1972) demonstrated the ecological importance of some of their physiological and morphological differences. His measurements were in lux (lx): sunlight measured in May 1991 in England as 100 000 lx was equivalent to approximately 600 W m^{-2} but, as explained in Table 2.1, it is not possible to make proportional conversions. **Light compensation point**, at which the amount of CO_2 fixed in photosynthesis is equal to that released in respiration, lay between 300 and 500 lx for *Oxalis acetosella*, *Athyrium filix-femina* (lady fern), *Luzula luzuloides*, and young beech plants. Amongst these four plants **light saturation**, the light intensity at which further light does not increase net photosynthesis, is attained at the lowest levels by *Athyrium* (2000–3000 lx) and by *Oxalis* (5000–6000 lx); these two plants also had higher maximum rates of photosynthesis than the others. *Deschampsia flexuosa* in contrast acted more like a sun plant, having a compensation point of 2000 lx and a net rate of photosynthesis which increased linearly up to 12 000 lx. As a consequence the existence of this plant on the forest floor is far more dependent on the occurrence of **sun flecks** than is that of the other species (Figure 2.11).

The **photosynthetic temperature optimum** of *Oxalis*, which is

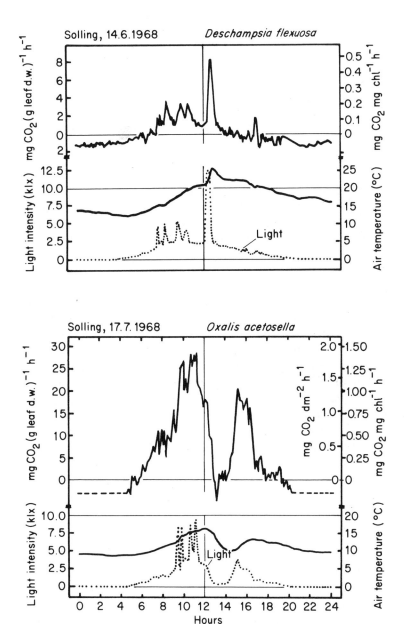

Figure 2.11 Apparent photosynthesis in the grass *Deschampsia flexuosa* and in *Oxalis acetosella* (a shade species) measured under natural conditions in a montane beech forest in West Germany. CO_2 exchange during a summer day is shown in terms of leaf dry weight (dw), leaf area and chlorophyll (chl) content of the leaves. (NB CO_2 exchange for *Oxalis* is plotted on a smaller scale than that for *Deschampsia*.) In contrast to the low amount of light reaching the forest floor, the crown top of the beech (*Fagus sylvatica*) may receive more than 80 klx at midday in summer. The shade leaves of the tree receive less than 15% of this and are light saturated at 6–7 klx during the summer, while the sun leaves are light saturated at 30–40 klx. (Redrawn from Schulze, 1972).

always low, was found to diminish from 13 to 18°C at 8000–12 000 lx to a range of even lower temperatures (9–12°C) at 1000 lx. This feature is of value in a wintergreen plant receiving little light and assists *Oxalis* to increase in weight longer into the autumn than most other species. Calculations of net photosynthetic gain are normally made with regard to the leaves rather than the whole plant. On this basis *Oxalis* has a higher net gain than *Luzula luzuloides* whether the reference system is unit dry weight, surface area or chlorophyll content. The greater competitive ability of *Luzula*, at least in the Solling Forest, is partially explained by the different **life forms** of the two species. *L. luzuloides* is a **hemi-cryptophyte** and some 63% of its total dry weight increment is partitioned to its aerial parts as against only 42% in *Oxalis*, which behaves as a **rhizome geophyte** here.

Shade-tolerant plants characteristically possess low metabolic rates which lead to low compensation points, a feature which contributes to their success in poorly lit habitats. In contrast, shade-intolerant species have higher compensation points, despite the fact that their rates of photosynthesis may exceed those of some tolerant species even at low radiation levels. Grime (1966) points out that the reserve of respirable materials, a good index of the ability to survive, is largely dependent upon conversion of photosynthate into plant structure. Intolerant species may fail to survive in shade because they quite quickly utilize a high proportion of the photosynthate produced during the period above the compensation point. Shade-tolerant plants retain larger reserves that enable them to survive relatively long periods when PAR is below the compensation point. Failure of seedlings in dense shade is often caused by fungal pathogens which are normally resisted if high sugar levels are maintained. Seedlings of common ash (*Fraxinus excelsior*) in woods with a well-developed herb layer often spend much of the year near the compensation point. Deaths usually occur in winter; after leaf fall new photosynthate is no longer produced and respiration soon depletes the small reserves of seedlings. Some disappear within a few days, while others first exhibit the 'wirestem' effect in which only the stele remains after the epidermis and cortex have been attacked by damping-off fungi. Only a few survive to grow above the herb layer. Rapid height increase does not enable herbs growing beneath forest trees to avoid shade, and stimulation of excessive extension growth at low PAR, as often occurs in shade-intolerant species, is disadvantageous as energy resources are dissipated and the seedling predisposed to fungal attack.

Leaf arrangement and leaf area index (LAI)

When poorly lit, plants tend to produce fewer layers of leaves than when unshaded; the measure employed to quantify this aspect of plant form is **leaf area index** (LAI), the total area of leaves growing above a given area of ground divided by the area of the ground itself. Horn (1971) found that the ground beneath an oak–hickory forest in New Jersey was covered, on average, by 2.7 layers of leaves of the canopy trees, while the LAI values of the understorey trees, shrubs and ground cover, which of course received successively less light, were 1.4, 1.1 and 1.0, respectively. Shade leaves are usually held horizontally where they absorb the maximum amount of radiation. The **sun twigs** of *Abies*, *Sequoia* and *Sequoiadendron*, however, are held at angles which reduce the heat load caused by absorbing long-wave radiation in situations where PAR is rarely limiting. The leaves of the seedlings and of the **shade twigs** of these three tree genera, on the other hand, are arranged horizontally and intercept the maximum amount of light. The pattern of **attenuation of radiation** passing through the forest is dependent on the cumulative LAI at a given level above the ground, the arrangement of leaves in **monolayers** or **multilayers** (section 5.4), and the angle at which the leaves are held to the horizontal. It is of great importance in influencing plant succession in forests, a subject considered in Chapter 5.

Foraging by modular herbs

Foraging, an activity common to both animals and plants, is a process in which an organism searches or ramifies within its habitat in order to acquire essential resources. Distribution of the resource-gathering structures of plants, mainly leaves and roots, in two- or three-dimensional space is partly controlled by genetical 'growth rules' that help determine many aspects of form. These include the distance between adjacent nodes on growing stems, the probability of branching at nodes, and the angle between branches (Hutchings and Slade, 1988). Figure 2.12 shows the modular structure of *Glechoma hederacea* (ground ivy); that of another woodland labiate (*Lamiastrum galeobdolon*) is very similar. **Ramets** consist of two horizontal laminas, each attached to a rooted node by an erect petiole. When planted such a parent ramet produces two primary stolons with new ramets at each node; a branched clonal system soon develops if conditions are favourable.

Slade and Hutchings (1987a–c) studied clonal integration and plasticity and the effects of different nutrient and light levels on foraging in *Glechoma*. Mean numbers of stolon branches, ramets per clone, and mean weight per unit stolon length were all highest in the high light, high nutrient treatment; they were least with low light and high nutrients. The latter plants had the lowest percentage dry weight allocation to roots and the greatest to petioles; the relatively light stolons and the few side branches allowed them to forage for irradiation extensively rather

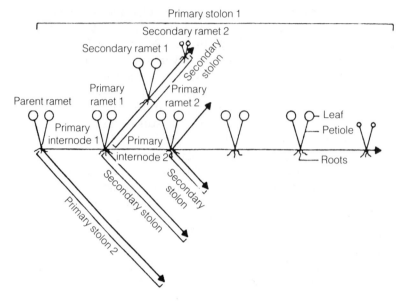

Figure 2.12   Schematic diagram of the basic vegetative morphology of *Glechoma hederacea*. (From Hutchings and Slade, 1988.)

than intensively as did plants in the high light, high nutrient treatment which would accordingly tend to stay in any favourable area they encountered naturally. Plants in the high light, low nutrient treatment also adapted effectively, possessing the highest root/shoot ratio of any of the three treatments.

2.5   Woodland hydrology and the water relations of forest plants

The amount and seasonal pattern of precipitation are among the most important factors influencing the growth and natural distribution of trees; they must also be carefully considered when creating new plantations. The extremely tall coastal redwood (*Sequoia sempervirens*) grows along the Pacific coast of North America in temperate rainforests which receive precipitation throughout the year. There is a cool maritime climate with abundant winter rain; in summer, clouds and fogs supply the tree foliage with water. In most of Britain this species grows less well than the big tree (*Sequoiadendron giganteum*), whose native habitat, at an altitude of 1200–2400 m in the Sierra Nevada, California, is much drier, and where most precipitation is as snow. Sitka spruce (*Picea sitchensis*), with a comparatively shallow root system, grows well when planted on the damp western side of Britain. In contrast, *Pinus sylvestris* and *Pinus nigra*, which have deep and effective root systems, grow well in Thetford forest, East Anglia, which has an average annual rainfall of only 23 inches (584 mm) and experiences severe drought at times.

The distribution of precipitation on the forest floor is locally very uneven; the ground becomes wetter under holes in the canopy, under the outer regions of the crown (from which it drips), and near the trunk. **Stem flow** is highest in trees with smooth bark whose branches are held at a steep angle to the ground. The amount of water infiltrating the soil at the bases of beech trunks can exceed that infiltrating in open sites by over 50%. Usually only a small amount of the water intercepted by tree foliage is absorbed by the plant; most evaporates. **Interception losses** vary in response to changes in meteorological conditions, while trees with open crowns and large smooth leaves retain less precipitation than those with dense crowns and small, easily wettable leaves. As a first approximation interception losses can be taken as 30% in evergreen coniferous forests and as 20% in broadleaved deciduous forest (Larcher, 1975).

Moisture often influences zonation in woodlands, as on the north-west face of the Ercall, Shropshire, which is largely covered by oak with a moderately open structure (Figure 4.3). Beneath is a heathy field layer and an extensive bryophyte cover in which *Plagiothecium undulatum* and the cushion form *Leucobryum glaucum* are common. The much smaller moss *Orthodontium lineare*, however, tends to form zones round tree bases. These are widest on the downhill side of the trunks down which water flows after rain; additional moisture clearly favours the moss, though it may also benefit from leachates in the water.

Figure 2.13 shows the hydrologic cycle of the Belgian mixed oakwood whose primary production is discussed in Chapter 9. In summer, losses caused by evapo-transpiration are high, those by drainage are low; the reverse is true in winter when stem flow is also greater. In sloping sites surface run-off and lateral drainage into streams are often large, particularly after winter rains.

Temperature influences both the metabolic state of plants, which are unable to absorb much water during very cold winters, and the potential evapo-transpiration rate. Trees frequently avoid much water loss by shedding their leaves. This is seen in northern temperate forests where deciduous trees lose their leaves at the beginning of winter, and in savanna where many trees are leafless during the dry season.

Drainage and **soil type** are very important with regard to water

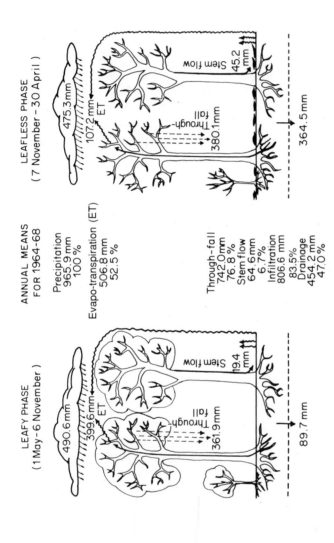

Figure 2.13 Mean annual and seasonal values for the main components of the hydrologic cycle in a mixed oakwood at Virelles-Blaimont, Belgium for the period 1964–68. The wood grows on a rendzina soil on a plateau without streams at an altitude of 245 m; all the water reaching the wood is of atmospheric origin. The tree canopy is unbroken and extends from 6 to 21 metres. Shrubs are sparse but the herb layer, in which *Mercurialis perennis*, *Lamiastrum galeobdolon* and *Hedera helix* are prominent, is well developed and varied.

Of the total precipitation in the leafless phase 89.5% infiltrated into the soil during the five-year period, while evapo-transpiration from the soil and the foliage of trees, shrubs and smaller plants reached only 22.5%. The corresponding figures in the leafy phase were 77.7% and 81.4%; as a result of the high losses by evapo-transpiration the water content of the fine earth was low in summer. (Redrawn from Schnock and from Froment *et al.* in Duvignead. © UNESCO 1971. Reproduced by kind permission of UNESCO.)

relations; a very high water table severely limits the soil volume which can be exploited by most trees. Even minor alterations in the depth of the water table may affect trees adversely, though roots of lodgepole pine (*Pinus contorta*) can adapt to very damp conditions by forming air cavities in the stele. Root death in waterlogged soils is sometimes caused by the presence of ferrous (iron II) ions, rather than by lack of oxygen (section 4.5). The **field capacity** of a soil, the amount of water which it contains (expressed as a percentage of the oven-dry weight of the soil) after it has been saturated by heavy rain and allowed to drain freely, is a most important character. Soils rich in clay and humus, colloids possessing a high affinity for water, have a high field capacity while that of soils containing a large proportion of the bigger mineral particles, such as sands and gravels, is low.

The shoot systems of trees and other land plants normally replace the water which they lose to the atmosphere when the roots absorb more from the soil. Thus water uptake, the conduction of water from the roots to the transpiring surfaces, and transpiration itself, are closely linked aspects of the water balance of the plant. **Potential evaporation** is proportional to the vapour-pressure deficit of the air (the saturated vapour pressure minus the actual vapour pressure), and transpiration by the tree generally follows the same trend as evaporation from a physical system until regulated by stomatal closure or until water content becomes severely limiting.

Water in the soil-plant-atmosphere system tends to move along a **water potential gradient**, travelling from where it is more readily available to where it is less so. This movement is now usually described in thermodynamic terms (Larcher, 1975; Meidner and Sheriff, 1976); the state of water in a plant cell or the soil is compared with that of pure water, the difference being expressed in terms of potential energy. **Water potential** is an effective measure of the availability of water; the larger the negative value of ψ the less available is the water. (Water potential, ψ, is defined as the difference between the free energy per unit volume of matrically bound, pressurized or osmotically constrained water and that of pure water. It is measured in bar or erg cm^{-3}: 10^6 erg cm^{-3} = 1 bar.) '**Readily available**' soil water can be easily absorbed by plants and represents the difference between the field capacity of the soil (where ψ soil is around -0.15 bar) and the **permanent wilting percentage** (PWP), which is the soil moisture content (% dry weight of soil) at which plants wilt and fail to recover overnight even when left beneath a bell jar to minimize transpiration. PWP varies with the species of plant, ranging from -6 bar for plants with a high moisture requirement to potentials of -30 bar in plants resistant to drought.

Plants can continue to absorb water from the soil as long as the water potential of their fine roots is more negative than that of the soil solution adjacent to them. The rate at which water is absorbed by roots is directly proportional to the active root area per volume of soil and the difference in water potential between the root and the soil. It is inversely proportional to the resistances to transport of water within the soil and from the soil to the plant. If the soil dries up in one place parts of the root system often desiccate and dry, while in damper regions the roots may grow rapidly and produce an abundance of absorptive rootlets.

Water travels from the roots to the leaves along a water potential gradient which becomes more negative, owing mainly to an increase in solute concentrations from the root cortex to the leaf mesophyll, as the transpiring surfaces are approached. Water changes from the liquid to the gaseous phase as it moves from the mesophyll cells to the substomatal cavities, from which it is lost by diffusion. It thus moves through the plant in response to a vapour pressure gradient. The maximum velocity of the transpiration stream is greatest in plants possessing wide vessels offering little resistance to flow. All estimates available (Daniel *et al.*, 1979) can be considered only as approximations, but commonly quoted velocities are conifers, $1-2$ m h^{-1}; diffuse porous hardwoods, $1-6$ m h^{-1}; and $4-44$ m h^{-1} in ring porous hardwoods in which large and conspicuous vessels are formed in spring. Velocities vary from 10 to 60 m h^{-1} in herbs but reach 150 m h^{-1} in certain lianas (Larcher, 1975).

The **water balance** of a plant reflects the difference between the amount of water which it absorbs and that lost in transpiration. In practice the water content of a plant varies constantly. Water balance becomes negative when transpiration exceeds absorption, often being restored, following a brief period when it is positive, after the reduction of transpiration accompanied by the continued absorption of water by the roots.

Alterations in the water balance are often observed by determining the **water saturation deficit** of the leaves. This is the percentage of water lacking from a tissue as compared with that present when it is 'fully saturated', i.e. turgid. As the water balance of plant leaves becomes more negative the osmotic pressure rises. Some species can withstand very considerable increases in osmotic pressure, whereas others are physiologically disturbed by much smaller increases. The ranges of osmotic pressures found in different groups of plants can be arranged in the form of an **osmotic spectrum**. Osmotic pressures of herbs in moist woodlands are lower and cover a much smaller range than those of dry woodlands. Similarly, maximum values for deciduous trees and shrubs are lower than those of conifers, and of woody plants growing in Mediterranean climates.

The **isohydric water balance mechanism**, in which stomatal regulation of transpiration prevents pronounced fluctuations in water content and osmotic pressure during the day, can be contrasted with the **anisohydric** type in which transpiration is not restricted until the water balance is strongly negative. The protoplasm of anisohydric plants, many of which grow in sunny habitats, can tolerate extensive fluctuations in osmotic pressure, and hence in water potential, over short periods. Isohydric plants have stomata which respond rapidly to lack of water and their water balance during the day usually stays close to zero; they normally possess large and efficient root systems. Many, but not all, of these species are physiologically disturbed by relatively small increases in osmotic pressure.

Trees and shade herbs are usually isohydric, but the degree to which different species can reduce transpiration from the leaves by closing the stomata varies considerably. **Cuticular transpiration** of leaves, which continues after the stomata are closed, can be expressed as a percentage of total transpiration when they are open. This has been estimated (Larcher, 1975) as 12% in *Betula pendula*, 21% in *Fagus sylvatica*, 12.5% in the herbaceous sciophyte *Oxalis acetosella*, and only 3% in *Picea abies* and 2.5% in *Pinus sylvestris*. Such values must vary with the seasons and within species, but the very low figures for the two conifers are noteworthy.

The total areas of the transpiring surfaces of full-grown trees are very large and distant from the roots. Even when there is abundant water in the soil, the rate at which water can be conducted up the trunk is insufficient to replace losses from the canopy in the absence of stomatal regulation at midday during clear sunny weather in summer. Even with such regulation, the tensions resulting from rapid transpiration may be sufficient to cause cavitation in the xylem, disrupting water movement between the roots and the leaves; water will usually continue to flow where cavitation has not occurred but the situation at the top of very tall trees seems likely to become critical during drought. In most trees the stomata are very sensitive to water deficit, reacting rapidly and preventing serious water loss; their closure causes the noon depression of transpiration characteristic of many trees on clear sunny days when the temperature, and consequently the vapour-pressure deficit, of the air are highest. **Water conservation** in various parts of the canopy follows an orderly sequence (Figure 2.14), occurring first in the shady parts of the crown. The leaves in sunny regions near the base of the canopy are the next to limit water loss, while those of the sunny top are the last. Water potential becomes negative most rapidly at the top of the tree, so the leaves here are preferentially replenished with water. Daily variations in osmotic pressure in trees rarely exceed 3 bar so the protoplasm is not subject to great changes in water potential, though the forces drawing water upward in the early afternoon are much greater than in the morning and evening. Conifers, sciophytes and a number of

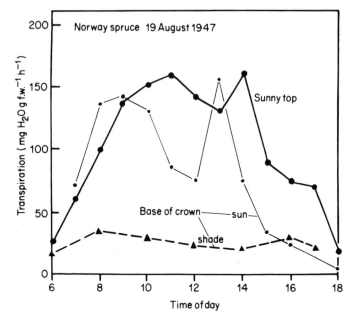

Figure 2.14 Fluctuations in the transpiration of shoots of *Picea abies* at 820 m near Innsbruck, Austria, during a sunny August day preceded by dry weather. (Redrawn after Pisek and Tranquillini, 1951; *Physiologia Plantarum*).

heliophytic trees such as the oaks can control their water contents to a considerable extent, but others, including some species of ash, cannot and their leaves often wither in dry weather.

When the soil becomes depleted of water for considerable periods in summer the stomata of trees may be closed almost continuously. During summer drought in regions of periodic dryness this results in the leaves of well-adapted trees and shrubs losing much less water per unit area than in the rainy season; a figure of 11% was estimated for *Quercus ilex* in the Mediterranean maquis. Many species of dry habitats even shed a considerable proportion of their foliage.

Trees near alpine or polar timberlines often suffer severe damage when their water relations are disturbed by cold. Water in the conducting elements often freezes at –2°C, while that in the surface soil remains frozen for months (Larcher, 1975). The most common damage occurring in isolated trees above the upper forest limit is caused by the slow desiccation of shoots projecting above the snow (Tranquillini, 1979), which lose water, often by direct evaporation, throughout the winter. This is a major influence in controlling growth near the timberline (section 4.4). In the European Alps the water content of *Rhododendron ferrugineum*, and of smaller individuals of Arolla pine (*Pinus cembra*), sometimes falls below the safety limit in early spring, when the sun warms the branches and transpiration increases before the ground has thawed. Larger trees are not so greatly endangered; their trunks contain sufficient water to tide them over until active absorption of water from the soil is resumed.

Climatic dieback of lodgepole pine (*Pinus contorta*) occurs in the Rocky Mountains of the USA, where periods of relatively warm weather alternating with severe cold result in reddening of needles. Lodgepole pine planted at altitudes of 300 m or more in Scotland suffered similarly in 1979, again apparently as a result of excessive transpiration during a period when frozen ground prevented a compensating uptake of water. In Scotland, trees of coastal provenances (Figure 1.5) suffered more severely than those of inland origin, another example of the need to secure a satisfactory match between genotype and climate when planting exotic trees (see also Figure 4.12). Needle reddening and dieback in *Pinus contorta* has since been shown to result from disease (section 6.3); climate may act directly or by predisposing infection.

Until recently prolonged drought has been relatively uncommon in Britain but many trees, particularly of birch and beech, died after the very dry summer of 1976. Though some were post-mature individuals whose end was merely hastened by drought, many young saplings of a range of species were also lost. Trees on shallow soil in sites with sharp drainage were particularly susceptible; in the Wyre Forest, England, there were considerable losses of western hemlock (*Tsuga heterophylla*) on a sharply sloping site overlooking the River Severn, whereas trees on the plateau above survived. Water relations interrelate with other factors; for example, insect attack can increase water loss, while trees stressed by lopping are more susceptible to infestation by wood wasps (section 6.6). Alterations in the forest canopy influence air and soil temperatures, soil water content, and shading of the forest floor simultaneously. Forests are also important in buffering the effects of heavy precipitation, acting effectively as 'sponges' which gradually release water into rivers.

2.6 Evergreen and deciduous strategies

Though the photosynthetic capacity of their leaves is generally low, evergreen conifers frequently have high net production. This has often been attributed to the prolonged growth period; photosynthetic gains can be made in spring, autumn and parts of the winter when deciduous trees are bare of leaves. On the other hand, the turnover rate of leaves is also important; if these have to be replaced frequently the amount of photosynthate which can be partitioned into the formation of wood and bark is greatly reduced. Whereas cold deciduous trees such as beech replace their major photosynthetic systems annually, evergreen conifers retain most of their leaves which continue to contribute to net production for many years, though their photosynthetic capacity gradually falls, as Clark (1961) showed for *Picea glauca* and *Abies balsamea*. Photosynthetic capacity and respiratory activity both tend, however, to be markedly higher in deciduous trees. When young leaves are forming they respire intensely, have a small surface area, and are usually low in chlorophyll. Consequently they are, for a short time, a cause of overall carbon loss, though their photosynthetic capacity is at its peak within a few days of full expansion.

The International Biological Programme investigations into the growth of *Fagus sylvatica* (Schulze, 1970) and *Picea abies* (Schulze *et al.*, 1977a,b) found within a kilometre of each other at 500 m on the Solling Plateau, Germany, afford an opportunity of comparing the production values of the conifer most commonly planted in German reafforestation and the deciduous tree which dominates many natural forests in central Europe (Figure 2.15). For the complex measurements of photosynthetic capacity a typical tree of each species, a 27 m high 100-year-old beech and a 25.6 m high 89-year-old spruce, was examined in 1968 and 1972, respectively.

The deciduous beech and the evergreen spruce were found to differ in four major respects with regard to production. Both the sun and the shade leaves of *Fagus* had a much higher photosynthetic capacity per unit dry weight than even the one-year-old needles of *Picea*. Beech had a shorter growing season than spruce; it showed positive CO_2 uptake during 176 days in the year, against 260. Beech had a higher annual production of leaves than spruce, but the latter had a much greater photosynthesizing biomass, because of the long life of its needles, some of which survived for as long as twelve years. In the last twenty years, however, increased levels of atmospheric pollution have reduced the useful life of conifer needles in many parts of the world (section 11.2), thus influencing competitive ability.

In the Solling it was the increased longevity of the foliage, rather than the longer growing season, which enabled the primary production of spruce (14.9 t C ha^{-1} y^{-1}) to be so much greater than that of beech (8.6 t C ha^{-1} y^{-1}). Though the dry matter which spruce 'invested' in its leaves every year was less than that of beech, the long-term return was greater because the leaves continued to fix carbon so much longer, albeit at a slower rate.

This analysis helps to elucidate the basis of the high productivity of evergreen conifers such as *Picea abies* in central Europe, where *Fagus sylvatica* is frequently the dominant of many natural forests despite its much lower productivity. Beech is more shade tolerant, being able to germinate and grow where spruce cannot. It is also less affected by storm and snow and ice breakage; the deciduous habit entails a reduc-

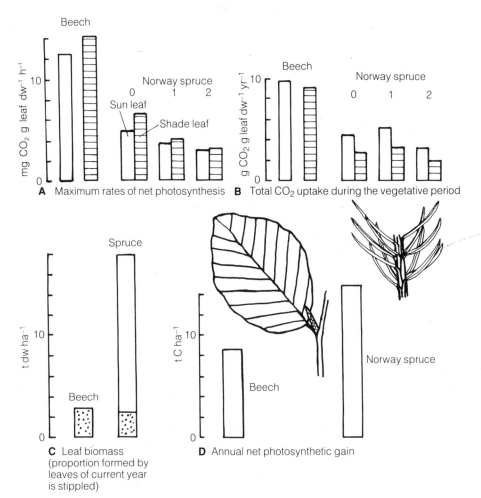

Figure 2.15 A comparison of photosynthetic activities of *Fagus sylvatica* and *Picea abies* growing on the Solling Plateau, Germany. In A and B columns for sun leaves are shown white, those for shade leaves are hatched. 0, 1 and 2 correspond, respectively, to spruce leaves developed in the current year, and 1 and 2 years previously. Carbon fixation per unit leaf dry weight is much greater in beech which, however, has a much smaller leaf biomass and lower annual photosynthetic gain than spruce. Drawn from the data of Schulze *et al.* (1977b, Table 1).

tion of the surfaces on which ice can accumulate in winter. The surface rooting of *Picea* means that gales may topple it relatively easily and often its life is shorter than that of beech, a powerful competitor with other trees, which when growing actively suffers markedly lower mortality from fungal diseases than spruce. Thus, relative growth rate, though very important, is only one aspect of competitive advantage whose balance is often swayed by climatic factors.

Picea abies flourishes in the north of Scandinavia and of Russia, where *Fagus sylvatica* does not ripen seed and is unknown as a natural forest tree. The deciduous habit and the production of strongly constructed, even xeromorphic, needles are both quite well adapted to the water stresses of northern winters, as the presence of *Populus tremula* and *Picea abies* in arctic Norway indicates. Rates of photosynthesis are still appreciable at low light intensities in stems of deciduous trees such as *Quercus petraea*, so stem photosynthesis must play some part in maintaining carbohydrate levels during the long leafless period. However, most boreal forests are dominated by evergreen conifers which clearly gain from the ability of their needles to photosynthesize for several years. In the very different climate of tropical rainforests, evergreen trees whose leaves have a low rate of CO_2 uptake can attain a high annual CO_2 gain because the active vegetative period is so long.

Competition between gymnosperms and angiosperms

Bond (1989) discusses the ecological importance of the different types of organization possessed by gymnosperms (such as *Picea*) and angiosperms (such as *Fagus*), pointing out that by accumulating several cohorts of leaves gymnosperm trees often achieve higher productivity than angiosperm trees. On the other hand, solute flow in gymnosperms is less efficient than in angiosperms, and their seedlings may be particularly uncompetitive since their initial growth depends on a single cohort of relatively stereotyped and ineffective leaves. The trend to angiosperm dominance of favourable habitats (warm, high light, high nutrients) is accentuated by their greater plasticity and initial rapid growth, frequently accompanied by use of insect pollination and seed dispersal by animals.

3
Reproductive strategies of woodland plants

3.1 Plant strategies

In recent years several approaches have been developed with respect to ecological 'strategies' displayed by living organisms. These strategies – involving such activities as resource capture, growth and reproduction – are not adopted by design, but have evolved as a result of the inexorable pressures of natural selection acting upon populations showing natural variation. Such pressures repeatedly result in distinct types of specialization associated with particular habitats or niches. The **r–K continuum** of MacArthur and Wilson (1967; section 7.1) is a two-strategy model which can be applied to both animals and plants. The **C–S–R model** (Grime, 1974) brings in a further strategy, the exploitation of continuously unproductive environments or niches by stress-tolerators, not recognized by two-strategy models. C–S–R theory also takes account of the separation of the established (adult) and regenerative (juvenile) strategies that is found in plants. It has been more thoroughly applied to a large and important set of plant species than any other strategic theory (Grime *et al.*, 1988), and is potentially a major tool for ecological prediction and the manipulation of vegetation.

Plant strategies in response to competition, disturbance and stress

Two main categories of external factors may be recognized as limiting the amount of living and dead plant material present in a habitat: stress and disturbance. '**Stress**', as it applies to plants, can be defined as the external constraints which limit the rate of dry matter production of all or part of the vegetation, and takes many diverse forms including shortages of light energy, water and mineral nutrients, or the influence of suboptimal temperatures. '**Disturbance**' may be regarded as consisting of the mechanisms which limit the plant biomass by causing its partial

or total destruction. It takes such forms as trampling, mowing, ploughing, the felling of trees, and the activities of pathogens and other herbivores; Grime (1979) also regards it as resulting from phenomena such
as wind damage, frosting, droughting, soil erosion and fire.

No plants can long survive both high stress and high disturbance.
There are, however, three combinations of stress and disturbance under
which plants can establish and continue to exist, and these have led to
the evolution of three primary ecological strategies. The plants possessing these extremes of evolutionary specialization are the '**competitors**',
which exploit conditions of low stress and low disturbance as do most
trees, the '**stress-tolerators**' (high stress–low disturbance) such as *Viola
riviniana*, and the '**ruderals**' (low stress–high disturbance), e.g. *Poa annua*
and *Funaria hygrometrica*.

Many attempts have been made to define competition; often authors
have used the word as a synonym for the 'struggle for existence' which
is part of the essential mechanism in Darwin's (1859) theory of evolution. In the definitions coined by Grime (1979) **competition** is 'the
tendency of neighbouring plants to utilize the same quantum of light,
ion of a mineral nutrient, molecule of water, or volume of space'.
Competition can occur both below and above ground, the latter being
especially important when the leaf canopy begins to close and the shoots
of one plant are substantially shaded by those of another. **Competitive
ability** is 'a function of the area, the activity, and the distribution in
space and time of the plant surfaces through which resources are absorbed
and as such it depends upon a combination of plant characteristics'.
While the presence or absence of any one of these characteristics is not
in itself diagnostic of high or low competitiveness, very many plants
possessing all the following four features are extremely competitive:

1. tall stature;
2. a growth form – such as a densely branched rhizome, as in stinging
   nettle (*Urtica dioica*) and creeping soft-grass (*Holcus mollis*), or an
   expanded tussock structure – which allows extensive and intensive
   exploitation of the environment above and below ground;
3. a high maximum potential relative growth rate (RGR); and
4. a tendency to deposit a dense layer of litter on the ground surface.

This three-strategy theory is best represented by means of a triangular
diagram whose points form maxima of competition, stress and disturbance, where the three **primary strategies** are situated (Figure 3.1),
and in which the position of a species is determined in relation to the
indices of competition (I_c), stress (I_s) and disturbance (I_d). Four main types
of **secondary strategy** are shown in intermediate positions. The three
small triangles beneath the main diagram show the strategic range of

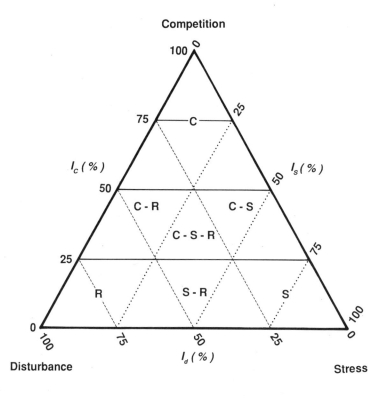

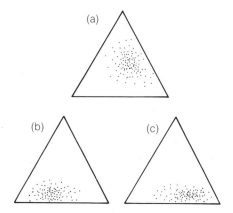

Figure 3.1 Model describing the various equilibria between competition, stress and disturbance in vegetation and the location of primary and secondary strategies. C, competitor; S, stress-tolerator; R, ruderal; C–R, competitive–ruderal; S–R, stress-tolerant ruderal; C–S, stress-tolerant competitor; C–S–R, 'C–S–R strategist'. I_c, relative importance of competition (——); I_s, relative importance of stress (. . . . .); I_d, relative importance of disturbance (-----). (Redrawn from Grime *et al.*, 1988; by permission). The strategic range of three life forms is also shown (a) trees and shrubs (b) bryophytes and (c) lichens.

trees and shrubs (a), none of which are ruderals, of bryophytes (b), which respond effectively to disturbance, and of lichens (c) which tend to be stress-tolerators.

Figure 3.2 is an ordination diagram illustrating the strategies employed by some common species of the woodland field layer. Of these only *Stellaria media*, a species occasionally encountered at disturbed woodland margins, is a ruderal. The relative stability of woodlands militates against the success of this strategy, though when herbs from all habitats are considered (inset) it is seen to be widely used elsewhere.

Grime *et al.* (1988) derive the C–S–R classification of a particular species by the use of a dichotomous key employing various morphological, behavioural and reproductive characteristics. The position of a species such as *Oxalis acetosella* (S to C–S–R) on the second triangular ordination (Figure 3.2) is, however, determined by the vegetational composition of the sample quadrats in which it occurred. As most quadrats contain species with a number of different strategies, agreement between the two methods of determination is broad rather than exact.

Regenerative strategies

All five of the regenerative strategies widespread in terrestrial vegetation occur in woodland species. **Vegetative expansion** (V) is common in habitats where disturbance is at a low level. **Seasonal regeneration** (S), where seeds or vegetative propagules are produced in a single co-hort, is common in habitats subjected to seasonally predictable disturbance by climatic or biotic factors. Species with **persistent seed or spore banks** (B_s) enjoy a selective advantage in places where the timing of disturbance is unpredictable. Numerous **widely dispersed seeds or spores** (W) are commonly found in species of habitats which are relatively inaccessible or subject to spatially unpredictable disturbance. Such species are often opportunists, indeed both fireweed (*Chamaenerion angustifolium*) and the moss *Funaria hygrometrica*, which frequently develop on woodland bonfire sites, are regarded as r-species (section 7.1) largely on this account. The last strategy, that of **persistent juveniles** (B_j), involves seedlings or sporelings derived from independent propagules persisting for long periods in a juvenile state in unproductive habitats subjected to low levels of disturbance. Dwarf trees of *Picea abies* (Figure 3.3) found in primeval Scandinavian forests exemplify this strategy (section 5.1).

Different regenerative strategies vary in degree of resource investment, mobility and dormancy. Moreover in many plants, and some animals,

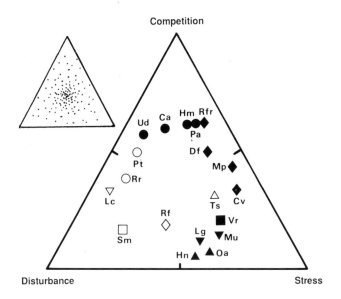

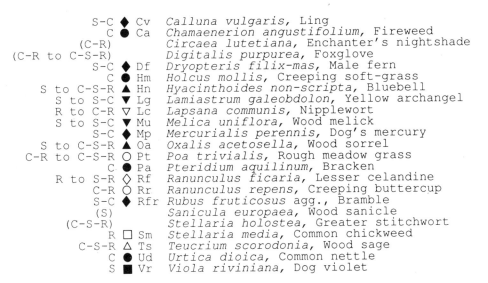

S-C	◆ Cv	*Calluna vulgaris*, Ling
C	● Ca	*Chamaenerion angustifolium*, Fireweed
(C-R)		*Circaea lutetiana*, Enchanter's nightshade
(C-R to C-S-R)		*Digitalis purpurea*, Foxglove
S-C	◆ Df	*Dryopteris filix-mas*, Male fern
C	● Hm	*Holcus mollis*, Creeping soft-grass
S to C-S-R	▲ Hn	*Hyacinthoides non-scripta*, Bluebell
S to S-C	▼ Lg	*Lamiastrum galeobdolon*, Yellow archangel
R to C-R	▽ Lc	*Lapsana communis*, Nipplewort
S to S-C	▼ Mu	*Melica uniflora*, Wood melick
S-C	◆ Mp	*Mercurialis perennis*, Dog's mercury
S to C-S-R	▲ Oa	*Oxalis acetosella*, Wood sorrel
C-R to C-S-R	○ Pt	*Poa trivialis*, Rough meadow grass
C	● Pa	*Pteridium aquilinum*, Bracken
R to S-R	◇ Rf	*Ranunculus ficaria*, Lesser celandine
C-R	○ Rr	*Ranunculus repens*, Creeping buttercup
S-C	◆ Rfr	*Rubus fruticosus* agg., Bramble
(S)		*Sanicula europaea*, Wood sanicle
(C-S-R)		*Stellaria holostea*, Greater stitchwort
R	□ Sm	*Stellaria media*, Common chickweed
C-S-R	△ Ts	*Teucrium scorodonia*, Wood sage
C	● Ud	*Urtica dioica*, Common nettle
S	■ Vr	*Viola riviniana*, Dog violet

Figure 3.2 C–S–R ordination diagram of woodland field layer species. This describes the various equilibria between competition, stress and disturbance in vegetation. The points shown for each species indicate the position of its maximum percentage occurrence in a matrix of vegetation types classified according to the strategies of the component species in Grime *et al.* (1988). Where points are not provided, the data for the species concerned are insufficiently clear. The small diagram shows the strategic range of herbs. (From Packham and Cohn, 1990; courtesy *Arboricultural Journal*.)

Figure 3.3 Dwarf tree of *Picea abies* from Fiby urskog, Sweden. Tree rings at ground level showed that this individual had an age of at least 43 years, despite a height considerably less than the 1.3 m at which age cores are normally taken. Note the weak leading shoot and the prominent lateral branches. (Photograph by Roland Moberg.)

the same genotypes may be able to regenerate in several different ways. *Galium aparine* has a single regenerative strategy (S), invariably developing from seed which is only briefly persistent in the soil. *Oxalis acetosella*, a plant better adapted to shade than glade margins, certainly uses two strategies and may use three (V, S, ?B$_s$), though most investigations have not discovered persistent seed banks in this species. It forms clonal patches whose rhizome extensions ultimately become isolated, regenerates from seeds which germinate in spring, and has been reported as forming a persistent seed bank. *Chamaenerion* (= *Epilobium*) *angustifolium* (W, V) develops adventitious shoots from its roots, forming large clonal patches. Numerous wind-dispersed seeds formed under unshaded conditions germinate mainly in autumn. Ungerminated seeds surviving until the following spring are not incorporated into a persistent seed bank. There is thus much to support the hypothesis (Grime, 1979) that **ecological amplitude** is determined by regenerative flexibility, as well as genetic variability and phenotypic plasticity.

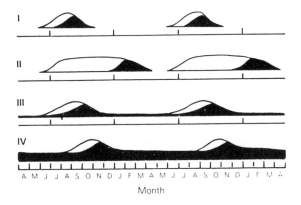

Figure 3.4   Four types of seed bank described by Thompson and Grime (1979). Shaded areas: seeds capable of germinating immediately after removal to suitable laboratory conditions. Unshaded areas: seeds viable but not capable of immediate germination. Type I, annual and perennial grasses of dry or disturbed habitats. Type II, annual and perennial herbs colonizing vegetation gaps in early spring. Type III, species mainly germinating in the autumn but maintaining a small persistent seed bank. Type IV, annual and perennial herbs and shrubs with large persistent seed banks. From Grime *et al.* (1988), courtesy of the authors.

Seed banks

Differences in germination patterns, dormancy and longevity of buried seeds (Salisbury, 1961), together with the development of seed banks have long been recognized as important features of the reproduction of vascular plants. Figure 3.4 shows the four types of seed bank detected by sampling the surface soil (0–3 cm depth) of 10 ecologically contrasted sites in the Sheffield region. The techniques adopted were designed to detect both the transient accumulation of germinable seeds during the summer and persistent seed banks; sampling occurred at 6 weekly intervals. Some species were in several types of habitat; the results suggested that seasonal variation in seed number is related to the species concerned rather than to the environment involved. The type IV seed bank is the most important in woodlands, where disturbance is likely to occur at long and very irregular intervals.

3.2   Reproduction and fruiting

In mature woodland most flowering plants are perennial, many being long-lived forms able to persist and propagate themselves vegetatively under conditions unfavourable to reproduction by seed. The formation of propagules (spores or seeds) which can be distributed by wind, water, animals, or explosive mechanisms, facilitates the invasion of habitats at appreciable distances from the parent plants. Three main **patterns of seed production** have been recognized (Grubb, 1977) in forest communities: (a) moderate production in most years; (b) fruiting rather irregular; and (c) abundant fruiting strongly periodic. Different species in group (b) will be favoured in different years. Three corresponding patterns of flowering seem also to occur, though unfavourable weather conditions may prevent a heavy seed set resulting from extensive flower production. High winds or rain may prevent pollination in woodland species; periods of drought and high temperature must also affect seed set. These variations are important because the distribution of viable seeds, in both space and time, greatly influences the developmental patterns of woodlands.

The effective sexual reproduction of tracheophytes in forests often demands very special conditions if young plants are to become established. The habitat must be open enough for seeds to be able to develop into mature plants without succumbing to competition, and sufficient seed must be produced to allow for loss resulting from plant pathogens and herbivorous animals, a situation which leads to the concept of **predator satiation** by masting (section 3.4). These constraints mean that in natural woodlands decades, or even centuries, may elapse between periods when substantial numbers of tree seedlings develop to maturity. **Mast years**, in which seed production is exceptionally high, are typically several years apart. Seedlings usually develop in large numbers after these, but are then often suppressed by competition for light, water and nutrients for many years and most of them die. Only if storm damage, felling or the death of a tree causes a gap in the forest canopy will some be able to mature.

Pollination mechanisms

Pollination of forest trees, notably of conifers, in temperate regions is mainly by wind (**anemophilous**), but towards the tropics many more species are pollinated by animals (section 3.4). **Entomophilous** (insect pollinated) flowers, whose features make them attractive to insects, often have bright colours, nectar and scent. The mechanisms may be very specific, as in a number of orchids in which pollination occurs after pseudo-copulation by male wasps, or the odour emitted by the cuckoo-pint (*Arum maculatum*) which attracts large numbers of small *Psychoda* flies. The pollen of insect-pollinated flowers is usually sticky, while that of anemophilous flowers is dry and produced in the enormous quantities necessary to the success of such a random method of distribution. Wind-pollinated flowers are usually well above the ground, the stamens are exserted and often with versatile anthers so that pollen is effectively dispersed. Male flowers are often in catkins as in alder, birch and hazel. Flowers or plants of **separate sexes**, strongly marked **dichogamy** in which the stamens and stigmas mature at different times, and **self-sterility** are common in **wind-pollinated plants**, helping to ensure effective gene-flow and the production of new genotypes which, together with phenotypic plasticity, are important if the population is to remain capable of making major adaptations to changing conditions.

Although many species show quite extreme adaptation to pollination by wind, animals or water, there is a balance between anemophily and entomophily in trees such as lime (*Tilia*), willows (*Salix*) and sweet chestnut (*Castanea sativa*), all of which are visited by numerous insects, but possess flowers showing some indications of the 'syndrome of anemophily'. Marked disparity in the effectiveness of two pollinating agencies in a particular species will influence the subsequent evolution of the floral mechanism. Most members of the olive family (*Oleaceae*) are entomophilous; indeed the Mediterranean 'manna ash' (*Fraxinus ornus*) possesses a white corolla and its fragrant flowers are pollinated by insects (Proctor and Yeo, 1973). Common ash, however, is anemophilous. Its flowers are simple and often bisexual (Figure 3.5), having two stamens and a long ovary (which later develops into a flat one-seeded 'ash key'), lack a corolla, and are borne in dark greenish masses on the naked twigs in March and April. Among the maples introduced to Britain, both sycamore (*Acer pseudoplatanus*) and Norway maple (*A. platanoides*) are pollinated mainly by bees. Temperate species of the Fagaceae such as *Fagus sylvatica*, *Quercus petraea* and *Q. robur* are predominantly wind-pollinated, while a number of subtropical oaks are entomophilous.

Similar examples of families possessing both wind- and insect-pollinated species can be found amongst the herbs: *Mercurialis perennis* (Euphorbiaceae) is an anemophilous member of a mainly insect-pollinated family. It is interesting that while the combination of features associated with anemophily has appeared independently in flowers with very varied basic structures, specialization to particular agents of pollination has often occurred within the evolutionarily recent past.

Most wind-pollinated deciduous forest trees flower while bare of leaves, when the numbers of active insects are low. The flowers are freely exposed to disperse and receive pollen (Proctor and Yeo, 1973) and the surrounding surfaces available to 'compete' with the stigmas for pollen are minimal. In the temperate zone most of the species which are dominant over large areas are wind-pollinated. Anemophilous plants tend to

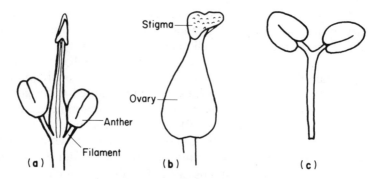

Figure 3.5 Common ash (*Fraxinus excelsior*). (a) Single hermaphrodite flower, (b) female flower, and (c) male flower. Some trees of this protandrous wind-pollinated species function exclusively as males, some as hermaphrodites and yet others as females. Stem diameter, height, and volume of female ash were found by Rohmeder (1967) to be considerably less than those of male trees, possibly because of the allocation of resources to seed production.

produce small numbers of ovules (grasses, sedges, *Corylus* and *Quercus* produce only one seed per flower), a distinct advantage in closed communities where the possession of large seeds confers higher competitive power.

Germination requirements

Dormancy and germination in seeds and bud dormancy and development are largely controlled by balances of the same growth promoters and growth inhibitors (section 3.5). Thiourea can be substituted for the natural germination stimulator which develops in *Fraxinus* seeds during chilling; it can also be used in place of cold temperature treatment in *Quercus*, *Larix*, *Picea* and other trees. The seeds of woodland species vary greatly in their germination requirements. The germination of fully imbibed seeds of ivy (*Hedera helix*) is inhibited by light, which acts as a dormancy-breaking agent in beech. Germination in *Anemone nemorosa* and *Theobroma cacao* is light-indifferent. Some seeds are affected not by short illumination, but by alternations between dark and light (photoperiod). Photoperiodic requirements may be altered by changes in other environmental conditions. *Betula* seeds at 15°C germinate only under long day (LD) conditions, requiring eight LD cycles to do so. If the temperature is raised to 20–25°C, a single exposure to light of 8–12 hours is sufficient to cause germination.

The seeds of *Quercus* (Figure 3.6) and *Viburnum* germinate in the

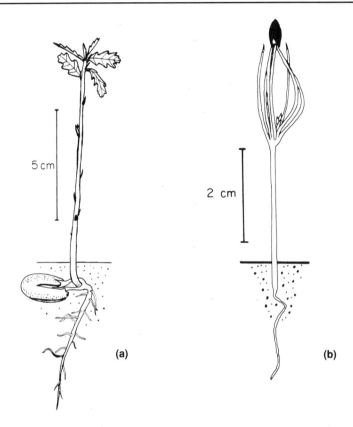

Figure 3.6 Tree seedlings of an angiosperm and a conifer. (a) English oak (*Quercus robur*). Germination is hypogeal and the cotyledons are largely concealed within the remains of the acorn. (b) Scots pine (*Pinus sylvestris*). The hypocotyl has elongated and the tips of several cotyledons are still trapped within the testa. Three single needles borne on the developing plumule can just be seen.

autumn at which time the radicle elongates. Both have been said to show 'epicotyl dormancy', as the epicotyl does not develop until after the seed is chilled during the ensuing winter. However, the epicotyls of ripe acorns planted in greenhouses in autumn germinate readily, so it is apparent that there is no absolute chilling requirement. Like many seeds, those of the herbs *Convallaria* and *Polygonatum* require a winter's chilling if the radicle is to grow; they are exceptional in that the epicotyl needs a further cold winter before it develops.

The seed is normally the first stage at which new genotypes are exposed to the processes of selection. Seeds also enable plants to perennate, multiply, and disperse, but they may serve a useful function by remaining dormant. *Digitalis purpurea* seeds can remain viable in brown earth soils for many years; they have a light requirement and so do not germinate while buried. When conifer plantations are felled the open conditions and freshly bared soil allow large populations of foxgloves to develop from seeds already in the soil. Populations of *Chamaenerion* (= *Epilobium*) *angustifolium* also build up quickly, but the seeds of this species are blown in by wind so the survival strategies of these two species, both of which have quite light seeds and flourish in well-lit sites, are different.

Seed size and distribution mechanisms

Species unadapted to vegetative existence in climax vegetation, such as mature woodland, are doomed to local extinction and are in this sense fugitive. The more frequently such a species undergoes the cycle *invasion-colonization-suppression-extinction*, the greater is the importance of a high intrinsic rate of natural increase and of dispersal as a means of escape to a more open habitat. Large seed number, small seed size, high powers of dispersal and a high proportion of net production partitioned to propagule production – all characteristics of r-species – can be expected in such plants.

There is a marked tendency for propagule weights to increase as the habitat becomes increasingly closed. Seed weight varies among species over a range of 10 orders of magnitude (Harper *et al.*, 1970). At one extreme are the dust seeds of the orchids, for example *Goodyera repens* (0.000002 g), while at the other are the huge 'seeds' of the double coconut palm (*Lodoicea maldivica*) which weigh 14 000–23 000 g. Angiosperms with really large seeds are restricted to the tropics or subtropics. The smallest known seeds are of plants with anomalous nutrition. Orchid seeds often lack any nutrient tissue and in nature the developing embryos are associated with mycorrhizal fungi. Total parasites frequently depend on host plants for early growth. Broomrapes (*Orobanche* spp.), of which *O. hederae* attacks ivy, are obligate parasites whose seeds require stimulus by a host root exudate if germination is to occur.

Seed dispersal mechanisms can be ecologically classified according to habitat: the mechanisms associated with the various forest strata frequently differ. In north temperate woodlands, for example, herbs very commonly produce hairy or hooked animal-dispersed propagules (*Arctium*, *Circaea*, *Galium*, *Geum* and *Mercurialis*), or even adopt a two-stage ballistic and ant-mediated dispersal (*Viola* and *Dicentra*), while shrub layer species frequently bear fleshy fruits (*Crataegus*, *Cornus*, *Hedera*, *Prunus*, *Rhamnus* and *Sambucus*). Similarly *Taxus* seeds, which are poisonous to mammals, are surrounded by brightly coloured, sweet-tasting, fleshy arils that are very attractive to birds, a fact which may explain, at least partially, the wide geographical range of this genus of nine species which occurs in North America, Europe and Asia, extending into Malaysia. Pioneer species of the tree layer commonly possess wind-dispersed propagules (*Betula*, *Fraxinus* and *Pinus*), while later successional and climax woodland species often bear heavy seeds, dependent on the specialized collecting habits of mammals or birds, as in *Fagus* and *Quercus*. However, there are exceptions: *Sequoia* and *Sequoiadendron* form climax forests of great antiquity yet have small winged seeds which grow well on seed beds left by fire, to which the mature trees are resistant. Certain species with heavy seeds, such as *Quercus robur*, do well as pioneer trees if their seeds reach relatively open habitats, as can occur when acorns roll downhill or are transported by squirrels, pigeons or jays.

Mellanby (1968) points out that birds in particular carry many acorns away from seed-bearing trees and drop them in the open where up to 5000 seedlings ha^{-1} may be found, far more than occur beneath oak woods. Though many of the vast surplus of acorns removed by animals are consumed, the germination of those which they bury in the ground and subsequently neglect may be an important factor in regeneration. In North America grey squirrels (*Sciurus carolinensis*) frequently bite the micropylar ends of acorns before hoarding them, thus preventing subsequent germination (Ratcliffe, 1991). Hoards of undamaged acorns and hazel nuts collected by wood mice (*Apodemus sylvaticus*), yellow-necked mice (*A. flavicollis*) and bank voles (*Clethrionomys glareolus*) may be 20–30 m from the seed source; these animals are also known to transfer acorns from large primary caches to smaller secondary ones.

3.3   Masting

Oak and beech are well known for their irregular periodic synchronous production of seed crops (Nilsson and Wastljung, 1987). It has been suggested that this masting habit is strongly developed amongst those tree species whose seed is subject to predation (Janzen, 1971; Silvertown, 1980). This will include those species, such as ash and pine, whose wind-dispersed seeds are commonly consumed by animals. The habit is less strongly developed in those species whose seed is dispersed by animals as a result of the consumption of fleshy fruits, though even in these there may be variable berry crops, as in holly.

The masting habit is well known in beech, where it may be associated with seed dispersal by birds. Good beech mast years in Europe attract large flocks of birds such as bramblings and great tits (Perrins, 1966; Jenni, 1987). Many seeds will remain uneaten through predator satiation, flocks not visiting every site, birds dropping seed when flying away to avoid other birds in the flock, and distraction caused by empty pericarps (Martin, 1988). Quantitative records of masting for many trees over long periods provide a factual basis for explaining the habit, but there are not many such records and some are difficult to interpret because beech sometimes produces cupules and perfect pericarps without seed within them if pollination has not taken place. However, an extensive series for Lower Franconia (southern Germany) has been collated by Maurer (1964). He singles out 1888 as 'Jahrhundertmast' the mast of the Century, resulting in many German beechwoods of that age: an excellent mast in 1946 has not had a similar afforestation effect as 80% of the nuts were collected for oil production. In his classic paper on the ecology of Swedish beechwoods, Lindquist (1931) gives records for four sites spanning 33 years. They show a clear pattern of masting every two or three years, broken once. He suggests an inherent periodicity which permits flowering every alternate year. In examining the failure of a regular biennial pattern, he studied weather records and noted the importance of high temperatures in June and July of the year preceding masting when flower buds are initiated, together with absence of severe frost in late April and May of the year of masting, when male and female flowers would be killed. Very hot summers result in good masting the following year unless they coincide with a year of good masting. The influence of weather on masting was further investigated by Matthews (1955). He established a significant regression of masting scale on daily mean difference from average July temperature: the higher the temperature is in July, the better the mast of the following year, providing it does not follow a good masting year.

The basic biennial masting pattern was very clear in a survey of English beech begun in 1980 by Hilton and Packham (1986). Each year mast was collected from the ground for a period of seven minutes below a number of beech trees at 11 sites across England. The mast was examined to see if it was full, attacked by mould or insects, or empty. The results are displayed in Figure 3.7: masting success was assessed on a scale of 1–5 on the basis of numbers of nuts collected, 1 indicating fewer than 10 nuts, 5 indicating more than 150 nuts. Observations in the seasons subsequent to 1985 have shown less regularity. Where sites mast well in consecutive years, individual trees which produced seed in successive years masted less heavily. Insect attack by the moth *Cydia fagiglandana* occurred while the nuts were developing in the cupule; its effects were most severe in the south of England and in years of poor masting. Empty pericarps are likely to be due to lack of cross-pollination. Nielsen and de Muckadeli (1954) enclosed branches on 28 beech to allow only self-pollination and recorded only 8% full nuts, compared with 58% full on control trees allowed to cross-pollinate. At the Himley site (Figure

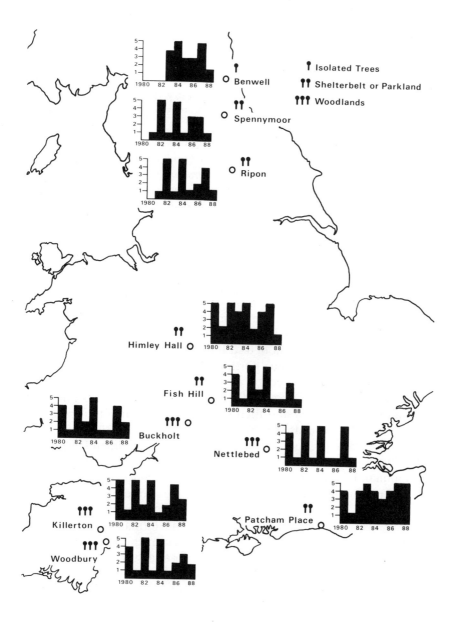

Figure 3.7 Positions of the main beech sampling sites, with masting performance for each group of trees expressed on a five-point scale, using the mean number of nuts collected in a seven-minute sample, for the years 1980 to 1988. 1. <10 nuts collected; 2. 10–50; 3. 51–100; 4. 101–150; 5. >150.

3.7) in 1987, 20 well-spaced parkland trees bore 32% full nuts, while 20 adjacent woodland trees bore 77% full nuts.

American beech (*Fagus grandifolia*) studied at a site in Michigan by Gysel (1971), also showed a basic biennial pattern, broken by failure in pollination leading to large numbers of empty pericarps. This study also revealed irregular insect predation, which was greatest in the poor mast years.

Oak is also a mast-seeding tree although its mast years do not necessarily coincide with those of beech and good mast years are generally more frequent. Moderate crops of *Quercus robur* and *Q. petraea* acorns occur at intervals of three to four years: even in years of general failure, there can be abundant seed in small areas. However, years in which there are uniformly heavy crops over considerable areas are not more frequent than every six or seven years in the south of England (Jones, 1959). The abundant crops produced by isolated trees suggest that cross-pollination is not essential to oak. In many continental districts of Europe 20–25 years may pass without appreciable production of acorns, though a few trees may fruit heavily.

As with many forest trees the amount of fruit produced by *Tilia cordata* varies widely from year to year. The direct influence of climate is most marked at the northern limit of distribution, which is reached in northern England but extends further north in Finland and Sweden and could well be modified by the effects of global warming. At its northern English limit the species flowers in July and August, but produces only small quantities of fertile seed and then only in very warm summers such as that of 1976. This is caused by failure of fertilization in the cool oceanic climate of Britain. In the more continental climate of Finland fertilization normally occurs but the seeds of northern trees fail to complete their development by autumn. Fruiting is prolific and much more frequent in central Europe and commences when the trees are more than 25 years old. In Russia *T. cordata* produces the best quality seed in the upper part of the crown; initiation of inflorescences and formation of fruit occur only on unshaded branches. This is true even in the Białowieża Forest where July and August are normally warm and sunny with mean air temperatures of 19–20°C. In the unbroken forest the production of fertile fruit during 1973 was almost confined to the emergent crowns of old trees, around which quite high densities of seedlings are visible even in August.

Since growth and behaviour vary considerably with environment and there are substantial genetic differences within species, studies of the flowering, fruiting, seed losses and germination of tree species in particular sites are of especial value. Gardner (1977), who made studies of this type on *Fraxinus excelsior* in a Derbyshire ashwood, found that fruit fall is evenly distributed throughout the wood, and continues from September to August. As Figure 3.8 shows, the amount of fruit produced varies greatly from year to year and many of the seeds are destroyed; only a very small percentage of the seed finally germinates. Each ovary contains four ovules of which one usually develops, though some fruits contain no seed; 1.1% of the ash fruits were found to contain two seeds, however, and in one example all four ovules developed.

A tree of common ash may produce 10 kg of winged fruits, containing over 100 000 seeds, every second year, and seedlings of this pioneer species have been recorded 125 m distant from the parent tree. Observations showed a basic alternation of good and bad years, supporting the view that fruiting in common ash is biennial. As *F. excelsior* flowers on the wood of the previous year, a lack of carbohydrate or other nutrient reserve during that previous year could prevent the production of flower initials. A good fruiting year was expected in 1968 but the flowers of some of the trees were destroyed by frost; fruit production was intermediate rather than good and the drain of the carbohydrate reserves was low. The following year differed from the expected pattern in being an exceptionally good fruiting year, presumably because the photosynthate for 1969 was supplemented with reserves saved from 1967 and 1968 by the loss of flowers in spring 1968. Ash fruit which falls in autumn is incapable of germinating immediately. A period of growth is required, during which the length of the embryo almost doubles. Embryo growth in ash ideally requires a temperature of 18–20°C and only after it takes place can prolonged **stratification** (exposure to low temperature), at about 5°C, overcome dormancy. Because of this, ash seeds usually germinate in April or early May two years after the flowering season in which they were formed, although a very small proportion germinate after the first winter. When foresters wish to germinate ash quickly they pick the seeds green in August and sow them immediately.

Masting in conifers is well seen in species of *Pinus*, *Pseudotsuga*, *Picea* and *Abies*. In Sweden the seeding of *Pinus sylvestris* is more irregular in the north than the south, where mast years are hardly discernible, whereas *Picea abies* is a much more erratic cone producer (Harper, 1977).

Many species of tropical Asian bamboos are **monocarpic** (cf. **semelparous**; Table 7.1), flowering and fruiting only once and then dying. The length of time between germination and flowering ranges from 3 to 120 years, according to species, so that synchronized flowering is observed not only in local populations but also in widely transplanted individuals of the same genetic stock.

Different species of trees in temperate woodlands may display masting at the same time, but mast fruiting at the community level is particularly

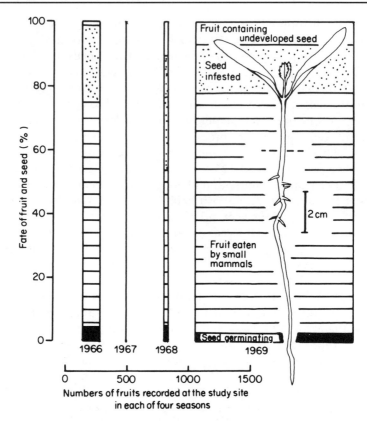

Figure 3.8 A young seedling of common ash (*Fraxinus excelsior*) and diagrams showing fates of ash fruit and seed up to the time of germination (numbers per m²) in Meadow Place Wood, Lathkilldale, Derbyshire, for the crops of 1966 to 1969. Most of the infested seed was spoilt by caterpillars of the moth *Pseudargyrotoza conwagana*. The largest seed loss was due to small mammals, notably wood mice (*Apodemus sylvaticus*) and bank voles (*Clethrionomys glareolus*). A very small crop was produced in the summer of 1967, of which no seedlings survived, and frosting diminished the crop of 1968. Following these two poor years, the crop of 1969 was very large. Seeds usually germinated in April or early May, about 20 months after falling in autumn. The values for seeds germinating are therefore based on the seedlings present in quadrats two years after flowering. (Drawn from the data of Gardner, 1977.)

spectacular in certain evergreen rainforests on nutrient-poor white sand soils in Borneo and Malaysia. Here it is especially the various species of dipterocarp which fruit synchronously over areas of several square kilometres at intervals of 5–13 years (Janzen, 1974).

Janzen (1976) suggests that the length of the inter-mast period is a genetic trait in 'semelparous' bamboos, whereas in masting trees this period seems to be determined by the environment. Many tree species, such as *Fagus sylvatica*, flower abundantly after a hot, sunny summer in which abundant carbohydrate reserves can accumulate. However, adverse weather at the time of fertilization or seed development, or the deprivations of herbivores, can easily upset correlations which have been sought between weather and masting, especially in those conifers which have a long period of cone development. On the other hand, vegetative growth is usually reduced in and immediately after mast years, so that, for example in *Fagus sylvatica*, annual ring increments may be half those of non-masting years.

Asian dipterocarps in sites of low primary productivity require a period of several years to build up sufficient photosynthate to produce a glut of fruit, and only then are they presumed to be sensitive to the environmental cue of several weeks without rain, as a break in an otherwise uniform climate. While Janzen (1971, 1974) stresses that the significance of masting lies in the satiation of seed predators (section 3.4), Harper (1977) suggests that reproductive advantage may also be gained by species typical of the later stages of succession producing large numbers of offspring as potential replacements should an aged parent die leaving a gap in which regeneration could occur.

3.4   Influence of animals

In temperate woodlands various herbs and shrubs, but very few trees, are pollinated by insects (Proctor and Yeo, 1973). Pollination by members of a wide range of animal groups becomes increasingly important towards the tropics, where the frequently wide separation of individual trees of the same species (**conspecifics**) by other species would render wind pollination inefficient. Among insects an example of extreme specificity is shown by the various species of fig-wasp (chalcids of the family Agaonidae), each mutualistically developing within the flowers of a fig; this relationship depends on asynchronous flowering within a fig population (Janzen, 1979). With few exceptions each of the 750 species of *Ficus* is pollinated by its own unique species of chalcid (Compton *et al.*, 1988). Krakatau (section 5.5) now has some 22 species of *Ficus*, of various distributions and ecologies. In order for the population of each species of tree to be capable of breeding locally, the relevant fig-wasp has to have reached the islands from the mainland and found the trees since the 1883 eruption. There must also be a sufficiently large tree population to provide, by asynchrony of fruiting, fruit for the chalcids to live in throughout the year. From existing observations it seems certain that these conditions are not met for all the *Ficus* species present;

regeneration of such species in the islands would be via a re-introduction of seeds by frugivorous bats and birds. *F. fulva* and *F. septica* were, however, observed fruiting on Anak Krakatau for the first time in 1985, the year in which cuckoo-doves (obligatory frugivores) appeared on the island.

Males of the oriental fruitfly (*Dacus dorsalis*) are attracted to their specific sex **pheromones** ('external hormones' concerned with co-ordination between individuals) which are also produced by flowers of *Cassia* trees; similar chemical lures are produced by certain orchids pollinated by neotropical bees, where active searching over long distances enables pollen to be transferred between sparsely distributed plants. Many bees in evergreen tropical rainforests are social, the colonies requiring large amounts of food throughout the year; these bees tend to be pollen and nectar scavengers and robbers, rather than pollinators.

Vertebrates are the primary pollinators of 20% or so of tropical species, their year-round food requirements precluding the one-to-one specificity seen in insects such as fig-wasps. Hummingbirds are the sole pollinators of thousands of New World tropical species, being attracted to odourless but brightly coloured flowers, often red or orange; honeyeaters are a comparable Old World example. Trees pollinated by bats include baobab, kapok and balsa; they tend to have large, dull-coloured flowers which keep open at night, producing copious nectar and a sour smell reminiscent of bats.

Contrasting **pollinator strategies**, observed by Janzen in two groups of Hymenoptera, also seem to occur within hawkmoths and hummingbirds, and can be exemplified by bats (Baker, 1973). 'Opportunistic' species concentrate on plants which display synchronized mass-flowering over short periods. Old World bats of the suborder Megachiroptera may migrate as flocks, following the sequential flowering of different species such as *Parkia clappertoniana*, each inflorescence of which yields up to 5 ml of nectar. 'Trapliners', on the other hand, include certain American Microchiroptera which pay fleeting visits to plants which may open only one or a very few small flowers per day over an extended period. Each flower yields very little nectar or pollen, but presumably sufficient to justify the expenditure of energy on complex feeding routes of 20 km or more.

Fruits and seeds may be sought as food by animals, some of which void or spit out viable seeds (e.g. mistletoe by birds) while others, such as agoutis and squirrels, practise scatter-hoarding, the successful germination of the seeds depending on the animal's forgetfulness. In extreme cases, partial abrasion of the endocarp in a bird's gizzard is a pre-requisite for germination; it has been suggested that the demise of the dodo (*Raphus cucullatus*) has left *Calvaria major* trees on Mauritius without a vital aid to regeneration for the past 300 years.

During its lifetime a tree produces immense numbers of seeds, of which only one is required to replace it. The dangers faced by individual seeds and seedlings are, however, very great, and a successful species requires a reproductive capacity sufficient to allow it to invade new habitats. Three basic strategies are shown by plants which achieve dispersal despite seed predation. Some produce extremely small seeds, others satiate predators with an over-abundance of seeds, while others are particularly rich in secondary compounds. Among Central American woody legumes, one species of *Indigofera* produces seeds weighing 3 mg, too small for the development of beetle larvae. Thirteen other species of legume average over 1000 seeds per m^3 of canopy, each of mean weight 260 mg, and even if all of an individual tree's crop is destroyed by larvae of bruchids ('pea-weevils'), predator satiation will be effective for the local population so long as a few seeds escape consumption. The remaining 23 species produce on average 14 large seeds per m^3, with a mean weight of 3 g, and these are completely protected from bruchids by toxic alkaloids. Bruchids are generally considered to provide a particularly clear example of co-evolution among insects and plants: 102 out of 111 species developing within seeds of deciduous forest trees in Costa Rica are apparently single-host-specific, other potential predators being excluded. The presence of up to 8% of L-dopa, a poisonous compound of great value in treating Parkinson's disease, protects the seeds of the leguminous 'vine' *Mucuna* from all insects, but certain rodents can include small amounts in a mixed diet. Larvae of *Caryedes brasiliensis* are able to metabolize canavanine, rather than incorporating this uncommon analogue of the amino acid arginine into their proteins, so breaking the defences of the large seeds of *Dioclea megacarpa* (Janzen, 1975).

Unlike plants which partition resources into chemical defence, masting species periodically produce an over-abundance of seeds which a wide range of predators finds highly edible. The essential characteristic of **satiation of seed predators** (Janzen, 1971) is that at least some seeds escape discovery (or recovery, if hoarded) and so retain the ability to develop. Over-abundance and satiation imply that predator populations cannot cope with the glut. In temperate forests, unfavourable weather and shortage of seed may reduce predator numbers between mast years. In the lowland tropical forests of Africa and Central America, with their great diversity of plants and animals, no examples are known of mast fruiting at the population or community level. However, the local animal communities in South-East Asian dipterocarp forests are low in numbers and species, and even when reinforced by migrant species of deer, ox, pigs and pigeons, could be satiated by community masting of dipterocarps (Janzen, 1974). There may be a similar explanation for synchronized fruiting in bamboos.

The influence of man on the distribution and abundance of these vertebrate predators makes these hypotheses untestable, while it is no longer possible to assess the role of passenger pigeons, which apparently were the chief seed predators of species of Fagaceae in North America. Beech mast (*Fagus sylvatica*) is totally lacking in toxins and in Denmark Nielsen (1977) found that on average 36% of the annual endosperm was destroyed by seed worms (*Cydia fagiglandana*, Lepidoptera); satiation of this species occurred in mast years. The data in Figure 3.8 possibly indicate satiation of caterpillars attacking seeds of *Fraxinus excelsior*. Gardner (1977) found that small mammals (*Apodemus sylvaticus* and *Clethrionomys glareolus*) were responsible for seed losses of up to 75%, removing them after seed-fall but before germination. Watt (1919, 1923) strongly implicated small mammals in similarly removing acorns and beech mast, and also in severing emerging radicles. Mice, rabbits, slugs, wireworms and fungi were shown to attack seedlings, whose photosynthetic efficiency was decreased by sap-sucking bugs especially in shade. Watt suggested that percentage mortality of beech seeds and seedlings was lowest in good mast years and that potential regeneration was virtually restricted to those years, supporting Janzen's ideas on predator satiation by masting.

It is frequently stated that, largely because so many acorns and seedlings are eaten by birds and mammals, the oak does not regenerate sufficiently to maintain satisfactory stands in British oakwoods. Shaw (1974) concluded that small mammals were the main predators of acorns, but when the crop is heavy (Tanton, 1965) mice and voles consume only a small fraction of it, whereas squirrels and pigeons remove most of the seeds, a few of which germinate after being dropped or buried. Defoliation of seedlings by caterpillars, which drop onto them from mature trees, may be as important as the influence of light in determining the survival of young oaks.

Environmental opposition may be so severe that regeneration occurs only under exceptional conditions. The lofty giant gum (*Eucalyptus regnans*) in South-East Australia typically produces 7.5×10^6 seeds $ha^{-1}y^{-1}$, but largely because of low germination and the deep burial of seeds by ants the seedling density is only *c.* 1500 ha^{-1}. Overheating of the surface soil, desiccation, attack by two species of litter fungi, and grazing by wallabies, together with their own poor competitive abilities, usually result in the death of all the seedlings; consequently saplings are conspicuously absent from the mature groves, which are characteristically of even age. When summers are dry enough, however, these groves are swept by fire as some were in 1939. A profusion of *Eucalyptus* seedlings then develops, which are able to grow before their enemies have recouped their losses (Evans, 1976). Though *E. regnans* reaches heights of well over 100 m, it is a relatively short-lived pioneer species whose forests are **perpetuated by fire**. With its flaky bark and characteristic leaf oils it burns readily during summer hot spells. In the rare cases where swamps, cliffs, or other barriers prevent the spread of fire for a century or more, these short-lived eucalypts tend to disappear and an understorey develops in which *Nothofagus*, *Atherosperma* and the tree fern *Dicksonia* are common. In the continued absence of fire these understorey trees gradually rise to dominance and a late successional southern hemisphere rainforest evolves.

3.5 Seasonal changes and aspect societies

Amongst the most significant environmental features of the temperate regions are the seasonal variations in temperature, light intensity, photoperiod, and often rainfall. This regular alternation of seasons favourable and unfavourable to growth is closely linked to the hormonally influenced **annual rhythms** of leaf production, dormancy, expansion and fall in deciduous forests. The duration and warmth of the **growth season** strongly influences primary production. Woodward (1989) defines the **heat sum** as the number of days in the year (n) when mean daily temperature exceeds the threshold temperature for growth (T_t), multiplied by the mean temperature for the n days concerned (T_m) minus the threshold temperature (which is about 0°C for plants of the temperate region). This calculation yields the **day-degree total** (D) with time being measured in days.

$$D = n\ (T_m - T_t)$$

Importance of the cold season

The effects of low temperature upon plants are profound; this environmental factor probably limits plant distribution more than any other, although the availability of water is the factor which seems to limit most greatly the productivity of world ecosystems. Even a sudden fall in temperature in which freezing is not involved can cause **chilling injury**; a number of tropical species are killed when cooled to between 0 and +5°C.

In winter, deciduous trees lose their leaves and, like the evergreens, undergo a period of dormancy in which there is a relative suspension of the overall metabolism. Even in regions with mild winters photosynthesis is distinctly depressed, and in *Picea abies* photosynthesis ceases entirely when temperature maxima fall below 0°C. Perennial plants growing in northern climates often have little tolerance of frost during the growing season; indeed one of the causes of canker and dieback in *Larix* is late frost after the trees have resumed active growth in spring (another major cause is infection by the ascomycete *Lachnellula wilkommii*). In much of Europe the exceptionally mild winter of 1989/90 was followed by very sharp frosts after the growing season had commenced. The plum crop was devastated and the foliage of many woodland plants including bilberry and a number of forest trees was damaged.

Once the growing season is over the plant undergoes a series of phases during which exposure to successively lower temperatures makes it increasingly tolerant or **frost-hard**. The buds of most dormant trees are covered by bud scales, which reduce water loss rather than protect against frost. If the water in active plant cells freezes intracellularly, mechanical damage from ice crystals often causes death of the cells. During frost-hardening there is usually an accumulation of sugars and other substances which depress the freezing point of the cell sap. Changes in the cell membranes, however, appear to be of major importance in frost-hardening.

Dormancy

Some seeds are unable to germinate because their coats are impermeable, but others remain dormant because they have a **chilling or light requirement** similar to that of buds or other organs. In *Betula pubescens* dormancy of both seeds and buds is removed by gibberellic acid, by chilling at 0–5°C, or by exposure to long days. Growth inhibiting hormones occur in both dormant buds and seeds: some seeds which fail to germinate if merely imbibed with water are capable of doing so if sufficient water is supplied to leach out the inhibitor.

In woody plants there is considerable variation in the time elapsing between the initiation of the flower and its complete development. The flower initials develop directly into mature flowers, as in herbs, in some late flowering trees (*Castanea*) and shrubs (*Hypericum*) where the flowers are formed on the current year's shoots. In blackcurrants (*Ribes nigrum*), gooseberries (*Ribes uva-crispa*), and many other genera including *Quercus, Fraxinus, Acer, Ulmus* and *Pinus*, however, there is a suspension of the development of the flower (or cone) initials, which are formed during the summer in resting buds produced earlier in the same year. Buds containing the flower primordia become dormant in the late summer or autumn with the onset of short days (SD). Dormancy is broken by winter chilling and the buds become capable of opening as soon as the temperature rises. In several species buds containing flowers can grow at lower temperatures than vegetative buds. As a result the flowers of such plants as hazel, willow, ash, elm, and oak are functional before the leaves expand, an advantage in wind pollination.

In most of the north temperate woody species which have been investigated, chilling is necessary to break bud dormancy when it has been fully induced by SD. Even so there are a few species which can be induced to resume growth under long days (LD) or continuous light. Buds of leafless seedlings of *Larix decidua*, *Betula* spp. and *Fagus sylvatica* placed under continuous illumination during the autumn will soon expand. There is also evidence that bud-break in *F. sylvatica* is dependent on increasing day length in spring; commencement of flowering in *Rhododendron* is controlled by temperature in some species and by day length in others. In nature the influences of temperature and photoperiod often lead to similar effects, as when temperature rises with increasing photoperiod in the spring. Experimentally induced environmental changes can be bizarre; bud dormancy can be induced in both *Betula* and *Acer* spp. by exposure to SD at quite high temperatures. In this case the leaves do not senesce but remain active on shoots with well-developed dormant buds.

The hormonal control of seed and bud dormancy is unlikely to be the same in all species. Individual hormones (endogenous growth regulators) may under various circumstances act to either inhibit or promote growth, for example the sesquiterpenoid abscisic acid (ABA) can promote growth in roots while, under other conditions, inhibiting it in shoots. In any particular instance it is the combination of the **hormonal balance** and the **sensitivity** of the tissue concerned which determines whether growth occurs (Trewavas, 1987). Insensitivity to a hormonal balance favouring growth often results from previously induced seasonal changes.

Annual rhythms

Phenology is concerned with the onset and duration of the activity phases of animals and plants throughout the year. As these are largely synchronized to long-term changes in the weather the dates on which they occur differ from year to year, although the order in which the various plant species unfold their buds, flower, fruit and senesce is much the same. Different individuals of the same tree species, however, often commence leafing out (**flushing**) earlier or later than their fellows; this can influence the extent to which they are defoliated by insects (section 6.6).

Salisbury (1916a) studied the periods of active vegetative growth and flowering in an oak-hornbeam wood in Hertfordshire. Here three species have pre-vernal flowering which is almost completed before the leaves of *Carpinus betulus* (hornbeam) are fully expanded in mid-May. The aerial parts of *Ranunculus ficaria* (lesser celandine) and *Anemone nemorosa* (wood anemone) die down completely by the end of June, but *Mercurialis perennis* remains active throughout the summer and shoots are commonly present during the winter in sheltered British woodlands. *Hyacinthoides non-scripta* continues to flower until mid-June (vernal flowering), but then dies down very abruptly, though its capsules are held high above the ground until early autumn. *Conopodium majus* commences vegetative growth at the beginning of March, has aestival (summer) flowering starting in May, and continues active growth throughout the summer. The actual times at which the various seasonal aspects occur alter with climate. This is reflected in the dates at which particular species begin flowering in various parts of their range. For *Ranunculus ficaria* this is usually late February in Hertfordshire, early April in Germany, and late April in the Ukraine where the climate is continental.

Figure 3.9 divides the species into broad **phenological groups**; it is based on central European woodlands, but the flowering sequence is similar to that in Britain. This diagram deals with far fewer species than the original version, which shows a number of plants not native to the UK. Hollow corydalis (*Corydalis cava*), for example, is common in beech forests on good soil in southern Sweden and central Europe but is not native in Britain where bluebell replaces it. *C. cava* flowers at the same time as the lesser celandine, but its foliage, though produced later, dies back slightly earlier. The conspicuous **chasmogamous** ('opening') flowers of *Viola* and *Oxalis acetosella*, which can be visited by insects, are developed in spring. Later in the year these plants form **cleistogamous** flowers which are closed, much reduced and self-pollinated. Those of *O. acetosella* have hooked peduncles and are often buried in the plant litter; most seeds of wood sorrel are produced by this type of flower. Cleistogamy is an efficient means of reproduction but its adoption reduces gene flow.

Studies of the phenology of such species as *Allium ursinum* (Ernst, 1979) and *Hyacinthoides non-scripta* (Figure 3.10) (Grabham and Packham, 1983) have investigated percentage dry matter allocation to the organs of plants growing in their native woodland habitats; this contrasts with studies of plants grown in cultivation (Packham and Willis, 1977, 1982). In both these geophytes the bulb is renewed annually, and there is considerable activity during the period spent underground. At the beginning of October bluebell bulbs are hidden well below the surface of the soil, and have sloughed off the roots which were active in the previous summer. The roots beginning to emerge through the sides of each bulb are new; they arise from the base of a newly initiated 'daughter' plant in the centre of the bulb. These roots make their way by enzyme action through the leaf and bud scales of the 'parent' bulb. By late autumn the bulb scales,

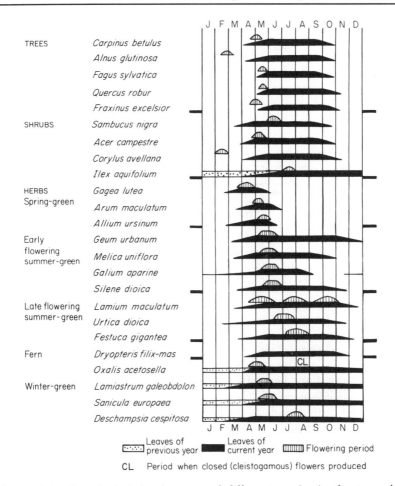

Figure 3.9 Phenological development of different species in the tree, shrub and herb layers of central European damp oak-beech (*Quercus–Fagus*) forests. (After Ellenberg, 1988.)

specialized tubular scale leaves, foliage leaves, scape and flowers of the new plant are clearly differentiated. They increase in size as the old bulb withers; by mid-February the shoot will often have emerged from the soil and begun to exploit the relatively high light intensities available on the forest floor before the tree canopy expands. The plant rapidly accumulates the food reserves needed for flowering and the formation of a new, and usually larger, bulb. In June flowering ends, capsules form, and the now flaccid leaves – which will soon decompose – come to rest on the soil surface. By autumn the cycle is complete and the seeds,

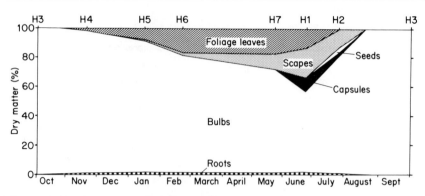

Figure 3.10   Percentage dry matter allocation to plant organs in *Hyacinthoides non-scripta* (bluebell) growing amongst *Pteridium aquilinum* (bracken) on an acidic sandy brown earth in an open region of Himley Wood, near Dudley, West Midlands. Note the high allocation to bulb weight. Each of the seven harvests (H1–H7) is based on the plants within a 0.1 m² area. Samples included plants of all the age classes present, but in this site there was a preponderance of mature bulbs, over 90% of bulb biomass occurring at a depth of 20 cm or more. (From Grabham and Packham, 1983.)

which under favourable conditions take five years to develop into flower-bearing plants, have been discharged from the capsules.

In *Hyacinthoides non-scripta*, as in *Allium ursinum*, there is usually a high rate of seed output and seedling establishment per unit area of stand; vegetative propagation is relatively rare in both species. The reverse of this situation is seen in wild daffodil (*Narcissus pseudonarcissus*) where establishment by daughter bulbs is high in comparison to that by seedlings.

4
Soils, climate and zonation

4.1 Soils and trees

The distinctive morphological characteristics of soils, with their often well-marked profiles (Figure 4.1), result from the integrated effects of **five soil-forming factors**: climate, parent rock, vegetation and associated organisms, relief of the land and time. These factors set the conditions under which the physical, chemical and biological processes operate to produce the horizons found in any profile. These processes (Crompton, 1962) fall into four main groups: weathering, translocation, the organic cycle, and the influence of erosion or deposition which can nullify the effects of the first three.

Many rocks contain all the nutrients required by plants apart from nitrogen, which is generally in short supply. Though the combined nitrogen received in rain will gradually build up, it is significant that the first plants to establish, and thus commence the **organic cycle**, very frequently possess a nitrogen-fixing mechanism. When plants die and decompose they release nitrogen for their successors, and also leave humus in the soil which provides a reserve supply of nitrogen, forms part of a base exchange complex, and helps retain soil moisture. There is a major difference between grasslands and forests in respect of the distribution of organic matter added to the soil. The enormous masses of roots produced by grasses die after a relatively short time so that organic matter is added directly to the profile throughout a considerable depth. In trees a higher proportion of the organic matter contributed to the soil arises from the foliage and tends to accumulate at the surface. Trees are often responsible for transporting mineral nutrients from deep in the profile to the surface, where they become available to herbs.

Soil reaction is measured on the **pH scale**, pH 7 being neutral, and soils with higher values alkaline. The average pH value of northern English soils under natural vegetation is about 5 (Willis, 1973). Such distinctly acidic soils often possess mull humus and may be quite rich in calcium ions adsorbed by colloids of the clay-humus complex, though lacking the calcium carbonate abundant in **rendzinas** (Figure 4.1g), alkaline or nearly neutral soils in which an organic A horizon rests

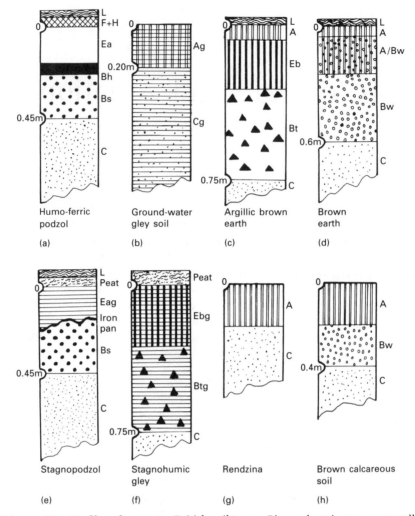

Figure 4.1 Profiles of common British soil types. *Pinus sylvestris* grows naturally and well on podzolic soils (a), but it is necessary to shatter the iron pan and improve the drainage of stagnopodzols (e) when establishing new plantations of conifers. Beech (*Fagus sylvatica*) and oak (*Quercus* spp.) establish naturally on a wide range of soils, but are not grown commercially on, for example, podzols. Most broad-leaved British deciduous trees (e.g. *Fagus, Quercus, Ulmus, Tilia, Carpinus*) are more exacting than conifers in their requirements, growing well on brown earths (c, d), especially those of higher pH. *Fraxinus excelsior* is widely established on brown earths and grows particularly well on rendzinas (g) and brown calcareous soils (h); the latter will support a wide range of tree species. *Betula pubescens* often grows amongst wet heath communities on surface-water gley soils, often with peat (f), but such soils need drainage, and often fertilization, to be of use in forestry. Ground-water gley soils with grey subsoils and rusty mottling near the surface are frequently wooded along minor streams. *Quercus robur* woodland with mull often occurs where water-logging is confined to the subsoil; in wetter sites *Q. robur* and *Alnus glutinosa* are often co-dominant and organic matter accumulates at the surface. *Salix* and *Alnus* are particularly common along river banks. (Redrawn from Burnham and Mackney 1964; by courtesy of the Field Studies Council.)

Soil horizon notation

Organic surface horizons
L   Plant litter, only slightly comminuted
F   Comminuted litter
H   Well decomposed humus with little mineral matter

Organo-mineral surface horizons
A   Dark brown, mainly mineral layer with humus admixture
Ap   Ploughed layer

Eluvial horizons that have lost clay and/or iron and aluminium
Ea   Bleached or pale horizon which has lost iron and/or aluminium
Eb   Relatively pale brown friable horizon which has lost some clay

Illuvial horizons enriched in clay or humus or iron and aluminium
Bt   Horizon enriched in clay
Bh   Dark brown or black horizon, enriched in humus
Bs   Orange or red-brown horizon, enriched in iron and/or aluminium
Bf   Thin iron pan

Other subsoil horizons
Bw   Weathered subsoil material, not appreciably enriched in clay, humus or
        iron, distinguished from overlying and underlying horizons by colour or
        structure or both
C   Little-altered parent material

Notes
g   The addition of 'g' denotes mottling or greying thought to be caused by
        water-logging
A/Bw   Indicates transitional horizon

Mull is a characteristic A horizon, which may be covered by a thin L horizon,
but F and H horizons are scanty or absent
Moder characteristically has an H horizon thicker than the L and F combined
Mor (raw humus) has thick L and F layers

---

directly on calcareous parent rock. Calcium is normally the dominant exchangeable cation in the soil. It causes clay particles to aggregate: Ca clays are flocculated, whereas Na clays, which are produced when the sea breaks through coastal defences, are highly dispersed and percolate down the profile in suspension. 'Clay shift' can occur under other conditions, especially in slightly acid or neutral soils. In more alkaline soils free bicarbonate ions inhibit movement; in more acid soils aluminium and dissociating ferric hydroxide have this effect.

Soil colloids, especially the clay minerals, are capable of ion exchange, absorbing cations of such bases as calcium, magnesium and potassium from solution and exchanging them for others. **Cation exchange capacity** or **base exchange capacity** is a measure of the extent to which soil cations can be exchanged in this way. **Base saturation** is a measure of the extent to which soil exchange sites are occupied by cations of bases rather than by hydrogen ions. In general, broadleaved trees require soils richer in bases than do conifers.

When substantial water is available for leaching, podzols develop relatively rapidly on sandstones, gravels and other permeable parent materials; they will also develop on boulder clay and other substrata under appropriate climatic conditions. Soil solutions from the surface layer of mor humus contain both carbonic and humic acids which leach bases from the eluvial Ea layer. The Ea layer often bleaches to an ashy grey colour, largely as a result of the downward translocation from it of iron and aluminium, probably in association with organic complexes. A siliceous residue of primary quartz grains remains: some secondary silica may also accumulate. Thus an increasing influence of mor humus will cause the Si/(Al + Fe) ratio of the Ea horizon to rise as leaching becomes stronger. Humus and the sesquioxides are subsequently deposited in the Bh and Bs (= Bfe) horizons of humus-iron podzols. **Podzols** are base-deficient, strongly acid and frequently support the growth of coniferous forest.

When waterlogged for long periods, soils become anaerobic and lose the bright red and ochre colours associated with the presence of ferric iron. Bacterial and chemical action reduce iron and manganese to ferrous and manganous forms which are soluble and more mobile. Such **gleyed soils**, very common in valleys and low-lying areas, may undergo seasonal drying, in which case re-oxidation causes the deposition of yellow and rusty spots and streaks of ferric iron, and of black manganese concretions. Gley (g) horizons are often neutral or mildly acid: they are also usually deficient in phosphate. Stagnopodzols (peaty-gleyed podzols, Figure 4.1e) are very common in upland Britain where podzolization and gleying result in a thin dense iron pan and a surface layer of peat.

Burnham (1970) found a strong correlation between the soil regions of Britain (Figure 4.2) and five different climatic regimes (Table 4.1), as defined by mean annual precipitation (over or under 1000 mm) and mean annual temperature. Under the **warm, dry regime** the characteristic soil process is clay eluviation within the profile resulting in **argillic (= leached) brown earths** of such high value for agriculture that forestry, being less profitable, is relatively uncommon. In south and east England the soils often dry out in summer. Washing out (lixivation) of the dry soil by the early winter rains brings clay particles near the surface of the profile into a suspension which passes into the subsoil, a process known as mechanical eluviation. Here the water is drawn into the dry lumps (**peds**) of soil leaving clay films on their external surfaces. Argillic brown earths have been termed grey podzolic soils by some authors. Soils thus mapped in northern England and south-east Scotland

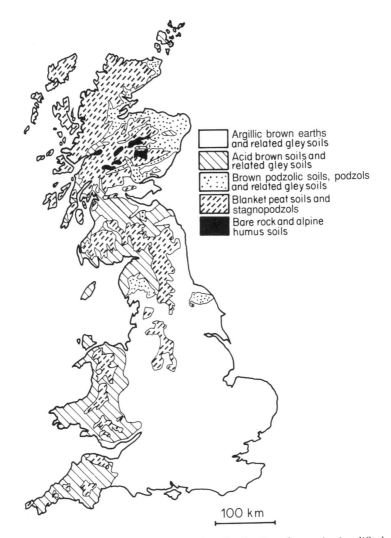

Figure 4.2 Soil regions of Great Britain. The distribution shown is simplified and very generalized; there are large local variations and there may be two or three different soil types within a single field or wood. (Redrawn from Burnham, 1970). For a more detailed map and account of the soils of England and Wales, see *Field Studies*, **5**, 349–63.

Table 4.1 Correlation of climatic regimes and soil types in Britain (from Burnham, 1970)

Climatic region	Mean annual temperature (°C)	Mean annual rainfall (mm)	Characteristic soils
1. Warm, dry	Over 8.3	Under 1000	Argillic brown earths
2. Cold, dry	4.0–8.3		Brown podzolic soils/podzols
3. Warm, wet	Over 8.3	Over 1000	Acid brown earths
4. Cold, wet	4.0–8.3		Stagnopodzols/blanket peat
5. Very cold, wet	under 4		Alpine humus soil

do not show clear evidence of clay eluviation, but brown earths of high base status develop on parent material which would give rise to argillic brown earths further south. Natural vegetation under this climate was almost everywhere broadleaved deciduous forest, predominantly oakwood, though lime (*Tilia*) is now known to have been more abundant than originally thought.

Parts of north and east Scotland, the North York Moors and certain areas on the eastern slopes of the Pennines have a **cold, dry regime** in which both rainfall and potential evaporation are low. Parent materials of low base content prevail in the area and this, together with the considerable volume of water available for leaching, causes most soils to be acid. Podzolization is widespread. Oakwood was the natural climax vegetation on most of the lower ground where the humus form was mull or moder. Such sites are now occupied by intergrades (**semi-podzols = brown podzolic soils**) between brown earths and podzols. On higher ground pinewoods frequently grew on podzols, but climatic change and human interference have caused the development of heathlands in many places. Except in the far north, forestry is normally a possible land use though yields are low. *Pinus* is the characteristic timber tree, though few native woods of Scots pine now remain (Bunce, 1977).

Under the **warm, wet regime** of south-west England, most of Wales and coastal areas in north-west England and south-west Scotland, subsoils are rarely dry and there is a high degree of leaching. Almost all sites bore deciduous woodland until cleared for cultivation and pozolization is weak, though it occurs in places. Forest yields on the **acid brown earths** are high and trees, especially *Picea sitchensis*, are planted quite extensively even on some of the better soils, modern plantations often employing very deep drainage ditches.

Much of the area having a **cold, wet regime** is too exposed for forestry. Use of other parts of the region for forestry often requires ploughing to break down the surface layers and improve surface drainage. Leaching is strong and weathering weak; recycling of nutrients is slow. Fertilization, particularly with phosphates, considerably improves the growth of trees. *Picea sitchensis*, sometimes planted with *Pinus contorta*, grows particularly well under damp conditions and is very successful in the less exposed parts of the region. On the other hand, British areas in which alpine humus soils have developed under **very cold, wet conditions** have never been forested.

The British Isles, being relatively small, do not show the diversity of climates which result in the even greater vegetational extremes found in other parts of the world, such as North America (Figure 1.5), in which variations in rainfall, potential evaporation, irradiation, temperature and length of growing season have major effects on both the soil type and the vegetation.

The above outline indicates the major features of soils and the way in which their regional distribution influences forestry in Britain. Some soils have been under woodland for very long periods, often for many hundreds of years; their characteristics are now considered.

Features of woodland soils

In Britain the most productive soils have normally been used for agriculture, so ancient woodlands (Rackham, 1980) are usually on soils which would, in medieval times, have been regarded as having low quality. Such soils differ substantially from adjacent agricultural soils which have been limed, fertilized and drained, practices which have sometimes influenced the soils of modern plantations. Woodland soils, on the other hand, have been penetrated by tree roots which often grow to a depth of up to 3 m though the majority are less than 0.6 m deep.

The soils of the ancient woodlands of eastern England are frequently derived from **drift deposits**, from 1 to 70 m thick, of boulder clay, sands and gravels, laid down over bedrock in glacial times. Drift deposits are very variable; boulder clay is usually a heavy clay with lumps of chalk in this area, yet woodland soils on it are often sandy or silty and include some of the most acid soils in Britain. In much of England the drift or solid geology is overlain by wind-blown deposits, rarely as much as 1 m thick, of silt and fine sand. These are almost certainly **loess** deposits dating back to great dust storms in late glacial times. Such deposits have often been mixed into agricultural soils by the homogenizing action of ploughing. In other instances they have been removed; the downhill edge of an ancient woodland may be a low cliff formed through the removal of agricultural soil down the slope under the action of ploughing and sheet erosion.

The soils of ancient woods are more stratified than those used for agriculture. They may, nevertheless, be disturbed when old trees are blown over causing the root plate, sometimes with tons of attached subsoil, to be pulled out of the ground. In other cases the trunks fall after major roots have decayed and the soil is not greatly disturbed. Many unfelled trees, however, merely rot above ground and fall to pieces leaving a stump. Superficial disturbance can be caused by rooting animals, particularly wild pigs. If grazing is heavy, or if many walkers use the area, the soil may become very compacted, reducing pore space and increasing the likelihood of surface run-off. Earthworms, moles and other animals which live in the soil have the opposite effect, and also tend to mix the soil horizons.

The **humus type** is of great importance. The litter of temperate woodlands on acid soils possessing few earthworms tends to accumulate as a thick organic deposit, in which three horizons may often be distinguished (Figure 4.1a). The L horizon at the surface consists of intact litter with little visible sign of decomposition. The F (fermentation) layer beneath is of well-comminuted (fragmented) litter; between it and the mineral soil is well-decomposed humus (H layer) containing little or no mineral matter. This **raw humus** or **mor** has a high C/N ratio (*c.* 20) and is of low fertility. It frequently develops under conifers such as *Pinus sylvestris*, especially those planted on former heathlands. It can also develop beneath oak or beech when these trees are on acid soils poor in bases. Mor has thick L and F layers, and its pH varies between 3.0 and 6.5. **Mull** humus has a pH between 4.5 and 8.0 and has passed at least once through the gut of one of the larger soil animals, usually an earthworm, while most mor has not. Mull typically forms under deciduous or mixed forests on moderately well-drained soils containing adequate calcium, but can develop in forests of cedars or of those spruces whose litter has a high calcium content. Its C/N ratio is frequently *c.* 10. Humus type is largely determined by the nature of the plant litter; in some areas planting Scots pine will cause podzolization to begin and mor humus to be deposited, whereas adjacent birchwoods on the same parent material have mull.

In the damp British climate most soils are leached by rain, and in old woodlands surface soils often have a lower pH and calcium carbonate content than does soil further down the profile. **Surface acidification** is especially obvious where the parent materials are uniform, as in Hayley Wood, Cambridgeshire, whose soils are derived from chalky boulder clay with a small admixture of loess. In this wood waterlogging on the central plateau, which is dominated by varying proportions of *Filipendula ulmaria* and *Primula elatior*, is frequent and severe in spring, and relatively slight on the steeper slopes (which never exceed 4°). Where the slope is slight there are many species in an area dominated by *Hyacinthoides non-scripta*; where it is steepest there is a closed community dominated by *Mercurialis perennis*. The dampest areas are in hollows on the plateau dominated by the large sedges *Carex acutiformis* and *C. riparia*, and the general pattern of zonation fits well with the resistance of the dominant species to ferrous iron. Surface acidification is greater, and extends deeper, on the ill-drained plateau than on the slopes, where mixing by earthworms and moles, which are uncommon in waterlogged soils, helps to counteract the loss of calcium from the topsoil by leaching.

Trees vary in their influence on soil acidity. In Connecticut old fields topsoil beneath *Juniperus virginiana* has a raised pH while the pH of the soil in the root zone is reduced; roots of this tree absorb considerable amounts of calcium and other bases which are returned in leaf litter. Foliage of *Juniperus communis*, which grows alongside, is of low base status and increases the acidity of the topsoil (Spurr and Barnes, 1980). The mean pH of the surface soil (0–3 cm) of 80 Coal measure woodland quadrats near Sheffield was higher than that at 9–12 cm in all six main polythetic divisive groups, and in one group the difference was statistically significant (Packham and Willis, 1976). Again bases contributed by tree litter are probably responsible for this effect.

In some woodland areas several rock types outcrop very close to each other. This is true of the Ercall, near Telford, Shropshire (Figure 4.3), where podzols occur above strongly acid, coarse-grained rocks, whether igneous (granophyre) or sedimentary (Cambrian quartzite), and acid brown soils on the more base-rich, finer-grained rocks. This is a fascinating area where variations in soil type, slope and water regime result in a complex pattern of vegetation types. As calcareous soils occur in Limekiln Wood, just over a kilometre away, all the major soil types of Shropshire are represented within two square kilometres.

Heathy oakwoods (*Quercus robur* and *Q. petraea* are both present), with a field layer in which *Calluna vulgaris*, *Deschampsia flexuosa* and *Vaccinium myrtillus* are prominent, occur on the north-west face of the Ercall. Oak grows in places on the south-east slope, but here birch is more common. Much of the birch was burnt in the early 1960s, but it soon regenerated from seed and by sprouting from stumps. Surface soils on the Cambrian siltstones and glacial deposits are damper and more base-rich; the calcicole *Sanicula europaea* is present on the boulder clay. The vegetation of this lower area is more species-rich and the alluvial deposits along the stream have a particularly diverse flora including *Athyrium filix-femina*, *Dryopteris dilatata*, *Caltha palustris*, *Carex remota*, *Filipendula ulmaria* and *Valeriana officinalis* as well as most of the species found in adjacent drier areas. A 5 × 5 m area beneath *Alnus glutinosa* here included 27 species of vascular plants and nine species of bryophytes, and there were many more growing under similar conditions within the immediate area. In this quadrat soil reaction varied greatly; five soil samples (0–5 cm depth), taken at the centre and near each of the corners, had pH values varying from 4.4 to

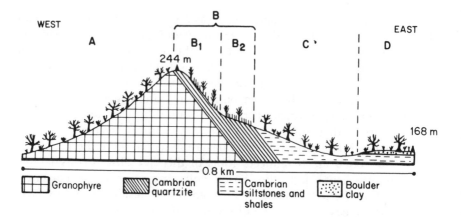

	Soils	Vegetation	pH	
			Topsoil	Subsoil
A	Podzols and brown podzolic soils	Oak-birch coppice; heather, bilberry, bracken or wavy hair grass locally dominant. The mosses *Leucobryum glaucum* and *Plagiothecium undulatum* are conspicuous	3.9	4.5
B₁	Humo-ferric podzols and podzolic rankers	Bracken, some birch scrub and heather	3.7	4.4
B₂	Gley-podzols	Bracken, some birch and bluebell	3.8	4.6
C	Typical brown earths	Oak-birch coppice, with rowan and holly, very variable field layer including male fern, ivy, honeysuckle, bluebell, bramble, creeping soft-grass, wood sorrel, yellow archangel	4.1	4.3
D	Gleyic argillic brown earths	Mixed deciduous woodland with ash, birch, elm (*Ulmus glabra*) and alder, with hazel, hawthorn (*Crataegus monogyna*), oak and ash regenerating; field layer includes dog's mercury, enchanter's nightshade, male fern, tufted hair grass, wood sanicle.	5.2	5.7

Figure 4.3 Soil–vegetation relationships of the Ercall, Shropshire. Leaching has decreased the pH of the topsoils. (Redrawn from Burnham and Mackney, 1964; by courtesy of the Field Studies Council.)

6.2 (mean value 5.7). Before the onset of Dutch elm disease, wych elm (*Ulmus glabra*) was common on the slope south of the stream. The death of these trees has allowed light into the woodland floor where many fallen trunks, together with the now abundant bramble, make the area difficult to traverse.

Various relationships between soil, climate and trees have been discussed above. The next section extends this theme to a consideration of the roles of root systems and the mineral nutrient requirements of various plants in competition.

4.2   Roots, foraging and competition

In addition to anchoring the plant, roots provide absorptive surfaces which forage for water and mineral salts as shoots do for light. Roots may have other functions (Jenik, 1979): the aerial roots of epiphytic tropical orchids carry out photosynthesis and also absorb water from the atmosphere by means of a velamen of dead cells. Breathing roots (pneumatophores) are common in mangroves and swamp cypresses (*Taxodium distichum*) growing in muddy swamps or damp soils. Beech trees growing on eroded banks are commonly held in place by massive exposed roots acting as buttresses while mangroves, many of whose seeds germinate viviparously, are supported over the oozing mud by aerial roots forming pillars and flying buttresses.

Differences during development may be illustrated by considering the root systems of three common European conifers. In its first year as a seedling *Pinus sylvestris* has the longest primary, and much the largest number of secondary and tertiary roots. The many small roots help the tree to flourish in dry barren soils, while the plasticity of the main root system, whose tap root can penetrate very deeply in suitable soils, but in which the laterals usually dominate and form a very widespread system, enables the tree to adapt to diverse habitats. The primary root of Norway spruce (*Picea abies*) stops growing after five years; as the mature root system is composed of shallow laterals the tree is commonly uprooted by the wind (Figure 4.4). As a seedling, the roots of the silver fir (*Abies alba*) are the least branched, and the adult tree has a deep root system with a dominant tap which does not adapt easily to shallow soils, unlike that of *Pinus*.

In highly competitive plants active spatial readjustment of the roots during the growing season involves high energy costs, but is an effective response in favourable environments. Stress-tolerant plants, in contrast, tend to have relatively long-lived roots which forage less effectively, but at lower cost, and survive periods of drought to resume absorption when the rains come. Evidence that tree species differ in such responses is

Figure 4.4 Root plate of Norway spruce (*Picea abies*) blown out of the peat at Sweat Mere, Shropshire. In this species growth of the tap root ceases after about 5 years and the lateral roots are very well developed. (Photograph by John R. Packham.)

provided by data on turnover in fine roots; it is also suggested that there may be cyclic replacement of main roots (Harris, 1981).

Another pattern to consider involves the extent of the root system. *Fraxinus excelsior* has very long, moderately branched lateral roots with long terminal branches which exploit a large volume of soil. Many other woody species also have an **extensive root system**, the coarse roots of which extend great distances into a large volume of soil. Such a system is well suited to stony soils in which the water is not uniformly distributed, and also to winter-rain regions where the roots of trees, unlike those of grasses, can draw water from great depths in the dry summers. In contrast, beech has an **intensive root system**: its shorter laterals with numerous short and extremely fine terminals enable the tree to utilize a smaller volume of soil more effectively. Both systems function suitably

Figure 4.5 (a) Forces acting on tree roots in windy conditions. (a) Thin roots, under tension; (b) moderately thick roots, under tension and some bending forces (if root plate tilts); (c) thick roots forming the 'root plate'. (After Helliwell, 1989; courtesy of *Arboricultural Journal*.) (**b**) Beech tree with shallow root system tilted during a gale several years previously. (Photograph by D. R. Helliwell.)

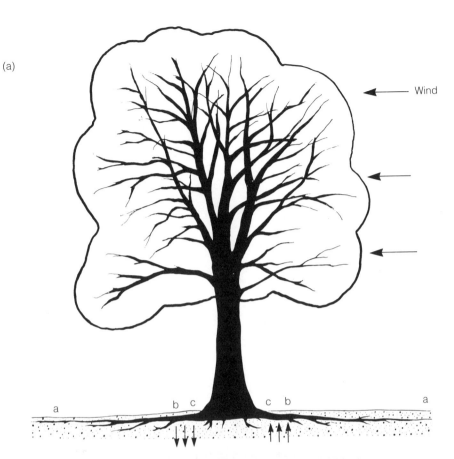

(a)

Wind

a   b c   c b   a

(b)

Countryside

TASK FORCE TREES

ROYAL
BOTANIC
GARDENS
KEW

FORM **A**	FOR SINGLE TREES
SPECIES	

WIND BLOWN TREE ROOT SURVEY - For single trees

PLEASE USE BLOCK CAPITALS THROUGHOUT; COMPLETE AS MANY BOXES AS POSSIBLE; USE ☑ TO MEAN YES

1. **Source of Information**

   Name: _____ Date: _____

   Occupation: _____

   Area of arboricultural interest: _____

   Address: _____

   _____ Post Code: _____ Tel. No. _____

2. **Tree Species:** Genus:_____ Species: _____

   Variety:_____ Common name if above not known:_____

3. **Are your observations based on:**
   Root plate only ☐   Root plate & trunk ☐   Entire tree ☐

9. **Roots exposed:** Root Plate, maximum radius as exposed, in m ☐ depth at centre, in m ☐
   Type:

   Lateral ☐   Tap and Lateral ☐   Lateral & Droppers ☐

   Vertical ☐   Sloping Roots ☐   Two or more tiered ☐

   Other, please draw: With droppers ☐
                                                          Without droppers ☐

   Diameter of main roots at 1m from where trunk entered the ground:
   Fibrous ☐   2-5 cm Diam ☐   5-10 cm Diam ☐   Above 10 cm Diam ☐

10. **Soil:**
    Topsoil Depth in cm:
    Sand ☐   Silt ☐   Clay ☐   Peat ☐

    Other, please specify: _____
    Subsoil:
    Sand ☐   Silt ☐   Clay ☐   Chalk ☐   Peat ☐

    Other, please specify:_____
    Soil Conditions:
    Dry ☐   Moist ☐   Wet ☐   Waterlogged ☐

Figure 4.6 Portion of the Kew Survey Form A used in assessing damage caused by the great gale of 16 October 1987. (By courtesy of the *Arboricultural Journal*.)

Other _____

Hard Pan ☐ Depth in cm below ground level: _____

Other impenetrable layer, please specify: _____

11. **Direction of Fall:** eg N, ENE, SW ☐

12. **Root Stability:**
Is there evidence of building/construction work that may have affected root stability:
Physical damage to roots ☐
Alteration of Water table ☐   Risen ☐   Fallen ☐

Restriction to root development ☐   Please specify if appropriate: _____

Other: _____

How long ago did the damage occur? _____ years

under normal conditions but *Fagus sylvatica* suffers more than most trees in drought, suggesting that it absorbs all the water near it and is then unable to extend its laterals towards damper soil sufficiently rapidly. Though the root systems of broadleaved trees can be so shallow as to compete with herbs, they frequently spread horizontally for great distances: Seeger (1930) records a radial spread of 18 m for 160-year-old *Quercus robur* growing on alluvial gravels and loams of the Rhine plain.

In humid regions large internally decaying yellow birch (*Betula lutea*) trees sometimes form roots that grow down through the decaying heartwood of the standing tree (Spurr and Barnes, 1980). Beech (*Fagus sylvatica*) can also form roots within damp, decayed and hollow trunks; a very large system of this type was revealed when a huge old tree blew down in the Wyre Forest a few years ago.

Root systems and the stability of trees

In practice, rooting pattern may be controlled more by soil depth and fissuring in the underlying rock than the species of tree. Helliwell (1989), in a discussion of the stability of trees, assumes that in the UK most trees have the majority of their roots in the fertile well-aerated upper 600 mm of soil, with few or no roots deeper than 1 m or perhaps 2 m. In drier parts of the country tree roots will penetrate to 4 or 5 m but this is exceptional.

The main forces acting on the root system are (a) the weight of the tree and (b) lateral forces caused by wind or the tree itself if it is leaning. Figure 4.5 shows how these forces are transmitted via the root system to the underlying soil. The main weight of the tree is borne by soil beneath the root plate, usually 2 m or so in radius, in which the main roots taper relatively rapidly. Beyond this plate, root taper is often very gradual. The stability of the tree is reduced if (a) the soil beneath the root plate is soft, (b) the roots are very shallow and easily lifted away from the ground on the windward side, or (c) the roots have been severed or are decayed.

Root and stem wood is strongest when its longitudinally arranged elements are in tension or compression, but relatively weak if subjected to forces lateral to these elements. A horizontal plank whose ends rest freely on solid supports will break under a relatively light load placed on its centre. This lack of resistance to lateral forces turns out to be the Achilles heel of many root systems.

Helliwell comments on storms where numerous apparently healthy trees have been blown down following heavy rain. The roots in these cases were not rotten but the very soft condition of the soil allowed the root plate to press into the ground on the lee side, while lifting away from the soil on the windward side where bending and tensile stresses at the edge of the root plate caused roots to break and allowed the tree to fall. This model works well when **dropper** ('sinker') roots are absent or very fine, as is commonly the case when trees grow on shallow chalk soils, but when they are present and intact (Figure 4.6, heading 9, top right) the tree will be more stable. Deciduous trees are more resistant to high winds when they are leafless and the soil is dry. Beech, unlike oak, can grow successfully on the shallow chalk soils of the South Downs; great numbers had their root plates pulled out of the ground when the great gale of 16 October 1987 struck southern England after a period of heavy rain. *Fagus sylvatica*, like many spruce (section 5.1), is a species whose root form and liability to wind blow is often along the lines outlined by Helliwell.

A more sophisticated picture of tree failure was revealed by the complementary investigations into the great gale of 1987 by the Forestry Commission (Gibbs and Greig, 1989) and the Jodrell Laboratory, Kew (Cutler *et al.*, 1989, 1990; Gasson and Cutler, 1990; Cutler, 1991). The latter discuss 15 factors commonly held to influence wind blow in the light of the statistics, which confirm that the majority of the overthrown trees had little or no disease. Interestingly, diseased trees with a thin canopy often survived, while surrounding and apparently healthy trees with large canopies blew down. Figure 4.6, from Form A used in the Kew survey, shows some of the main types of root system encountered. Mean root depth was a little greater than assumed by Helliwell (1989) with oak rather deeper than beech (Figure 4.7). Though many beech on chalk soils blew down, the stability of this species was less on clay and even worse on sand.

Giant redwood (*Sequoiadendron giganteum*) is remarkably wind-stable; none were reported to have blown down in the Kew survey, while one at Wakehurst Place supported a huge coastal redwood which fell against it. *Aesculus* tended to remain standing after branches had blown off the trees, thus reducing wind resistance. Even when wet, clay and loam soils may provide firmer root anchorage than sand. Many trees, including *Tilia*, are very vulnerable on gravel. Isolated trees with their well-developed roots were less susceptible than those in groups, of which only the perimeter trees were usually stable.

Soil pH, nutrients and microorganisms

Just as various trees and other plants show different tolerances of drought or waterlogging, so with soil pH, which directly affects the viability of plants as well as influencing the nutrient supply (Figure 4.8). Outside the range pH 3–9 the protoplasm of most tracheophyte roots is severely damaged (Larcher, 1975). Further, in very acid soils roots are poisoned by increased concentrations of Al^{3+}, whereas borate poisoning is more likely at higher pH. The availability of iron and manganese is reduced in alkaline soils, often to the point where it limits the growth or performance of some species.

Most vascular plants can exist in single-species cultures, and possess a broad central optimum, between soil pH 3.5 and 8.5. Such **amphitolerant** species have an ecological distribution optimum corresponding to their physiological development optimum where they can compete successfully with others in the mid-range of the soil pH which tracheophytes can tolerate. A number of plants are unable to do this and in nature are often forced into habitats where the competition is less intense. *Deschampsia flexuosa* resembles *Calluna vulgaris* in occurring on acid soils, although it is **acidophilic-basitolerant**, while *Tussilago farfara*, a ruderal plant sometimes found along woodland rides, has an ecological optimum at the opposite end of the spectrum and is **basiphilic acidotolerant**. The bearberry (*Arctostaphylos uva-ursi*), uncommon on soils of intermediate pH, competes best on acid and basic soils.

Dry weights of pure pot cultures of *Holcus mollis* show its physiological pH optimum to be about 6, but in central European woods it is frequently crowded out by other species above pH 4.5 (Ellenberg, 1988). Below pH 4.0 creeping soft-grass is unable to compete effectively with *Deschampsia flexuosa* in English acid oakwoods, so competition from other species tends to restrict it to a rather narrow pH band.

Mycorrhizas are extremely important in forest communities; fungal associations with tree roots are almost universal amongst trees in woodlands of long standing. The fungi of ectotrophic mycorrhizae form a compact sheath of hyphae over the roots, which are stimulated to form numerous stubby branches of the kind commonly seen in beech, pine, oak and the Australian eucalypts. These fungi receive simple sugars from

Figure 4.7 Exposed roots of beech (*Fagus sylvatica*) on the Chalk Downs at Arundel, West Sussex, where its shallow root plates are sometimes pulled completely out of the ground in gales. (Photograph by J. R. Packham.)

the trees, while mineral nutrients and water absorbed from the soil by the fungi are passed on to the trees. Growth of trees is often limited by the level of available phosphate (section 9.6), and much improved by the association with mycorrhizal fungi. Inoculation of young trees with mycorrhizal fungi can be well worthwhile, particularly in establishing and maintaining agroforestry systems in difficult semi-arid areas.

Soil microorganisms are strongly influenced by soil pH. Though soil bacteria commonly tolerate a range between pH 4 and 10 the optimum is slightly on the alkaline side of neutrality (Clark, 1967), and some species have relatively narrow tolerance to soil reaction. Soil fungi are better adapted to distinctly acid soils than are bacteria. Even if fungi are restricted experimentally by use of a fungicide, the low pH will itself prevent any vigorous colonization by bacteria of a strongly acid forest soil. As a result decay in acid soils is relatively slow and mineralization leads to an accumulation of ammonium rather than nitrate nitrogen. Similarly, decomposition patterns may also be influenced by direct and indirect effects of soil pH on soil fauna (section 8.3).

So far this chapter has been largely concerned with soils, climate and competition. The wide variation in such environmental conditions within woodlands is of great importance to the mechanisms causing zonation and maintaining diversity, which are now discussed.

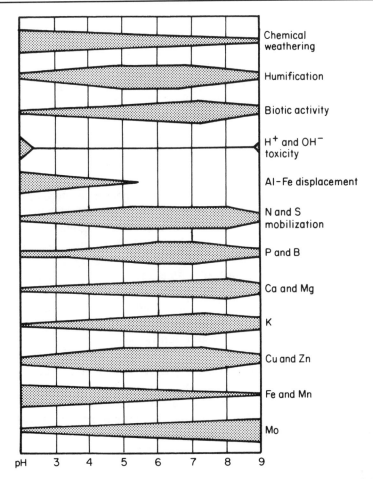

Figure 4.8   Influence of soil pH on soil formation, mobilization and availability of mineral nutrients, and the conditions for life in the soil. The width of the bands indicates the intensity of the process or the availability of the nutrients. (Redrawn from Larcher, 1975; after Shroeder, 1969, *Bodenkunde in Stichworten*, F. Hirt, Kiel.)

4.3   Zonation: distribution in space

Zonation, the segregation of various species and communities in space, and **succession** the changes which occur in communities with time (Chapter 5), are two of the great related themes of ecology. The major types of woods and forests are described in sections 1.4 and 1.5 (which includes a vegetation map of North America; such maps are often primarily concerned with the plants which grew before the destructive influence of man). If climax vegetation is to develop, environmental conditions must remain stable for long periods, as they often did before human interference. The world now has only few areas of natural climax forest remaining; most forests have been planted, are still undergoing succession as a result of the influence of man, introduced pests or pathogens, or exist in a state of semi-equilibrium different from that formerly found in nature.

Zonation within the primeval forest of Fiby urskog

Fiby urskog, one of the finest examples of southern coniferous forest in Sweden, is 16 km to the west of Uppsala, in an area formed by the mainly Pre-Cambrian Baltic shield of granitic and gneissic crystalline rocks. The reserve (Figure 4.9) consists of 65 ha of forest and 13 ha of lake, and is a grade 1 primitive forest (*urskogar*). *Picea abies* is dominant on the boulder moraine, glacial till and later sedimentary deposits of the hill slopes and lower areas. *Pinus sylvestris* is the most important tree in the relatively open forest on Goat Ridge and the other granite ridges of the central area, where depressions in the granite contain shallow soils of low nutrient quality which frequently bear mire vegetation with *Eriophorum vaginatum*, *Ledum palustre*, *Vaccinium uliginosum* and *Sphagnum* spp. The jointed granite is usually covered by a cryptogamic mat of mosses and lichens where trampling is not excessive. Though several broadleaved tree and shrub species are present, including some (e.g. *Tilia cordata* and *Ulmus glabra*) characteristic of the **boreo-nemoral zone**, the area they occupy in the mature forest is relatively low. The woodland possesses many features, such as standing dead trunks, rotting logs, and an uneven age structure, characteristic of primeval forest (Hytteborn and Packham, 1985, 1987).

Dwarf shrubs are a feature of the forest, *Calluna vulgaris* and *Vaccinium myrtillus* being common in drier places. *Deschampsia flexuosa* occurs in the drier, reasonably open areas of the reserve and in the spruce forest is often found with its tillers well dispersed amidst a ground cover of mosses in which *Hylocomium splendens*, *Plagiothecium curvifolium*, *Pleurozium schreberi*, *Ptilium crista-castrensis* and various species of *Dicranum* are prominent. Amongst the hepatics *Ptilidium ciliare* is a fairly common member of the bryophyte mat, while the smaller *P. pulcherrimum* is frequent on wood and fallen trunks, occasionally occurring on rock.

The highest parts of the reserve are now about 60–65 m above mean sea level, and were exposed to wave action when the land was first raised above the sea so that drift and other sediments were washed down from the present hilly areas. The coarser down-washed material is now present amongst the huge granite boulders and finer material,

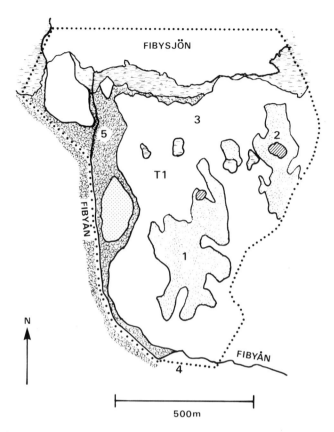

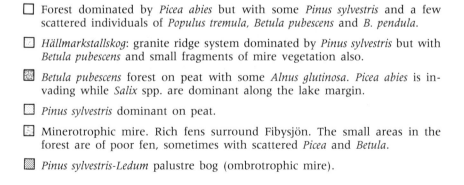

Figure 4.9 Fiby urskog forest reserve, central Sweden. The surface of the lake is normally some 40 m above mean sea level. 1–5, positions of forest relevés referred to in section 2.2; T1, midpoint of the transect shown in Fig. 4.10.

Key

☐ Forest dominated by *Picea abies* but with some *Pinus sylvestris* and a few scattered individuals of *Populus tremula, Betula pubescens* and *B. pendula*.

☐ *Hällmarkstallskog*: granite ridge system dominated by *Pinus sylvestris* but with *Betula pubescens* and small fragments of mire vegetation also.

▨ *Betula pubescens* forest on peat with some *Alnus glutinosa*. *Picea abies* is invading while *Salix* spp. are dominant along the lake margin.

☐ *Pinus sylvestris* dominant on peat.

☐ Minerotrophic mire. Rich fens surround Fibysjön. The small areas in the forest are of poor fen, sometimes with scattered *Picea* and *Betula*.

▨ *Pinus sylvestris-Ledum* palustre bog (ombrotrophic mire).

including clay and silt, is found in still lower places. Later deposits include peat.

Immediately to the north of Fiby urskog lies the dystrophic lake of Fibysjon, whose acid humic waters are fringed by communities of helophytes and hydrophytes. Various willows are present including *Salix cinerea*, *S. pentandra* and *S. repens*, while the northern species *S. lapponum* occurs on a quaking mire. The western margin of the reserve is formed by a small stream which now has a fairly extensive tract of seral birchwood growing on peat along its eastern edge. This birchwood developed after the dam at the southern end of the reserve fell into disuse in the early 1930s, allowing the water level of the stream to fall.

Fiby urskog shows both mosaic structure and zonation on several different scales, from the broad sweep of areas dominated by pine, spruce or birch to the fine scale of variation within the cryptogamic mat where adjacent 0.01 m² quadrats often show considerable differences. Successional processes are also of great interest: the seral birchwood will eventually be dominated by spruce and the mature spruce forest is the type area from which storm gap structure, produced by a characteristic form of cyclic change, was described (section 5.1).

Distribution patterns of herbs and dwarf shrubs are strongly influenced by variations in topography, soil conditions and the tree canopy. Figure 4.10 shows biotic variation along a rather clear-cut environmental gradient whose position within the primeval spruce forest is shown in Figure 4.11. When the 59 quadrats involved were subjected to indicator species analysis (Hill *et al.*, 1975) the six major polythetic divisive groups showed remarkably little overlap, with a smooth transition from group 010 through groups 011, 001, 000 and 11 to group 10. Thirty-one species are involved and mean species densities per m² for the six groups, listed in the order given above, are 5.8, 10.8, 9.4, 5.6, 3.3 and 4.4.

The indicator species for group 0 are *Anemone nemorosa*, *Gymnocarpium dryopteris*, *Linnaea borealis*, *Luzula pilosa*, *Maianthemum bifolium*, *Trientalis europaea* and *Vaccinium myrtillus*. *Anemone nemorosa*, which is restricted to group 0, is a much better indicator than *Vaccinium myrtillus*, although this species is still considerably more likely to occur in a stand belonging to group 0 than group 1. *Melampyrum pratense*, *M. sylvaticum* and *Rubus idaeus* are the indicators for group 1, where the field and bryophyte layers are rather open. The lowest nine quadrats (A1–A9) have a shallow peat layer resting on clay, with clay continuing until about A14. Low boulders are prominent from A24 to the top of the transect. The moister conditions of the lower slopes evidently favour higher species density, though the extremely heavy shade beneath the small spruces at the very bottom of the hill decreases it.

In the last few decades great strides have been made in the field of vegetation classification and mathematical modelling. Ecological physi-

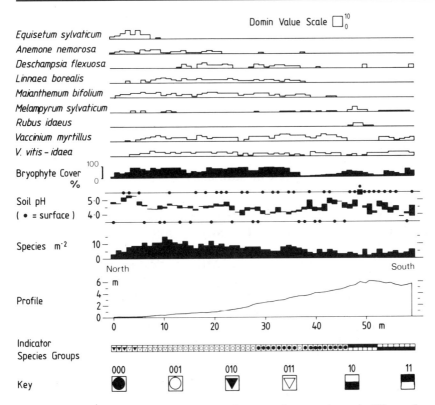

Figure 4.10 Transect 1, a sloping area of primeval spruce forest in Fiby urskog. The diagram shows; (a) Domin values for a number of field layer species; (b) the percentage of the ground covered by bryophytes within each m² quadrat; (c) the difference between the soil pH at 0–3 and 10 cm depth. The dot for each quadrat is placed at the value for the surface soil. (d) Species density per square metre; (e) topographic profile, (f) the indicator species groups to which the stand belong. (Unpublished data of Packham, Hytteborn and Moberg.)

ology has already made a contribution to the understanding of distributions such as those just discussed; increasing effort in this area seems certain to reap rich rewards.

Indices of similarity and gradient analysis

In some respects North America (Figure 1.5) offers better examples of forest zonation than Europe, where the mountains are not so high, the climatic differences not so extreme, and man has changed the natural vegetation more extensively. Although authorities often disagree as to exactly where boundaries should be drawn, there is widespread agreement as to the existence of **floristic provinces** separated by **tension zones** which coincide with the distributional limits of many species. The floras of stands near such a boundary, or within a single province, can be compared as follows, using species presence only to obtain an objective measure of similarity. Where the number of species found in one community is a, the other contains b species and the species common to both is c:

$$\textbf{index of similarity} = \frac{2c}{a + b}$$

(The index of similarity is sometimes called the **coefficient of community** or the **Sørensen coefficient**.) The forests of Wisconsin, to the south of Lake Superior, afford an example of relatively rapid transition. The northern mesic forests contain 27 tree species and are dominated by sugar maple, eastern hemlock, beech, yellow birch, and basswood (*Tilia americana*). They have 17 species in common with the southern mesic forests which possess 26 tree species of which the most important are sugar maple, basswood, beech and northern red oak (*Quercus borealis*). The index of similarity for the trees of these two areas is

$$\frac{2 \times 17}{26 + 27} = 0.64$$

The boreal forest, or taiga, of North America forms a broad crescent extending from Alaska to Newfoundland. In undisturbed sites it is usually dominated by three species, white spruce (*Picea glauca*), black spruce (*Picea mariana*), and balsam fir (*Abies balsamea*), which often occur together. La Roi (1967) sampled 34 taiga sites, all dominated by white spruce and/or balsam fir, arranged in an arc from near the arctic circle in Alaska, where balsam fir did not occur, to Newfoundland. While simple presence lines are sufficient to demonstrate differences in the distributions of the major tree species, an index of similarity is a useful

Figure 4.11 Map of the forest area surrounding Transect 1, Fiby urskog, showing fallen trunks, tree and shrub canopy. Marks at the north and south ends of the map indicate the position of the metre-wide area recorded for herbs and dwarf shrubs. *, mature spruce for which a radial increment diagram is given in Figure 5.2. Numbers 1–8 refer to spruce whose diameters at breast height (cm), heights (m) and estimated age in years are as follows **1**: 4.6, 6, 37; **2**: 8, 12, 110; **3**: 50, 30, 132+; **4**: 42.6, 28, 167; **5**: 46.6, 30, 181; **6**: 35.5, 28, 170+; **7**: 35, 26, 166+; **8**: 35, 30, 144+; **9**: 64, 30, 160 (aspen). (Unpublished data of Packham, Hytteborn, Claessen and Leemans.)

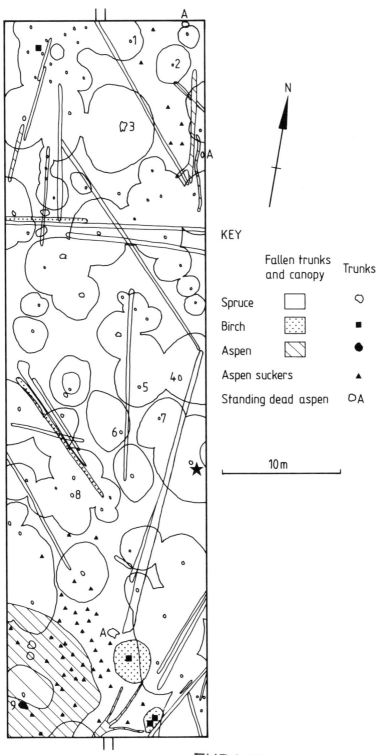

KEY

Fallen trunks
and canopy   Trunks

Spruce

Birch

Aspen

Aspen suckers

Standing dead aspen   ◯A

10 m

comparative measure when all the flowering plants are considered (Packham and Harding, 1982, pp. 74, 75).

The extent to which vegetation can be regarded as a complex **continuum** of populations which gradually blend with each other, rather than a series of integrated communities with discrete boundaries, has been much discussed. Brown and Curtis (1952) supported the continuum concept and used the statistical technique of **gradient analysis** to illustrate the gradual nature of the transitions between species populations in the upland conifer-hardwood forests of northern Wisconsin.

However, some consider that if the hypothesis that discrete associations occur in nature is to be tested, the vegetation should be in equilibrium; a number of stands studied by Curtis and his co-workers had been clear-cut only 26 years before and others had been selectively logged. Daubenmire (1966) also considers that the failure of gradient analysis to recognize that some species are *ecologically* dominant to others is a serious deficiency. He recognizes that there is a continuum in the distribution of coniferous trees with respect to altitude in the Rocky Mountains (Figure 4.15), but takes this to be a **floristic** rather than an **ecological continuum**; the competitive dominance of one species of tree over the others present can, on such a view, be sufficient justification for recognizing discrete communities in an altitudinal transect.

In detailed investigations of forest communities Daubenmire (1952) names the groups (**unions**) of overstorey and understorey species after the dominants. Different habitat types are distinguished by specific combinations of these unions, for example, the *Thuja plicata/Pachistima myrsinites* association, so that the influence of both the subsoil and the topsoil is taken into account. Habitat-type classifications based on this approach are now widely used for commercial forestry in western USA. Late successional trees are used to define tree-species **series**. Within each of these there is one or more characteristic understorey plant union which, together with the tree overstorey, defines the **habitat type**, which is based on potential climax vegetation. A third level (**phase**) is sometimes used to subdivide the habitat type.

So far in this section we have considered the ways in which plant species are distributed in relation to various environmental or historical factors; our next example considers variation within a single widely distributed species.

Climate and variation in Douglas Fir (*Pseudotsuga menziesii*)

When the natural distribution of a species covers a wide area, as does Douglas fir in North America, the various populations within its geo-

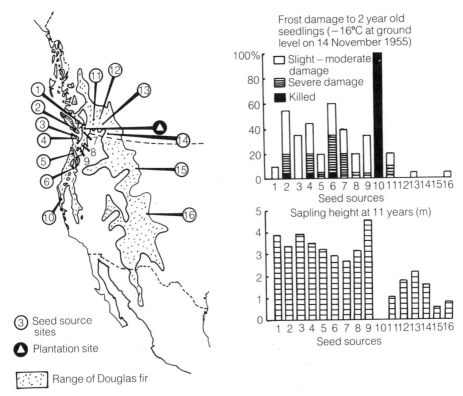

Figure 4.12 Ecotypic differences in Douglas fir (*Pseudotsuga menziesii*) of different provenances. Note the vigour of saplings grown from coastal seed and the extreme frost sensitivity of the California provenance (seed source 10). In addition to their frost hardiness trees of interior provenances (11–16) have considerable resistance to drought. (After Haddock *et al.*, 1967.)

graphical range often show variation in their appearance, in their physiology, or in both. Many species form **clines** in which there is a continuous gradation of character differences, often involving external form, within the species population that is related to its geographical or ecological distribution. Such differences may be environmentally induced, in which case the various forms which may be recognized are known as **ecads**, i.e. phenotypic modifications of the same genotype. (A **phenotype** is defined by its observable characteristics, whereas a **genotype** is defined by its genetical constitution.) Frequently, however, individual populations of a species with a wide geographical range, or growing in a variety of habitats, tend to evolve differently because the selection pressures to which they are subject are not the same. This often results in the production of ecotypes. Each **ecotype** is a group of plants within a species that is genetically adapted to a particular habitat, but able to cross freely with other ecotypes of the same species.

Douglas fir grows in a wide variety of sites in the Pacific northwest. It is very vigorous in coastal areas of British Columbia and Washington State where it can reach a height of 116 m and a girth of 16 m. It reaches maturity at an age of 400 years or more and at this stage has deeply creviced thick brown bark. The timber, often sold as 'Oregon pine', is very strong with a well-developed band of summer wood in every annual ring. Figure 4.12 shows the native range of Douglas fir, frost damage to young plants collected from 16 different **provenances** (places of origin) and grown together at a single experimental site, and mean sapling heights of the plants which survived the frost. Two varieties of *Pseudotsuga menziesii* are commonly recognized: a coastal form, var. *menziesii*, and an interior form, var. *glauca*. The young leaves of var. *glauca*, which are thick and blue-grey, emit little scent when crushed whereas the green leaves of var. *menziesii* have a strong fruity-resinous aroma when pressed. Plants from the interior have a much greater resistance to frost and a lower growth rate than those from the coast; their seedlings also survive longer under soil drought. Var. *glauca* seldom exceeds 16 m when planted as an ornamental in Britain but has reached 25 m in Surrey. Though var. *glauca* can be grown in areas subject to severe frosting and tolerates lime quite well, the greater growth rate of var. *menziesii* makes it far more suitable for general forestry in Europe, where the slow growth of the interior form makes it susceptible to fungal attack.

4.4   Altitudinal zonation and alpine timberlines

In the European Alps and other mountainous regions the lower zones are frequently occupied by broadleaved trees with conifers dominating the upper zones. A clear example of this type of **altitudinal zonation** is shown by the forests of the north-eastern United States which at heights of up to 750 m are usually of the 'northern hardwood' type, dominated by sugar maple (*Acer saccharum*), beech (*Fagus grandifolia*), yellow birch (*Betula lutea*), and sometimes eastern hemlock (*Tsuga canadensis*). Red spruce (*Picea rubens*) is important at higher elevations and above 850 m the forest consists largely of this species and balsam fir (*Abies balsamea*). Fir gains in importance with increasing altitude; the zone below the timberline is almost pure fir forest in many areas, particularly where disturbance is common. The upper limit for forest growth is in the 1350–1700 m range, with the topmost trees dwarfed to prostrate or shrubby 'krummholz' forms. Alpine tundra is present above this level, although small pockets of shrubs or shrubby trees occur occasionally in protected sites.

The upper forests of this region frequently show **wave regeneration** (section 5.3) which, because of the short cycle time between disturbances, favours balsam fir at the expense of red spruce, whose overall survival strategy is well suited to long disturbance-free periods. Sprugel (1976) found only one north-eastern wave system in a spruce–fir mixture, the rest were in pure fir forests. Red spruce lives for up to 300 years, persisting after balsam fir, whose life span is 80–100 years, has died out of a mixed stand. The fir produces large crops of seeds frequently, while red spruce gives smaller crops at irregular intervals. Established seedlings of both species are very tolerant of dense shade in which they can persist for many years, but the initial advantage of more seeds and more vigorous seedlings causes the understorey of most spruce-fir forests to have a higher percentage of fir than the overstorey. Other types of disturbance, such as very heavy cuttings or severe spruce budworm outbreaks, which affect both species (section 6.6), frequently result in regeneration forests with a higher proportion of fir than was formerly present.

Timberlines, and vegetational zones in general, are usually at greater altitudes in regions of large mountain masses than on isolated mountains. It is possible that wind velocities are lower, or snowfall greater, in large mountainous areas such as the Rocky Mountains or the European Alps; either of these factors would result in more soil water being available for summer growth. Within the forest, winter desiccation is much less severe than on a deforested slope (Figure 4.13 and section 2.5), but above the sharp timberline conditions rapidly deteriorate and trees of normal stature occur only on unusually favourable sites. Above the alpine timberline (or forest limit) is a timberline ecotone (**kampfzone**, struggle zone). An **ecotone** is a zone of transition between two ecosystems, in this case between forest and open mountainside. At the **krummholz** ('bentwood') limit, which marks the top of this ecotone, such trees as exist are very contorted, and often have such a low structure that they are almost completely covered and protected by snow in winter. Many species take a multi-stemmed low bush form in the kampfzone; this may well result from the conditions of growth though genetic factors may also be involved. The squat habit of these trees is paralleled by that of **elfin woods** near the timberline of high altitude forests much less affected by man. Such elfin woods are well seen at 40°S in the Argentinean Andes where the southern beech (*Nothofagus pumilio*) becomes more and more stunted as it approaches the timberline, but even here they form dense stands with an almost closed front against the alpine level (Ellenberg, 1988).

In some parts of the Austrian Tyrol (Tranquillini, 1979) high altitude forests appear to have been felled during the Middle Ages to provide timber for mining. The forest zone has since gradually spread up the slopes, but the kampfzone is still unusually wide, whereas in undisturbed regions close by the transition at the upper forest limit is very abrupt,

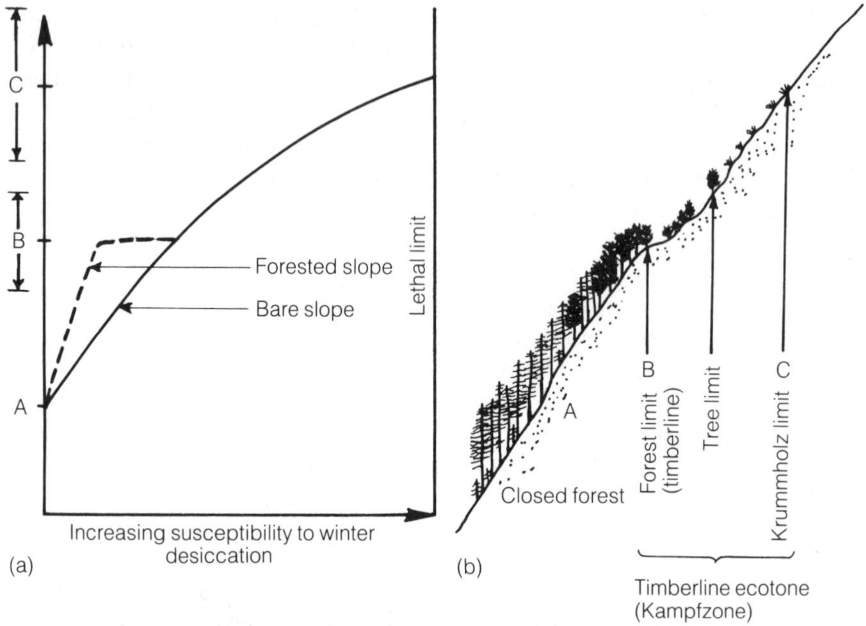

Figure 4.13 Diagrammatic representation of the susceptibility of trees to winter-desiccation on a mountain slope. Risk of damage begins at altitude A; above this height conditions become increasingly extreme and the growing season is shortened so that there is less time for needles to form and develop cuticular protection adequate for drought resistance. Up to the **forest limit** (altitude B) a closed stand can gradually develop by natural regeneration. Within the forest winter-desiccation is much less severe than on a deforested slope, but above the sharp **timberline** (B) conditions rapidly deteriorate and trees of normal stature occur only on unusually favourable sites. C, the **krummholz** limit, is determined by the resistance to desiccation of the trees growing at the highest point; all trees near to this point are stunted and deformed.

In diagram (a) the arrows above and below B and C indicate the height ranges within which these two limits are found after a series of warm (upward movement) or of cold (downward movement) seasons. The trees shown in this diagram of the European Alps are *Picea abies* and *Pinus cembra*, of which the latter has its lethal limit for winter-desiccation at the higher altitude. At Obergurgl, Austria, the timberline (*P. cembra*) is at 2070 m. (Modified from Tranquillini, 1979.)

as it is in the South American example just quoted. Trees such as *Picea abies* and *Larix decidua* may form adventitious roots on older branches weighed down by snow, and reproduce vegetatively by layering, often forming oval patches extending away from the prevailing wind which may cause branches that project above the winter snowline to have a strikingly flagged appearance. Islands of *Pinus cembra* in the kampfzone do not arise from layering but from the activities of the nutcracker, the corvid *Nucifraga caryocatactes*, which buries heaps of 10–30 seeds.

Tree root systems at the timberline show intensive development of mycorrhizas. Almost all the short roots of *Pinus cembra* at central European alpine timberlines are mycorrhizal, and mycorrhizal fungi are known to have evolved high altitude strains adapted to low temperatures. Mycorrhizas must be present if such diverse species as *Nothofagus solandri*, *Eucalyptus pauciflora*, *Picea engelmannii*, *Pinus contorta*, *P. flexilis* and *P. hartwegii* are to establish near timberlines.

Natural treelines are usually sharp; these upper limits to distribution are set by temperature, moisture, soil and wind conditions. As a mountain is ascended conditions become, on average, windier, wetter and colder – but above all more variable. In exposed parts of the northern Rocky Mountains seedlings of *Picea glauca* are killed by a few hours of high temperature at ground level which causes stem girdling, though at night the temperature drops sharply. Partial shading prevents most heat deaths in this tree and drought is then the most likely cause of mortality, especially as growth at high altitudes is so slow that seedlings take several seasons to form tap roots long enough to enable them to survive a drought.

In New Zealand, evergreen *Nothofagus* forests end abruptly at between 1000 and 1500 m above sea level. Near this treeline *Nothofagus* seldom has good seed years and the seed shows poor germination (0–3%). Those seedlings which do develop often die, their tops having dried out. When such small seedlings were planted higher up the mountain, all those in the open died in the first year, but shaded seedlings survived and established 200 m above the treeline.

Variations in forest dynamics

The behaviour of various tree species changes with increasing altitude. In the Vallibäcken forest, northern Sweden, for example, the population dynamics of trees growing in the high altitude sites differ from those at lower levels, which possess storm gaps of a similar type but smaller size than those described for the boreo–nemoral forest of Fiby urskog (section 5.1). Figure 4.14 shows cumulative age distributions for *Picea abies* and *Betula pubescens* in two forest plots as a semi-logarithmic diagram with

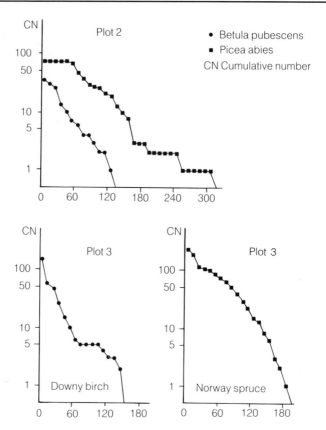

Figure 4.14 Cumulative age distributions of all individuals of *Betula pubescens* and *Picea abies* in two forest plots, of similar area in the Vallibäcken forest, N. Sweden. Plot 2 is at an altitude of 460 m and plot 3 at 335 m. The coniferous forest limit is at 580 m in this area. (From Hylteborn, Packham and Verwijst, 1987, courtesy of *Vegetatio*.)

age classes of 10 years. If there is at least one good seed year with acceptable germinability during each 10-year period, and the mortality in each age class is constant, such curves can be interpreted as survivorship curves (Deevey, 1947). These criteria are met in the case of plot 3, of which some 140 m² was a storm gap, while the remaining 260 m² were covered by tree canopy. The mortality of both spruce and birch is relatively high in the first two age classes. After 70 years birch reaches a plateau which continues to 110 years. Following initial high mortality in the very first age classes spruce died off more slowly, increasing its mortality rate later on, but not as fast as birch in the same higher age classes. It is suggested that the form of this curve reflects the ability of spruce to survive as **dwarf trees**, with poorly developed leading shoots and well-formed lateral branches. This effect is shown even more clearly

by the survivorship curve for spruce in a plot containing
from Fiby urskog (Hytteborn and Packham, 1987).

In plot 2 at 460 m, and in its surroundings, there was
typical storm gaps though the plot was open with a crow..
15%. The cumulative age distribution of *Picea* here cannot be int...
as a survivorship curve because the number of cones produced at th..
altitude and latitude (67°N) is both low and irregular while mature
viable seeds are only produced in any quantity in the rare years when
mean temperature between June and September exceeds 10°C. In plot
2, therefore, tree growth and regeneration is dominated by environmental factors, whereas in plot 3 the major influence is exerted by the biotic
influence of the mature trees.

Temperature–moisture gradients

Figure 4.15 shows the order in which a number of conifer species are
found at increasing altitude in a northwestern region of the USA, where
there is a well-marked temperature–moisture gradient from the warm,
dry lowlands to the cold, wet mountain peaks. The lower limit of distribution for these trees usually shows a gradual transition, which is
apparently set by soil moisture levels. Ponderosa pine (*Pinus ponderosa*),
Douglas fir (*Pseudotsuga menziesii*), Engelmann spruce (*Picea engelmannii*: not
shown in Figure 4.15) and alpine fir (*Abies lasiocarpa*) are encountered
at successively greater altitudes and their roots will usually grow in
successively moister soils. This tallies with their drought resistance, which
is highest in ponderosa pine and least in alpine fir. Temperature seems
not to be the major factor; all the species can be grown at low altitude
if watered, while dry atmospheres produce little effect as long as there
is sufficient soil moisture.

Climate (especially the wind, water and temperature regimes), soil
type, altitude and biotic factors including the influence of man, are thus
major influences that have caused the zonation of the world's forests
and woodlands. These include the most diverse plant communities known;
the mechanisms which maintain this diversity are now considered.

4.5   Diversity in communities of woodland plants

In this section we are largely concerned with **species diversity** within
a single community, and the factors which either allow many species to
grow together, or alternatively tend to result in simplification, often
with large numbers of individuals belonging to relatively few species. On
a world scale, however, the severe and predominantly cold climates of

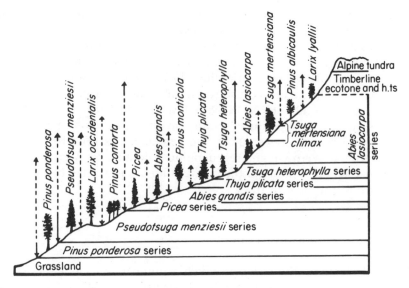

Figure 4.15 Coniferous trees in the Rocky Mountains of north-western Montana arranged to show the usual order in which the species occur with increasing altitude. The relative altitudinal range of each species is shown by arrows. The dashed portion of each arrow shows where a species is seral (early successional) and the continuous portion where it is the potential climax dominant (late successional). The temperature–moisture climatic gradient runs from the lowlands where the conditions are warm and dry to the timberline where they are cold and wet. There are several timberline types (h. ts): all have a timberline ecotone with krummholz above them. (After Pfister *et al.*, 1977.)

the northern taiga provide niches for far fewer species than do forests where the climate is normally warm and moist.

Assessment of diversity necessarily involves the use of quantitative data. At its simplest the number of species per unit area may be taken as a measure of species richness; for more complex measures diversity may be established in relation not just to area but to the numbers of individuals of each species present in that area. A **diversity index** considers the number of species present in relation to the size of the sample area or the total number of individuals of all species. The more complex **Shannon-Wiener** or **information index** takes into account the **equitability** (evenness) of the contributions made by different species; an overwhelming preponderance of a single species is itself a loss of diversity. **Dominance concentration** measured by the Simpson index gives values which are inversely related to both equitability and diversity. Species diversity is discussed by Whittaker (1975, Ch. 3) which should be consulted for details of the relevant indices.

Relative abundance is an important factor involved in diversity: various relative abundances of the same plant species can result in woodland areas of quite different appearance, some of which may favour particular species of animal. **Structural diversity**, as between the successive phases of a coppice cycle or a number of regeneration gaps, favours a wide range of herb, insect and bird associates. In practice experienced ecologists can quite rapidly form a preliminary view of the **conservation value** of a woodland by reading a species list and mentally assessing the diversity of form, phenology and rarity of the species involved and taking particular note of ruderals and indicators of ancient woodland.

Sources of diversity

Factors associated with the regeneration gap (section 5.1) are of major importance in helping to maintain species diversity in woodlands, but other mechanisms (Grubb, 1977) are also involved and are discussed here. One of the most important concerns the difference in **life forms**, whose complementarity is shown in a particularly marked way by woodland plants (sections 1.2 and 2.1). In a mature woodland the dominant trees do not utilize the resources of the area completely, indeed they often bear epiphytes themselves. One or more species of shade-tolerant shrubs or herbs (section 2.4) can usually co-exist, often in places where the tree canopy is thin so that the shoots of smaller plants receive sufficient light to persist or even flourish. Trees, shrubs and herbs frequently compete for water and mineral nutrients, but the different sizes of these plants helps the various species to fit together in a diverse mosaic which allows the minimum needs of them all to be satisfied. Heterogeneity is further enhanced by **dependence relationships** between trees and their epiphytes, parasites, hemi-parasites and the saprophytes which can grow on their dead products in very dense shade.

Differences in phenology are of great importance in enabling several species to grow vigorously in the same place but at different times of year. The phenology of the herb sequences of north temperate deciduous woodland (section 3.5) has been extensively investigated with regard to changing light and temperature conditions; its importance with regard to the transmission of mineral nutrients is also becoming apparent (section 9.6). Equally interesting is the non-coincidence of flowering, fruiting and flushing periods found in trees of at least some tropical forests.

Temporal fluctuations in the environment, such as winters in which minimum temperatures are exceptionally low, may have important effects, in this instance favouring species whose distributions are predominantly in higher latitudes. Such fluctuations may involve biotic factors. **Host-specific parasites** and **herbivores** may play an important role in maintaining species richness. When their activities are concentrated on adult plants they may inflict particularly high mortality on offspring near the parents. When pairs of species of the same age are grown together one will usually gradually oust the other, but the two species may sometimes move towards a position of balance which, once achieved at a given density and under particular environmental conditions, can persist indefinitely. **Balanced mixtures** may result when two species are limited by different factors, so that the balance of competitive advantage can change, as when mineral nutrient concentrations or soil pore-size vary locally. They may alternatively represent a balance of intraspecific versus interspecific competition; a possible cause of this is the production of autotoxins, substances more toxic to the species producing them than to others.

Many natural and semi-natural woodlands are of **uneven age**, which can result in the developmental (pioneer) phase of one species being pitted against the pioneer, building, mature or degenerate phase of another (section 5.1) with various results. Plants of different physiological age have very different competitive abilities, so it is possible for two species to persist in mixture indefinitely when the plants are of various ages.

Soil type may substantially affect species diversity in grasslands and woodlands. In north-west Europe, grasslands on shallow calcareous soils which are short of nitrogen, phosphorus and other mineral nutrients, have far more species than grasslands on deeper soils of pH 5–6 with a better supply of nitrogen and phosphorus, which are dominated by relatively few species of high RGR. If species-rich, nutrient-poor grasslands are fertilized, standing crop is increased and diversity decreased because many of the species with low RGR, formerly favoured by their diverse regeneration requirements, are suppressed by the vigorous growth of their taller competitors in the mature phase. In contrast with natural grasslands in the same regions the field layers of beechwoods in central and northern Europe on deep, fertile soil of pH 5–6 (Asperulo-Fagion) are generally more species-rich than those on shallow calcareous soils (Cephalanthero-Fagion). Several factors are involved here. Heavy shade prevents rapid growth of shrubs and herbs of potentially high RGR even on fertile soils, while relatively few strict calcicoles grow in woodlands, though many do in grasslands. It is also significant that many species common in soils with a pH of less than 5 can grow at pH 5–6, but do not occur on highly calcareous soils.

Effects of tree litter upon plant diversity

Tree litter is an important constituent of all woodland ecosystems; information on its rate of decomposition is needed in attempts to understand more fully the processes of mineral nutrient cycling and energy flow. The litter on the forest floor affects the humus type and mineral nutrient status of the soil, while its physical presence often influences species differentially.

The direct effects of litter upon vegetation are particularly marked in many temperate and boreal forests, where tree litter often accumulates in large quantities on the woodland floor and individual leaves persist for years. In contrast, leaves are known to disappear completely in 2–7 months under the warm moist conditions of tropical forest. Persistent tree litter is often distributed very unevenly within woodlands, the highest densities occurring in hollows, which may be devoid of herbaceous plants. Litter is also trapped by robust ground-flora species including *Rubus fruticosus* agg. (bramble), *Pteridium aquilinum* (bracken), and *Vaccinium myrtillus* (bilberry).

Investigations of woodland floors beneath canopies of sycamore and oak near Sheffield (Sydes and Grime, 1981) showed that the main ground flora constituents could be arranged in the series *Lamiastrum galeobdolon*, *Hyacinthoides non-scripta, Anemone nemorosa > Milium effusum, Holcus mollis, Poa trivialis > Mnium hornum* in terms of their tendency to occur in areas with a high density of litter. The emergence of aerial organs is a critical phase in the lives of woodland herbs (Salisbury, 1916b); species most frequently associated with persistent tree litter have shoots which can penetrate it effectively. This is well seen in bluebell (*Hyacinthoides non-scripta*), whose spear-shoots can puncture the tough leaves of oak and force their way through the coherent heavy litter commonly present in early spring during wet weather. The evergreen overwintering shoots of *Lamiastrum galeobdolon* connect to rather wiry stems ramifying through the litter which produce new shoots in early spring. The young shoots are robust, erect and with rather narrow apices well adapted to penetrate weak points in the litter, and this plant often flourishes on woodland floors where much litter is trapped by a low growth of bramble. *Anemone nemorosa* has its young stems folded in the 'penknife' mode and these also can push through weak areas of the litter. Because of their small size, *Mnium hornum* and other mosses have poor powers of emergence through litter. Similarly many woodland grasses, particularly *Poa trivialis*, are not suited to resist burial by litter, having weak leaves and low growth when vegetative.

On the other hand, in the woodlands mentioned above, young seedlings of *Fraxinus excelsior* were most abundant where deep litter was present; perhaps fruits dispersed in litter are less conspicuous to predators. In the same region, slower rates of predation and smaller losses in viability have been observed where acorns were immersed in tree litter. Again *Oxalis acetosella* grows very much better where there is an abundance of broadleaved litter in which its roots and rhizomes often run; it was found (Watt, 1925) to disappear gradually from plots beneath *Fagus sylvatica* into which the continued fall of leaf litter was prevented.

Soil conditions and vegetational mosaics

Local variations in edaphic conditions involving features such as soil depth, particle size, soil reaction and exchangeable calcium levels, strongly influence the distributions of many woodland species and frequently result in the development of vegetational mosaics. *Fagus sylvatica*, *Daphne laureola* (spurge laurel) and *Allium ursinum* (ramsons), for example, all grow well in soils rich in exchangeable calcium, whereas *Vaccinium myrtillus* (bilberry) is a strong **calcifuge** in all respects, growing in distinctly acid soils in which calcium carbonate is absent and exchangeable calcium is low. *Calluna vulgaris* (ling, section 4.2), normally grows in soil devoid of calcium carbonate and its seedlings are markedly calcifuge. Species which in nature usually behave as **calcicoles**, such as *Sanicula europaea* (wood sanicle), or **calcifuges**, such as ling and bilberry tend to have much narrower ranges in respect of soil reaction and exchangeable calcium than do, for example, *Pteridium aquilinum* (bracken), *Mercurialis perennis* (dog's mercury) and *Teucrium scorodonia* (wood sage).

Salisbury (1920), in early studies of the calcicole/calcifuge problem, noted that 'in woods on acid soils the calcicolous *Mercurialis perennis* occurs either in dry areas with a high calcium content (or low acidity) or in damp areas where the lime requirement is high (considered in terms of unit weight of dry soil) but is apparently ameliorated as a consequence of the high water content.' *Mercurialis* is known to grow in soil with pH values from 4.5 to 8.2, but its main roots are seldom in the more acid regions of soil profiles. When grown in a number of acid boulder clay soils with initial pH values varying from 5.7 to 6.7 there was a positive correlation between exchangeable calcium in the soil and dry weight accretion, although the most alkaline soil did not have the most exchangeable calcium (De Silva, 1934).

Mercurialis perennis, *Brachypodium sylvaticum* (slender false-brome) and *Deschampsia cespitosa* (tufted hairgrass) grow well on many woodland soils (mulls and rendzinas) containing quite moderate amounts of phosphorus. In some instances where *Urtica dioica* (stinging nettle) grows in the vicinity it has been found that its seedlings cease growth at an early stage if transferred to the *Mercurialis* soils. *Urtica* grows vigorously on these soils if phosphate is supplied, but in its absence the addition of other nutrients, including nitrogen, has little effect (Pigott and Taylor, 1964). The seedlings of *Mercurialis perennis*, *Brachypodium sylvaticum* and *Deschampsia cespitosa*, however, scarcely respond to additional phosphate and grow successfully on the natural soils. *Chamaenerion angustifolium* and *Sambucus nigra* (elder) respond very positively to additions of phosphate, but when nitrogen alone was added fireweed showed virtually no response and dry weights of elder seedlings were actually less than those of the control plants. *Galium aparine* (cleavers), a large fruited annual herb, was found to be checked at an early stage when grown on soil beneath *Mercurialis* in Buff Wood, Cambridgeshire, and resumed vigorous growth only when phosphate and nitrogen were added together; either alone produced little response.

Lamiastrum galeobdolon on nutrient-poor soils responds positively to additions of phosphate, or phosphate and nitrogen (Packham, 1983). This ability promotes the survival of yellow archangel as an underlayer to *Urtica dioica* in eutrophicated woodland sites such as parts of Cantreyn Wood, Bridgnorth, which have been subject to fertilizer drift from cereal fields.

On the chalky boulder clay of Cambridgeshire woodlands, such as Buff Wood and Hayley Wood, *M. perennis* is confined to the well drained and better aerated soils, whereas *Primula elatior* (oxlip) grows on soils with poor drainage which become surface waterlogged during the spring months of most years. In pot experiments *Mercurialis* grew as well, and sometimes better, in soil from the *Primula elatior* areas (mean pH 5.86) of Buff Wood as in soil from *Mercurialis* areas (mean pH 7.17), thus eliminating soil acidity and mineral nutrients as the direct causes of the distributional pattern shown here by *M. perennis*.

Though *M. perennis* is confined to areas where soil oxygen diffusion rates are generally high, Martin (1968) has shown that it is the presence of ferrous ions (Fe^{2+}) in damp soils which causes root death in this species. Sand culture experiments in which waterlogging occurred in the absence of ferrous ions did not cause injury. There is a clear correlation between the extent of intercellular airspaces in the roots and the ability of species to grow in parts of these Cambridgeshire woodlands that are waterlogged in spring. *M. perennis* has small intercellular airspaces at the angles of the cortical cells and does not form enlarged airspaces in response to poorly aerated root environments. In contrast, *Primula elatior* has larger airspaces and occasional small lacunae, *Filipendula ulmaria* (meadowsweet) has a spongy root cortex of small, loosely packed cells, while in *Deschampsia cespitosa* well-developed aerenchyma forms a lacunar system round the central stele. Measurements of rates of oxygen diffusion from the roots of *Deschampsia cespitosa* and *Mercurialis perennis* in deoxygenated water showed that the former had by far the greater power to oxidize the surrounding root medium, a characteristic of plants which grow success-fully in bogs. When *M. perennis* survives on soils liable to poor aeration in the wet season, it does so by producing an entirely superficial root system absent from regions of the soil profile in which reducing con-ditions occur. Plants with such a growth form are very susceptible to summer drought.

Martin (1968) showed an increasing tolerance to ferrous ions in the series *Mercurialis perennis, Hyacinthoides non-scripta, Brachypodium sylvaticum, Geum urbanum, Circaea lutetiana, Primula vulgaris, P. elatior, Carex sylvatica, Deschampsia cespitosa* when species were grown in well-aerated water cultures. The main symptom of ferrous ion toxicity was root death and the order given above corresponded with the ecological behaviour of these species towards waterlogging, with *D. cespitosa* being tolerant of both waterlogging and ferrous ions, while *M. perennis* is very sensitive to both. Ferrous ion solubility is mainly determined by redox potential and pH, being greater in more acid soils. (**Redox potential** is a quantitative measure of ability to gain or lose electrons, and hence, respectively, to be reduced or oxidized.)

Despite the similarities of their phenology and gross morphology the cowslip (*Primula veris*), primrose and oxlip have distinct habitat re-quirements. The exclusion of *P. veris* from woods appears to result from intolerance to shade; this species is more tolerant of drought than *P. vulgaris*. *P. elatior* has the least tolerance of drought; Whale (1984), who examined the responses of these species to perturbation when subject to competition, ranked tolerance of waterlogging in the reverse order. Shading reduces seed production in all three species but particularly in *P. veris*. *P. elatior* has a shorter life span than the other species so the reduced seed set and increased seedling mortality resulting from grazing tends to cause accelerated decline in numbers in established populations.

Competitive ability of clonal plants

Clonal plants have recently been a focus of attention in that some of them, notably *Glechoma hederacea*, display remarkable plasticity in their foraging behaviour (section 2.4). *Clintonia borealis* (Liliaceae), which is restricted to the hardwood forests of eastern North America, is also of interest in that its rhizomes persist for up to 15 years allowing the number and position of the ramets to be determined precisely. Angevine and Handel (1986) reported that only 14.4% of ramets produced two new rhizome segments in a year. Single rhizomes tended to diverge from the direction of the previous year's growth by a mean of 22.9°, but where two rhizomes were produced they diverged from the previous line of growth by an average of 48.1° each. High resource availability in

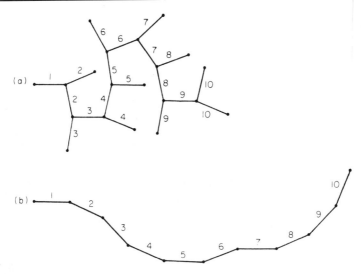

Figure 4.16 Diagrams of two rhizome systems produced by a simple model of (a) branching and (b) unbranched growth in *Clintonia borealis* over ten years. Internode lengths are proportional to actual internode lengths before (6.1 cm) and after (5.5 cm) branching in rhizomes. The angle of divergence for unbranched rhizomes is 22°. Branched rhizomes have a major angle of 21° and a minor angle of 77°. Left or right orientation was determined by random simulation. In the branched system, one branch (chosen at random) was killed each year, to maintain a single growing point. The unbranched system had travelled 62% further from the origin by the end of year ten. (From Angevine and Handel, 1986; courtesy *Journal of Ecology*.)

a local site leads to more frequent branching so that net movement of a clonal fragment is reduced and a favourable patch of forest floor is exploited more intensively (Figure 4.16). The areas occupied by clones as they invade the forest floor are broadly fan-shaped but the above-ground ramets are present only at the front edge of the rhizome system, quite unlike the distribution in species such as *Oxalis acetosella* where leaves are present throughout the area occupied by a clone.

Hutchings and Barkham (1976) studied colonies of *Mercurialis* in Foxley Wood, Norfolk, in attempts to determine the extent to which the shoots were comparable to discrete units in populations. Their results suggested that the connected shoots of *M. perennis* (Figure 4.17) are part of an integrated system, being mutually dependent rather than competitive in many activities. Rhizomatous perennials such as *Pteridium* and dog's mercury probably persist in pure stands because their dense shoot, rhizome and root systems exploit the resources of the habitat thoroughly and efficiently, preventing the entry of competitors, while not developing intense competition between individual shoots and other organs of

Figure 4.17   Leaf mosaic of female colony of dog's mercury (*Mercurialis perennis*) in Cantreyn Wood, Bridgnorth, Shropshire. (Photograph by John R. Packham.)

the same plant. Populations of clonal plants, though typically made up of a limited number of genotypes – each represented by large numbers of copies or ramets – can thus compete very efficiently with other species.

Dog's mercury grows as a woodland herb in many parts of Europe, where, as indicated above, its success in a variety of habitats is influenced substantially by several environmental factors, especially light, calcium levels, the nutrient status of the soil, and soil waterlogging. It is a hairy perennial hemicryptophyte with erect stems and long creeping rhizomes which bear roots whose 'working depth' varies according to soil type. In soils in which water and humus contents are both high the roots branch profusely, but are mainly confined to the upper 10 cm of soil. Root growth is poor in sandy soils and even poorer in peat. In calcareous clays the greatest development is at 15–20 cm and the plant grows well on this soil type.

Plants associated with vegetation types that are seldom disturbed and of moderate productivity usually have characteristics intermediate between those of competitors and stress-tolerators. *Mercurialis perennis*, which has a marked capacity for local vegetative spread, is a woodland and hedgerow example of such a **stress-tolerant competitor**, with a pronounced ability to endure low light levels while itself, particularly the female colonies, casting a dense shade. Although shade-loving, it can also tolerate high light levels, i.e. it is a **sciophyte** capable of growing as a **facultative semi-heliophyte**. There are a number of habitat forms (Mukerji, 1936), but the leaves tend to be larger, thinner, relatively wider, less hairy and a deeper green when the plants are in shade. Female plants grow in shady habitats; the development of male colonies is favoured by higher light intensities. The leaf canopy of dense stands of female plants in many woods where light levels are low is very complete (Figure 4.17), a major factor in the ability of this plant to suppress competitors. Tree seedlings, including those of common ash, are frequently shaded out or held in check until their shoots arise above the canopy of *M. perennis*. Light levels and water regimes influence the distribution of dog's mercury in non-woodland situations also; in the grasslands of the Sheffield area Grime and Lloyd (1973) found it, together with *Anemone nemorosa* and *Oxalis acetosella*, amongst the familiar woodland plants restricted to, or concentrated upon, the comparatively cool and moist north-facing slopes. The distribution of *0. acetosella* within the heathland, grassland and flush communities of the Long Mynd, Shropshire, appears to be largely controlled by the water regimes of the various microhabitats (Packham, 1978, 1979).

Part of the complex mosaic of a climax woodland. Black 'bootlaces' (rhizomorphs) of honey fungus (*Armillaria mellea*) run over the trunk of an elm killed by Dutch Elm disease (*Ophiostoma ulmi*), which lies surrounded by dog's mercury (Mercurialis perennis), stinging nettle (*Urtica dioica*), lesser celandine (*Ranunculus ficaria*) and a little cleavers (*Galium aparine*). Suckers derived from the dead tree grew nearby.
Buff Wood, Cambridgeshire, May 1982. (Photograph by John R. Packham.)

5
Forest change

5.1 Succession, cyclic change and gap regeneration

Foresters, landowners and others have for centuries observed changes in vegetation with the passing of time. These are examples of **succession**, which can be defined as non-seasonal directional change in the types and numbers of organisms present in a particular habitat over a period of time. **Cyclic change** is a related process in which there are repeated and similar sequences of change, frequently associated with particular patches in the general vegetation. In the first few decades of this century many ecologists considered succession to follow a broad general pattern with distinct phases. Moreover, Clements (1916) considered succession to involve 'a sequence of plant communities marked by the change from lower to higher life forms', and did not recognize the possibility of regression. This is now known to occur; taiga dominated by conifers can revert to moss-dominated muskeg for example. Miles (1979, 1987), whose reviews of the subject are of great interest, now considers, with others, that the only fixed attribute of plant succession is floristic change, and that the quest to find a unifying theory of succession is a pointless one. On the other hand, an understanding of how succession operates in particular habitats is of great value to vegetation managers, especially those concerned with forestry or woodland conservation. In such cases computer modelling is often valuable in predicting future standing crop, trends in age cohort and species composition.

Primary succession involving trees frequently occurs where regions of completely bare ground (landslip scars, bottoms of drained lakes, areas exposed along rising coastlines or cleared by fire) are colonized by organisms within reach of the bare substratum. As the vegetation develops it provides shade, enriches the soil with humus and, together with the associated animal communities, generally modifies the often harsh conditions originally present. The original (pioneer) plants tend to be suppressed by competition from the mesophytes, which under the changed conditions tend to grow more vigorously than the pioneers. Successional changes within plant communities provide a variety of niches for heterotrophs so that, for example, successions of insects, birds and mammals and of soil fungi have been recognized as accompanying plant successions.

The classical concept of succession was developed by a number of workers, notably by Clements (1916), who emphasized both the re-action of organisms with the environment and the importance of com-petition between members of the changing populations which come to occupy the original site. This is often termed the **facilitation** or **relay floristics** model of succession, the latter term being a reference to the sequential change in the major species. It culminates in the final stabilization of the community when it reaches a **climax** at which it is held to be in equilibrium with its environment. The views of Clements (1916, 1936) on the nature of the climax are now out of tune with modern opinion, which holds that as the environment is constantly changing, no matter how slowly, the vegetation is eventually bound to do the same. Nevertheless the idea of climax vegetation remains a useful concept, provided it is realized that its permanence is only relative and that apparently similar successions may result in more than one endpoint.

Krebs (1985) provides a concise description of the three major models of succession, not one of which can by itself account for the complete range of floristic replacements found in nature. In the **facilitation** model species replacement is assisted by environmental changes brought about earlier in the succession, whereas in the **inhibition model** it is pre-vented by the present occupants of the site, perhaps through heavy shading or allelopathic mechanisms (section 1.1), and replacement will only occur when these are damaged or killed. The species which colonizes the site first thus gains a major advantage; succession on this model involves the replacement of short-lived species by long-lived species. The third, or **tolerance** model is intermediate between the first two. It does not depend on the initial presence of early successional species. Any species can start the succession but those which establish first are replaced by others which are more tolerant of limiting resources.

Climax vegetation

The climatic climax is the only climax recognized by Clements (1936), a view often referred to as the **monoclimax theory**, though he noted the way in which soil reaction, competition, migration barriers, and man could prevent the full development of the 'true' succession and so cause a number of 'sub-climaxes'. He thought of succession as a deterministic process and of a patch of vegetation as being a highly organized society of plants, whereas Miles (1979) considers it to be 'an aggregation, essen-tially temporary in nature, of independently behaving plant species occurring on a site where each can grow and happens to have arrived through the chance processes of dispersal.' Clements also viewed a plant formation as a supra-organism which arose, grew, matured and died. Tansley (1939) placed greater emphasis on factors which could differentiate vegetation from the climatic climax, stabilizing it in corresponding edaphic, physiographic, biotic or anthropogenic climaxes (the **polyclimax theory**). Tansley drew attention to the dynamic nature of vegetation: 'each plant community has had an origin and will have a fate'. He realized that 'even climatic climax communities, though seemingly permanent so long as the existing climatic complex persists, may contain within them the seeds of their own decay', as when they cause changes in the soil which will eventually prevent the regeneration of the dominant species.

Primary succession consists of all the phases involved in the development of mature (or climax) vegetation on an area of bare ground. This sequence of phases constitutes a **primary sere**. The Glacier Bay and Krakatau sequences (sections 5.4 and 5.5) are striking examples of primary succession accompanied by parallel changes in the development of soil. Frequently, however, vegetation undergoes succession after being damaged but not destroyed, as when forest is clear felled and then abandoned in the hope that it will regenerate naturally. **Secondary succession** occurs on disturbed ground with an existing soil, as in American old field sequences where secondary forest has developed on abandoned farms (section 5.4). Almost all the species involved in such successions are present at its outset as buried seeds, roots, rhizomes, etc., or invade shortly afterwards. Egler (1954) emphasizes the importance of this **initial floristic composition** factor: as each successive group drops out of dominance another, there from the start, takes its place. Annual weeds are replaced by perennial grasses which are later supplanted by shrubs and trees, but viable propagules of all the species involved tend to remain in the general area throughout the succession. The first part of this succession fits the inhibition model; in North Carolina the horseweed (*Erigeron canadensis*) inhibits the aster which eventually succeeds it by both shading and influence of chemicals exuded from its decaying roots. On the other hand, the oaks which replace the pines much later in the succession cannot develop until enough pine litter has accumulated to protect the acorns from desiccation and enable the soil to hold more moisture.

A **plagioclimax** develops when the vegetation is subject to some influence which deflects the progression of a sere in an unnatural way, such as the repeated selective felling of the most economically valuable species in a forest. The roles played by deer and other animals in deflecting forest successions can be demonstrated by **exclosure experiments**, as in northern Wisconsin, where protective legislation has led to high populations of white-tail deer (*Odocoileus virginianus*). These animals

browse saplings of both sugar maple (*Acer saccharum*) and eastern hemlock (*Tsuga canadensis*), but the latter is more severely damaged. The
rapid replacement of hemlock seedlings and saplings by sugar maple was
reversed in trial plots from which the deer were excluded.

It is now clear that **natural disturbance** occurs far more frequently,
and is much more important in forest dynamics (Pickett and White,
1985), than was allowed for in the classic self-reproducing equilibrium-
type climax (cf. Jones, 1945). Indeed, in the cold damp climate of northern
Finland old spruce stands become moribund; reproduction is poor even
when gaps appear in the original canopy. Decomposition here is so slow
that nutrients are not released in quantity until the site is burnt and
normal regeneration becomes possible. In such an area a stable self-
reproducing climax forest will never be achieved. Moreover, many North
American forests previously thought to be examples of 'climax vegetation'
are in fact large, wholly or partially even-aged stands which regenerated
after fire (evidence for which is provided by charcoal in the soil), so they
must now be considered in an entirely different light. Even the forested
areas of the tropics are no longer believed to be as unchanging as previously thought (section 5.2).

Pickett and McDonnell (1989) claim that **community dynamics**
'emphasizes process rather than the end point, accommodates the richness of causes of succession, and motivates diverse research approaches.'
They further point out that the dynamics which lead to succession are
often **allogenic** (controlled by outside factors) rather than **autogenic**
(governed by biotic interactions within the community).

Gales often modify the structure of forests. As long ago as 1913 Cooper
found 'a mosaic or patchwork which is in a state of constant change' on
the Isle Royale, in the northwest of Lake Superior, where old wind-
throws regenerated as new ones developed. A similar structure is found
in many forests; Sernander (1936) described the mosaic found in old-
growth Swedish spruce forests on unstable morainic soils as the 'the
storm-gap structure'. Figure 5.1 is a map of a typical area of Fiby urskog,
near Uppsala, which still possesses a structure very similar to that described by Sernander, whose efforts preserved it from felling. The radial
increment diagram for a mature spruce shown in Figure 5.2 illustrates
the marked increase in width of the annual rings of this tree immediately after 1795, when very many mature trees were blown down.
Figure 5.3 shows a typical storm gap.

Watt (1947) investigated the relationships between the distribution
patterns found in various plant communities and the processes by which
they were produced. In many such instances there is a mosaic of patches
or phases which are dynamically related to each other and the vegetation
is in a state of **cyclic change**, with the dominant species exhibiting the
phasic series pioneer, building, mature and degenerate during which its

(a)

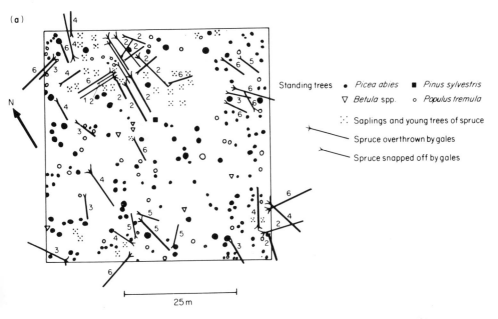

Standing trees • *Picea abies* ■ *Pinus sylvestris*

▽ *Betula* spp. ○ *Populus tremula*

∴ Saplings and young trees of spruce

Spruce overthrown by gales

Spruce snapped off by gales

25 m

Figure 5.1 Diagram of an area of the Fiby Forest, near Uppsala, Sweden, showing 'storm gaps' resulting from the windthrow of Norway spruce (*Picea abies*) on boulder moraine. Storm gaps also develop on marly moraines. Seeds of spruce often germinate on the fallen trunks whose degree of decay (necrotization 1–6) is indicated. In stage 6 the area is usually entirely covered by a growth of moss, but a row of young trees may mark the line where the trunk fell and subsequently decayed. (Redrawn from Sernander, 1936.)

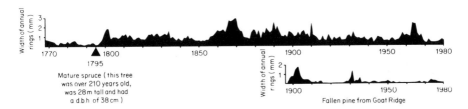

Mature spruce (this tree was over 210 years old, was 28 m tall and had a d b h of 38 cm)

Fallen pine from Goat Ridge

Figure 5.2 Radial increment diagrams showing variation in the width of the annual rings of a mature spruce (*Picea abies*) and a small Scots pine (*Pinus sylvestris*) from Fiby Urskog, near Uppsala, Sweden. The annual rings of the spruce show a marked increase in width immediately after 1795; it appears that this tree was one of those whose growth was released when storm gaps were torn in the forest canopy by the great storm of that year. The pine was only 3.9 m high. Its roots were entirely superficial, being developed in a mat of humus and vegetation (in which lichens were conspicuous) overlying solid granite. Though stem diameter where the tree was bored (10 cm above the ground) was only 6.5 cm, the increment core showed over 80 annual rings, some of them paper thin. (Unpublished data of H. Hytteborn and J. R. Packham.)

Figure 5.3   Storm gap and two standing dead trees in spruce forest on boulder moraine, Fiby urskog. The spruce were blown down a relatively short time ago and most of their trunks are still covered by bark. Regrowth of spruce and aspen occurs in the foreground, the aspen have been grazed by elk. (Photograph by Håkan Hytteborn.)

competitive ability changes. This series develops naturally in unmanaged woodlands, which frequently possess mosaic structure, but the same sequence is also exhibited by plantations, where it can be termed the **stand cycle**. During the **pioneer phase** of a woodland, tree seedlings frequently grow in an open area in which many herb species flourish. The canopy is closed by lateral contact between saplings early in the **building phase**, during which intra-specific competition is so intense that many trees are suppressed and die (Cousens, 1974). Accumulation of woody biomass reaches its peak in this phase during which, particularly in plantations of evergreen conifers, the ground flora is greatly diminished. In the **mature phase** competition between the trees is reduced and they bear their heaviest crops of seed. The canopy becomes less dense and shade-tolerant herbs are able to establish. The **degenerate phase**, in which biomass accumulation is small and eventually negative, seldom occurs in commercial forests. Heart rots and other decay organisms reduce the quality and quantity of the wood. Branches die, trees often become stag-headed, dead organic matter accumulates in the ecosystem and herbs receive far more light (Figure 9.3).

Watt describes how in the 1920s there was intense shade and root competition between the young trees of even-aged *Fagus sylvatica* on the South Downs. Their initial high density decreased markedly as the woods aged and a ground flora dominated by *Oxalis acetosella* developed where the soil, which rested on chalk, was not too shallow. Bramble later succeeds wood sorrel as the ground layer dominant and Watt used this bare ground–*Oxalis acetosella*–*Rubus fruticosus* sequence to date stands of beechwoods growing on different soils. The ground layer succession proceeds most rapidly in beechwoods on soils of the greatest depth and surface acidity, normally correlated with the highest humus content, lowest calcium carbonate level and closest texture.

Figure 5.4 shows the spatial distribution of bare, *Oxalis* and *Rubus* phases in an old unmanaged beechwood, together with the gap phase in which regeneration occurs following the death of old beech trees. The gaps may at first be dominated by another kind of tree, such as ash, oak or birch, whose lighter shade allows young beech to grow more successfully than beneath *Fagus* itself. Such ancient woodlands with an uneven age structure give an idea of what primaeval wildwood was like and show the importance of cyclic change, often involving other species in gap regeneration, in maintaining temperate climax forests dominated by a single species. Though the mosaic pattern constantly changes, the overall species composition of a large area of such a forest remains broadly similar. However, as in many natural populations, particular age groups may be represented much more strongly than others. The events responsible for the existence of a large proportion of ancient trees in a semi-natural wood may have happened 200 or 300 years ago, when heavy masting and weather conducive to germination followed the death of many old trees from some general cause such as water stress.

In the montane coniferous Vallibäcken forest, northern Sweden, at lower altitudes a cyclic change occurs that is governed by the occurrence of storm gaps. These are created mainly by winds from the north-west, so that logs fall in a southeasterly direction (Hytteborn *et al.*, 1987). *Picea abies* and *Betula pubescens*, the dominant tree species, differ in their population dynamics due to their differences in total life-span and in

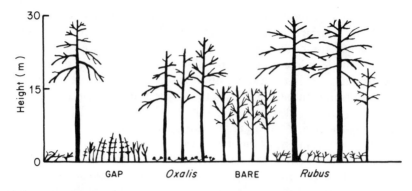

Figure 5.4 The distribution of the phases in space when an old beechwood has been left to itself and has trees of all ages. Beech (*Fagus sylvatica*) is forming a reproduction circle in the gap (see Figure 5.5). This typically has young ash (*Fraxinus excelsior*) in the centre. (Redrawn from Watt, 1947; by courtesy of *Journal of Ecology*.)

shade tolerance. The process of cyclic change can best be described in a simple model which allows spruce to regenerate both inside and outside the gaps, in the latter case being able to survive for a long period in the form of dwarf trees. Birch regenerates everywhere but only yields survivors in the tree layer if they regenerated in a period when a gap was formed (see also section 4.4). Establishment of both species is good at lower altitudes and takes place primarily on logs. Gap regeneration is equally important in tropical forests as Martinez-Rames *et al.* (1988) point out.

Gap regeneration

What allows so many plant species to grow side by side in tropical forests and ancient temperate deciduous woodlands? Competition between animals is often seen to be very direct; these mobile organisms often engage in physical combat or compete for the same food. This relates closely to the principle of **competitive exclusion** which was defined by Gause (1934) '. . . as a result of competition two similar species scarcely ever occupy similar niches, but displace each other in such a manner that each takes possession of certain peculiar kinds of food and modes of life in which it has an advantage over its competitor'. In so writing he was using the definition given by Elton (1927): 'the **niche** of an animal means its place in the biotic environment, its relations to food and enemies'. Zoologists generally accept the idea that every animal must occupy a different niche. The full requirements of an animal are complex and it is easy to envisage many types of animal niche, whereas all photo-autotrophic plants need light, carbon dioxide, water and the same mineral nutrients, though their tolerances of various environmental conditions differ.

Woodland autotrophs, in contrast to animals, are sedentary when adult, so many of the most critical aspects of competition between them must concern the struggle to occupy any areas of ground which become bare. Mobility in green plants is usually restricted to the seeds or other propagules, which are dispersed in a variety of ways (section 3.2). Mechanisms which contribute to plant diversity in woodlands without involving the **regeneration gap** are reviewed in section 4.5. Grubb (1977) shows that these alone are insufficient to provide an adequate explanation of why species richness can persist in plant communities; it is a failure to consider the nature of regeneration that previously caused many to reject 'Gause's hypothesis' with respect to plants. Heterogeneity of the environment (physico-chemical and biotic) ensures that the replacement plant, or plants, which come to occupy the regeneration gap left when a plant dies may sometimes be of a species different from that present formerly. The replacement stage is thus crucial to our understanding of species diversity in plant communities, and to the basic processes of evolutionary divergence in plants.

Only when differences in their seeding patterns, germination and early life are considered is it possible to understand why so many species of essentially similar life form, adult phenology and habitat range persist together in species-rich communities. The size of the gap is itself important; when a large tree falls the gap produced is likely to be dominated next by a light-demanding species such as pine or ash, whereas shade-tolerant species such as beech or western hemlock are more likely to be the next occupants where the gap in the canopy is small.

The successful invasion of a gap by a given species will be influenced by the timing of the processes leading to the production of the means of vegetative reproduction (e.g. stolons, rhizomes, suckers, bulbils) or, more usually, of viable seed (flowering, pollination, setting of seed), its dispersal in space and time, requirements for germination (section 3.2), establishment and onward growth. The orientation, size, shape, and time of formation of gaps help to provide heterogeneity in the environment, as do the nature of the soil surface, the litter, other plants, animals, fungi, bacteria and viruses present. All stages and processes leading to the replacement of a mature tree or other plant by another are important, though as Grubb remarks 'The idea that all these stages are important in the maintenance of species-richness has all too often been ignored'.

Foresters realized early the importance of differences in regeneration requirements; at first tolerance to shade received emphasis but later differences were also recognized in the ability to tolerate root competition. Aubreville (1938) suggested that any one tree species never regenerated under itself in the African rainforests that he had studied. This is not true of forest trees in general, though direct regeneration beneath a complete canopy of the parent tree does not occur in some species.

Three factors are particularly significant in controlling the sequence of events during colonization of a gap developed in *Fagus sylvatica* woodland.

1. The age of the woodland. If the wood is not completely mature but still has much ash and oak there will be plenty of shrubs and herbs to colonize the gap. Later in the succession, when the beech is almost pure, the lower strata may be largely suppressed or even absent. If bramble (*Rubus fruticosus* agg.) colonizes the woodland floor at about the time the beech begins to bear seed it may grow so luxuriantly in the gap as to prevent regeneration, at least for a time. In a natural forest, however, the passage of large mammals would probably damage the bramble cover sufficiently for ash or oak seedlings to grow up and at least partially suppress the bramble, thus enabling beech to develop again.

2. The size of the gap is important because beech mast formed in high forest drops almost vertically and is largely confined to the margins of wide gaps, the centres of which are often occupied by a dense growth of common ash (Figure 5.5) whose seed is wind-dispersed. The occasional beech seedling has little chance of establishing in the central core of ash until the taller growth and deeper shade of the young peripheral beech, together with the gradual extension of the parent beech canopy, tend to suppress the more light-demanding ash. The outcome varies, but in at least some cases the ash in the centre of the gap grows upwards fast enough to avoid suppression. In narrow gaps the whole floor can be seeded by the beech whose seedlings and saplings may not meet with competition from any other tree.

3. The time when the gap is formed in relation to a full-mast year. If the beeches have just seeded heavily their seedlings can establish themselves before a serious competitor, such as bramble, can fill the

---

Figure 5.5   (a) Reproduction circle of beech (*Fagus sylvatica*) from Figure 5.4. The gap is in beech associes (sere 2) and has ash in the centre and a peripheral zone of young beech. Beech has epigeal germination; the two semi-circular cotyledons of the seedling (b) are photosynthetically active for a considerable period. ((a) Redrawn from Watt, 1925; by courtesy of *Journal of Ecology*.)

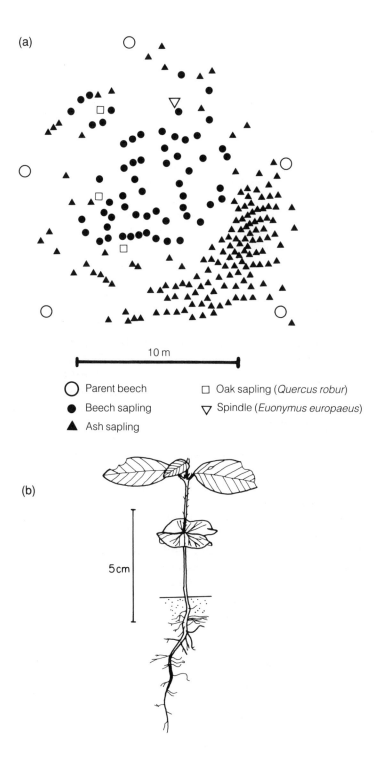

(a)

10 m

○ Parent beech   □ Oak sapling (*Quercus robur*)

● Beech sapling   ▽ Spindle (*Euonymus europaeus*)

▲ Ash sapling

(b)

5 cm

gap. When beechwoods are managed on the selection system it is important to fell and extract mature trees in winters when there is plenty of viable mast in the soil. Even when this operation was carried out 30 years previously, mapping of the surviving, relatively young, trees often provides evidence of the original reproduction circles.

Seed banks

Seed availability is crucial to the subsequent development of the re-generation gap. In practice early successional species usually appear soon after woodlands are disturbed by fire, cultivation, severe thinning or windthrow. There is strong evidence that these plants often develop from seeds which have been long buried in the soil. For example, in-vestigations of mineral soil cores (taken after careful removal of up to 8 cm of litter and humus) from 16 sites in successional field and forest stands in Harvard Forest, Massachusetts, showed that seeds of many field species germinated in virtually all the soils tested. The number of species present as viable seed was not greatly different whether the soil came from recently abandoned cultivated fields, dense *Pinus strobus* plantations devoid of ground cover, or an 80-year-old *P. strobus* stand with a hardwood understorey which represented the mature phase of the secondary succession from old field. Seed populations beneath *Pinus resinosa* and *Picea rubens* were similar to those under *Pinus strobus* (Livingston and Allessio, 1968).

However, an old growth coniferous forest in British Columbia had a far smaller seed bank than an adjacent recently cutover area (Kellman, 1974). The seed bank beneath old growth forest here seemed to be insufficient to account for the large herb populations which soon develop when fresh areas are logged.

Seeds of late successional and climax species, together with those of most trees, tend to be deposited in surface litter and humus. Many are carried into forest stands by wind, while birds and mammals transport others. Nevertheless, woodland floors frequently contain viable seeds of many successional species from pioneer to climax. This can be of com-mercial importance as in the control of blister rust in the western white pine (*Pinus monticola*) in North America where *Ribes* spp., alternate hosts of the fungal pathogen, developed again after systematic eradication, cutting of old growth forests and burning of slash. Viable seeds present in the soil were favoured by humus reduction and the increased light and temperature (Olmsted and Curtis, 1947). Various species of *Rubus* frequently establish rapidly when European and North American wood-lands are disturbed. The fact that forest soils contain so many of their seeds, which are widely dispersed and remain viable after passing through the gut of many birds and mammals, is a major contribution to the success of this genus.

5.2   The changing mosaic: a comprehensive theory of forest maintenance

Whitmore (1982) builds on the ideas of A. S. Watt to develop a theory applicable to forests everywhere believing, like Shugart (1984), that they are fundamentally similar despite great differences in structural complexity and floristic richness. The basic similarity of successional patterns in space and time is the product of the same processes. Though the species may differ, all are subject to the forest growth cycle with its gap, building and mature phases. As we have already seen, gap size is crucial: it may vary from a metre or two to several km across. Within gaps, whose shape and orientation are also important, new tree shoots grow up from seeds, seedlings, roots, stumps or fallen trees.

Foresters have long been aware of the **forest growth cycle,** or stand cycle in the case of plantations, and manipulate it to produce the desired product. Productivity is greatest in the building phase, fuelwood and chemical feedstock are often harvested early in this phase, while the crop is economically mature with regard to poles, and chipwood later when current annual increment begins to drop (Figure 10.5). High quality timber should be harvested early in the mature phase of the cycle; if left longer shakes and other faults may develop in the timber. If the forest is to be maintained for conservation mature and degenerate phases, together with a range of patches of all sizes and ages, will enhance ecosystem and hence niche diversity amongst both plants and animals.

A natural forest is in continual flux, being a mosaic of structural patches or phases which constantly alter as one phase changes to the next. The scale of the mosaic depends on what creates the gaps; an individual tree may die or huge numbers may be destroyed by gale or fire. Light-demanding species establish where gaps are large, while shade-tolerating forms flourish where gaps are small.

Foresters have commonly placed trees in **tolerance classes,** based largely on their ability to survive shade while regenerating in the understorey. Spurr and Barnes (1980) give five such classes ranging from very tolerant, e.g. *Tsuga heterophylla*, where growth can occur with irradiance as little as 1–3% of full sunlight, to very intolerant, e.g. *Pinus contorta*, where the young tree requires at least 60% of full sunlight. In a Chilean forest co-dominant tree species were found to have a continuum of responses to different scales of disturbance (Veblen *et al.*, 1979). The main canopy species, *Aexticon punctatum*, regenerated under continuous shade or in small openings, while at the other extreme *Nothofagus obliqua* was only able to do so in the open sites left when forest stands were destroyed.

Ashton (1977) supported the view that the great richness of tree species in tropical rainforests is based on niche differentiation, but though tree species do differ greatly in their ecology the evidence suggests that in these complex communities many species have largely overlapping or almost identical niches. Thus any one of several species may be well-fitted to grow in a gap of a particular size, so it is largely a matter of chance, perhaps dependent on the presence of a seed or seedling when the gap is formed, that determines which species is successful (Whitmore, 1975).

Finally, multiple gap sizes must be considered. If small gaps alternate in time with large ones, floristic fluctuation results. Thus Whitmore (1982) postulates that gap-forming processes drive the forest growth cycle and greatly influence forest floristics. It seems reasonable to include within climax forest those areas undergoing cyclic regeneration by gap phase replacement. Large areas undergoing floristic succession in their recovery from catastrophic disruption by wind, fire or volcanic action are another matter.

Floristic richness in European temperate deciduous forests was, in contrast to the forests of East Asia and North America, greatly reduced by extinctions during the glacial maxima of the Pleistocene; British forests are now especially impoverished (section 5.6). It is now believed that tropical forests were also affected by Pleistocene climatic change. Tropical rainforests were, until the present continuing destruction by man, at or near their maximum extent and have attained these dimensions for only a small fraction of the last two million years (Flenley, 1979). Factors which influence **species richness** at successively smaller scales are given below.

1. **Availability of flora** which depends on past extinctions and the way evolution has progressed. Pleistocene climatic fluctuations are important here. An extreme example of how capriciously these processes can operate is provided by the tropical rainforests of west Malesia most of which are, apparently uniquely, dominated by a single tree family, the Dipterocarpaceae.
2. **Site differences** like those between the heath forest, peatswamp forest and lowland evergreen forests in Borneo.
3. **Variation within each forest formation** such as the structural and floristic differences between the evergreen rainforest on ridges, hillsides and valleys at Andulau, Brunei.
4. **Forest dynamics** largely determine species composition within particular patches because some species are adapted to regenerate in large gaps and others in small.

Thus species richness is maintained by continuing disturbance at several scales, indeed the 'intermediate disturbance hypothesis' (Connell, 1979) postulates that the most species-rich forests are those which, because they are still recovering from major disturbance, contain both pioneer trees and their successors. This is supported by observations on old field successions in eastern USA (Spurr and Barnes, 1980, p. 445) and by a number of tropical examples. It does not always hold, however; richness is limited by the available flora and the ability of the individual species to adapt to the available regeneration niches.

Computer modelling and the relative abundances of forest trees

As many trees live for hundreds of years and some for millennia, real data on long-term forest dynamics are scarce. The use of computer modelling to make maximum use of the information available is a valuable approach provided its limitations are recognized. Shugart and Urban (1989) use this technique in investigations which start from the premise that relative abundances of tree species in a forest vary along environmental gradients, and change at a given site-type as a result of gap creation. The forest is seen as a mosaic whose dynamics are driven by the demographic mechanisms of tree establishment, growth and mortality. The first approach employed is to use a set of forest simulation models based on the fates of individual trees, thus exploring long-term implications of tree-to-tree as well as tree–environment interactions. The second involves using a functional classification of tree species, with species roles being defined according to the coupling of mortality and regeneration.

In addition to biotic factors, the relative abundances of forest trees are influenced by:

(a) **environmental filtering**: the sorting of trees by available light, water regime, temperature, mineral nutrient supply and other abiotic factors;
(b) **competition** between trees.

Both types of process were investigated in the context of a model conceptually similar to that put forward by Whitmore (1982). It involved changes in species composition along various environmental gradients, when subject to different disturbance regimes, and in response to disease and 'stress' over periods as long as 18 000 years.

The system shown in Figure 5.6 involves a basic 2 × 2 categorization in which two of the four roles (1, 4) are self-reinforcing. Role 1 species create the gaps they need for regeneration, while role 4 species can regenerate in shade and do not markedly open the canopy when they die. When trees of role 2 and role 3 species die they tend to relinquish

(a)

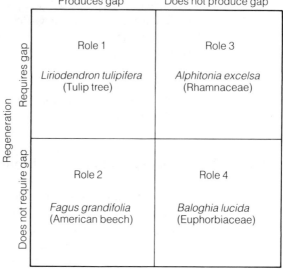

Mature tree mortality

	Produces gap	Does not produce gap
Requires gap	Role 1 *Liriodendron tulipifera* (Tulip tree)	Role 3 *Alphitonia excelsa* (Rhamnaceae)
Does not require gap	Role 2 *Fagus grandifolia* (American beech)	Role 4 *Baloghia lucida* (Euphorbiaceae)

Regeneration

(b)

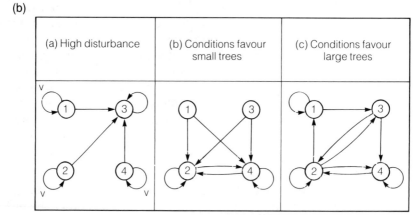

(a) High disturbance	(b) Conditions favour small trees	(c) Conditions favour large trees

Figure 5.6 (a) A categorization of four roles for forest trees, dependent on whether or not they produce a gap when they die, and whether or not they require a gap for regeneration (based on Shugart, 1984.) Information for the examples named is used in the simulations shown in Figure 5.7. (b) The effects of the conditions under which a forest may develop on the patterns of interrelationships found among trees with the four roles defined above (based on Shugart, 1987). Arrows show the most likely transfers of space following the death of a tree.

space. When their individuals die role 2 species tend to let in both role 1 and role 3 species, both of which need a gap to regenerate. The death of role 3 species does not create the gap they need to regenerate, so entry of role 2 and role 4 species is favoured. Figure 5.6 shows the effects of three sets of conditions on the patterns of interrelationships found among trees with the four roles envisaged. This example is evidence of the potential value of modelling techniques, which should act as a spur to further fieldwork.

The ecology of the species used to exemplify the four regeneration roles is of considerable interest (Shugart, 1984). Models showing total biomass and stem numbers when grown in single-species stands are shown in Figure 5.7. *Liriodendron tulipifera* (Magnoliaceae, role 1) regenerates best on mineral soil with adequate light and moisture. It regularly produces large numbers of wind-dispersed seeds which can survive up to 7 years in soil. The species is shade-intolerant, grows rapidly and can reach a height of 50–55 m. Large trees are often windthrown. It was common in the original southern Appalachian forests where it was potentially the largest tree. There is a strongly accentuated saw-toothed biomass curve with very high biomass maxima. The associated number of stems give a curve saw-toothed in the opposite direction, with heavy recruitment (followed by gradual thinning) as the last of the previous cohort die.

Fagus grandifolia (Fagaceae), a smaller tree, is shade-tolerant and long-lived. In this species, as in many other role 2 trees, vegetative reproduction may be as important as seedling regeneration. The relatively large edible seeds need to be sheltered in leaf litter if they are not to be consumed by squirrels and other forest mammals. The saw-toothed biomass curve has lower maxima and larger minima than that for *Liriodendron*. Stem numbers again show increases following decreases in biomass.

Alphitonia excelsa (Rhamnaceae, role 3) is a soft-wooded, fast-growing, and shade-intolerant tree of the Australian rainforest whose life is usually short. The seeds of this relatively small tree are widely distributed in the soils of the forests in which it grows. Regeneration follows death of a group of trees, the gap left by a single tree is too small. Stem numbers show a regular series of peaks with time. Role 3 species are termed 'nomad' or 'pioneer' trees by some authors.

Baloghia lucida (Euphorbiaceae, role 4) reaches *c.* 20 m in height and has relatively hard wood, it can be quite abundant as a subcanopy tree in mature Australian rainforests. In the model, numbers and biomass tend to be more stable after the first hundred years than in the three species just discussed.

If numbers of species can be grouped together on a functional basis it simplifies discussion of their behaviour. Shugart (1984) draws attention to the way in which Whitmore (1975, pp. 69–73) uses regeneration

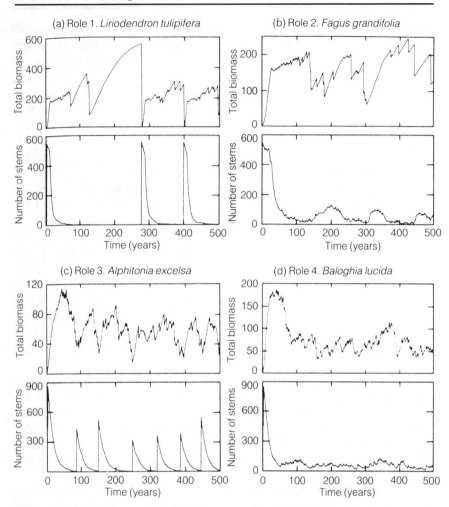

Figure 5.7   Number and biomass dynamics of hypothetical patches in forests of a single species that conform to the roles shown in Figure 5.6. (Redrawn from Shugart, 1984.)

strategy, shade tolerance, and tree size to create essentially the same categories as the four roles considered above.

Effects of changing climate

Shugart *et al.* (1980) investigated the behaviour of a hypothetical two-species mixture of American beech (role 2 *Fagus grandifolia*) and yellow-poplar (role 1 *Liriodendron tulipifera*) as climate slowly changed. The growth rates of both species increased under warmer conditions, but both could regenerate under all the climatic conditions considered. In cool climates growth rates were slow and all trees tended to be smaller with *Fagus* dominating at the lowest degree day values. Under warmer conditions trees became larger and *Liriodendron* dominated the stand. The effect was very clear-cut at either end of the climatic range considered, but the types of forest found at intermediate values showed a lag, or hysteresis. At 4750 day-degrees, for example (section 3.5), *Liriodendron* accounted for around 20% of stand biomass if the climate had been gradually warming, but was over 80% if the climate had been cooling from a much warmer level.

5.3   Patch dynamics and regeneration waves

Picea abies is best thought of as a role 1 tree, needing gaps and growing large enough to produce them; though shade-tolerant its saplings grow very slowly until released by a gap in the canopy above them. In Fiby urskog, and Scandinavian boreal forests in general, *Pinus*, *Betula* and *Populus tremula* are role 3 trees which do not themselves produce the large gaps they need to regenerate. With patch sizes from 1000 m² down to 50 m² the revised FORSKA computer model (Prentice and Leemans, 1990) produced an accurate simulation of the dynamics which Sernander (1936) described for Fiby (section 5.1). *Pinus*, *Betula* and *Populus*, all light-demanding species, are important early in the simulated succession but are eventually replaced by *Picea*.

Prentice and Leemans (1990) conjecture that two types of forest dynamics can be distinguished because the trees involved have different ratios of maximum crown size to patch size. Forests of narrow-crowned trees typically show '**Sernander-type dynamics**' with directional succession towards a self-replacing climax of shade-tolerant trees. Competitive exclusion of light-demanding species from such forests is prevented only by agencies that create multiple-tree gaps. Various authors agree that a patch size of around 1000 m² is critical. Such a size typically allows 80% of incoming diffuse light to reach the forest floor; a proportion which falls to 20% with a patch size of 100 m². In Fiby urskog storm gaps often reach 1000 m² while crown sizes of the largest canopy *Picea abies* have an area of only 60 m²; indeed in Scandinavian boreal forests canopies of this tree do not exceed 100 m². The prevalence of Sernander-type dynamics in the boreal zone may be partly due to low sun angles at high latitudes, but tree morphology seems to be the main factor concerned.

In temperate forests with '**Watt-type dynamics**' crown size is much larger in relation to patch size and the fall of a single tree can create an opening suitable for colonization by a light-demanding tree.

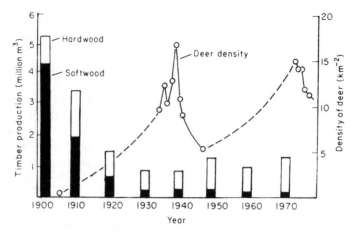

Figure 5.8   The relationship between yearly timber production by volume in Pennsylvania since 1900 and estimated deer population densities in the Allegheny National Forest. Broken lines represent periods for which the data are sparse or inadequate. The decline in the 1940s was due to a lack of browse and high winter mortality. (From Whitney, 1990; courtesy of *Journal of Ecology*.)

Disturbance regimes: response of a hemlock–hardwood forest to exploitation

The Allegheny National Forest (ANF) was established in 1923 on the Allegheny High Plateau in Pennsylvania, where *Fagus grandifolia* (American beech, *c.* 43%) and *Tsuga canadensis* (eastern hemlock, *c.* 20% of the whole ANF) originally dominated the moister regions. Outliers of the more southern Appalachian oak forest occupied the more xerophytic sites such as those on upper slopes or with stonier soils. Considerable amounts of very large *Pinus strobus* were felled in the early post-European settlement period (1797–1880) and floated out down river, but the most intensive felling was between 1880 and 1930 (Figure 5.8). Leather tanning demanded huge amounts of eastern hemlock bark, generally six times the weight of the dry hides. Once the logging railroads had been installed softwoods and hardwoods of almost all species were exploited, and an extensive hardwood furniture industry developed. Before the petro-chemical industry rose to prominence in the 1920s almost all important organic chemicals were derived from the distillation of wood. This was the cause of the next major assault on the trees of the High Plateau, clear cutting for chemical wood which by 1922 was using trunks down to a 5–8 cm diameter limit.

The most notable change in the history of the ANF has been the drop in extent of the hemlock–northern hardwood type (*Tsuga canadensis, Fagus grandifolia, Betula* spp., *Acer rubrum, A. saccharum* and other hardwoods) from 83.4% of a total land area of 207 000 ha in 1800 to 15.8% in 1986, when it consisted predominantly of maple, beech and birch. During the same period the area covered by the Allegheny hardwood type (black cherry and maple) has risen from 6.8 to 53.0% with black cherry (*Prunus serotina*) increasing from 1% to over 22%.

These changes, though partly caused by selective logging, can largely be attributed to the nature of the disturbance regimes imposed by man. Whitney (1990) considers that in many ways the primary forest of the High Plateau region approximated to the stable, self-reproducing climax community envisaged by Clements. Large-scale destruction, by windthrow and fire, occurred perhaps once in a thousand years. Small-scale disturbances involving the occasional loss of single trees due to old age, drought or disease favoured the very shade-tolerant beech and hemlock, whose individuals in some cases went through a number of release–suppression cycles before reaching the overstorey. Clear cutting, fire and overbrowsing destroyed the suppressed saplings of *Tsuga*, while heavy cutting reduced the proportion of *Fagus* in second- and third-growth stands despite the ability of this species to reproduce from suckers. In contrast shade-intolerant species such as black cherry, and trees which resprout vigorously from cut-over seedlings and saplings (red and sugar maple), have greatly increased.

Market hunting almost eliminated white-tailed deer (*Odocoileus virginianus*) from Pennsylvania in the late 1800s, but its numbers have since greatly increased (Figure 5.8). The density of this animal should be less than 8 deer km² if the forest is to regenerate successfully; today browsing in the ANF is so intensive that most regeneration growth below 2 m high has been eliminated.

Wave regeneration

Sprugel (1976) describes a particularly regular type of cyclic change which occurs in balsam fir (*Abies balsamea*) forests at altitudes of over 1000 m in north-eastern USA. Here aerial photographs show crescentic waves of dead trees which were first thought to represent windthrown areas. Closer examination showed almost all the dead trees at the fronts of the crescents to be still standing; they did not fall until some time after death. In **wave regeneration** mature trees continually die off at the front edge of a wave, which lies behind an opening in the forest canopy and is exposed to the prevailing wind. Young trees spring up

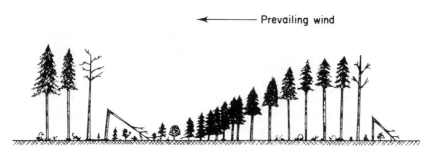

Figure 5.9 Diagrammatic cross-section through a regeneration wave in an *Abies balsamea* forest, Whiteface Mountain, New York State. (From Sprugel, 1976; by courtesy of *Journal of Ecology*.)

again after the wave has passed on, at a speed varying between 1 and 3 m y^{-1} according to the degree of exposure (Figure 5.9). Endogenous factors play a part in this rhythmic pattern. Balsam fir is moderately short-lived and becomes increasingly susceptible to stresses, particularly those caused by pathogens, at the age of 50–60 years. The trees bearing the brunt of the prevailing wind are very commonly partially senescent but are finally killed by environmental stresses which cause a loss of branches and needles in winter due to accumulations of rime-ice (Foster, 1988), death of needles owing to winter desiccation, and decreased primary productivity because of cooling of the leaves in summer. Their death releases the young saplings previously held in check by heavy shade and also exposes the trees immediately downwind to much greater environmental stress. Initiation of the individual waves is likely to occur when one, or a small group of trees, is killed by localized stress (e.g. wind-throw or butt rot) exposing the arc of trees standing in the lee.

Regeneration waves in the *Abies balsamea* forests of New York, New Hampshire and Maine tend to follow each other at 60-year intervals. Very similar moving regeneration waves occur in *Abies vietchii* and *A. mariesii* forests at 2000–2700 m on several mountain ranges in Japan, but these firs grow more slowly and live longer (to about 100 years) than balsam fir so the waves tend to be further apart. In wave-regenerated forests there is an endless cycle in which the trees degenerate, die, regenerate, mature and degenerate again at any one point. The composition of the forest as a whole, however, remains relatively constant with time and the ecosystem is in a **steady state**.

Robertson (1987), who made a detailed study of the aerodynamics, suggests that a series of helical roll vortices (Goertler vortices) are mainly responsible for the formation of wave forests. Sinusoidal regeneration waves in the *Abies balsamea* forest of Spirity Cove, Newfoundland, the largest wave forest known, are characterized by dead tree strips 100–150

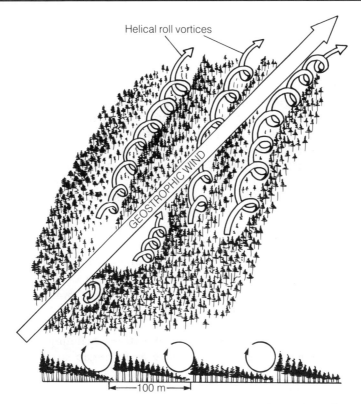

Figure 5.10 Hypothetical diagram illustrating the aerodynamic mechanisms by which helical roll vortices form wave forests. The angle of the major axis of the helical roll vortices in relation to the geostrophic wind direction implies the classic Eckman Spiral. (From Robertson, 1987; courtesy of *Canadian Journal of Forest Research*.)

m apart which move in 55-year cycles (Figure 5.10). Crest-shaped wave fronts are concave on the windward side, occur randomly, and leave a trailing edge to the right of the prevailing wind which becomes a sinusoidal wave front. On the 'vortex theory' the dead tree strips are roughly parallel to the direction of the geostrophic wind. Only in crest-shaped waves will the dead trees be arranged in crescents whose concave centres point downwind.

The paper birch (*Betula papyrifera*) forests of the wilderness area in south-east Labrador range from less than 1 ha to several km² in extent. All occupy steep slopes that have been burned in the past 110 years and have sharp borders with adjoining conifer forests. *B. papyrifera* regenerates rapidly on open sites with moist well-drained soils and nearby seed sources. Canopy openings in the resulting even-aged overstorey ultimately enable a gradual invasion by *Picea mariana* and *Abies balsamea* and a progressive conversion to conifer forest ensues. Thus, as Foster and King (1986) point out, fire maintains *B. papyrifera* and other early post-fire communities in a mosaic pattern and increases vegetational diversity. The pattern is not random but controlled by interactions between the autecology of the species involved, the relief of the ground and the way in which lightning-induced fires develop in space and time. The disturbing factor of fire thus helps to maintain the forest vegetation of south-east Labrador in a state of dynamic equilibrium, in the same manner as does wind in areas where regeneration waves occur.

In a large steady-state forest (Sprugel, 1976) features of the ecosystem such as net primary productivity, soil nitrate and species richness would show little or no variation with time, apart from seasonal changes. The 'steady-state' theory takes into account the large-scale natural disturbances which occur in many forests, rather than assuming, as is implicit in the classical theory of climax, that tree-for-tree replacement is the normal pattern of ecosystem maintenance in all forest systems. It is unlikely that the maintenance mechanisms of all the varied types of forest are the same; there may indeed still be some forests where large-scale disturbances are so rare that tree-for-tree replacement is the normal pattern of re-generation. The steady-state concept, however, applies very well to forests in which such disturbances are a normal feature, including those which are fire-controlled, fire-susceptible, have a storm-gap structure, or are wave-regenerated. Such a system could also be expected to adapt well in the face of long-term changes in the weather.

5.4   Forest change and seral woodlands

Substantial areas of vegetation seldom remain static; change is ever present. This section is concerned with the changes which have occurred in recent times in an ancient European forest, and also with the development of seral woodlands. All the examples considered involve change, usually of management but in one case of climate. Even the mechanisms of change vary; future forests may well evolve differently even if left undisturbed.

Change in an ancient lowland Polish forest

Changes in the proportions of the dominant trees within a forest can often be detected by measuring stem diameters of the trees and their saplings. A good example is afforded by measurements made soon after the First World War in a typical area of the Białowieża National Park (Figure 5.11). At about this time (Pigott, 1975) there were many large

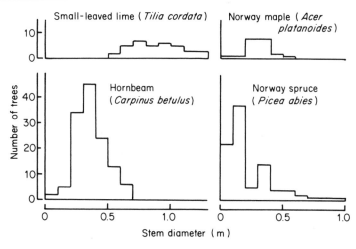

Figure 5.11 Frequency of trees in successive diameter-classes for the principal species in a typical plot of one hectare in the Białowieża Forest, Poland, before 1928. (Histograms redrawn from Pigott, 1975; by courtesy of The Royal Society.)

old trees of *Tilia cordata* (small-leaved lime) but virtually no young ones. In contrast, hornbeam and Norway maple were regenerating, though not as strongly as Norway spruce. The species present vary with the soils but almost half the wooded area of the Park is now covered by Querceto-Carpinetum in which the main tree species are *Carpinus betulus*, *Tilia cordata*, *Quercus robur*, *Acer platanoides*, *Ulmus glabra* and *Picea abies*. The forest is primary in the sense that most of it has had a virtually uninterrupted tree-cover since prehistoric times, indeed Falinski (1986) refers to it as being primeval. This complex boreo-nemoral forest has, however, been damaged and exploited by man for thousands of years, as Falinski demonstrates. Despite these depradations it still possesses some of the tallest and oldest individual trees in Europe. *Picea abies* (55 m, 300 years), *Pinus sylvestris* (over 40 m, 350 years), *Quercus robur* (43 m, 400–500 years), *Fraxinus excelsior* (42 m), *Tilia cordata* (40 m) and *Acer platanoides* (38 m) are particularly notable.

A survey of the Park in 1973 showed an almost continuous canopy of hornbeam with tall emergent trees of small-leaved lime, common oak and Norway spruce (Figure 5.12). Small-leaved lime now regenerates freely. Groups of seedlings, saplings and young trees are common both in gaps and beneath the main canopy; those of lime probably all date from later than 1923 so there is a discontinuity in the distribution of the diameter classes of this tree (Figure 5.13).

The forest is noted for its herbivores – European bison (*Bison bonasus*), red deer, roe deer and wild pigs – which graze and debark trees and

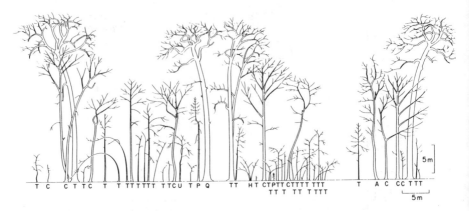

Figure 5.12 Vertical section of trees in the Białowieża Forest, Poland, along a strip of width 2 m passing through 2 main groups of saplings of *Tilia cordata*. **T**, *Tilia cordata*; **C**, *Carpinus betulus*; **A**, *Acer platanoides*; **U**, *Ulmus glabra*; **Q**, *Quercus robur*; **H**, *Corylus avellana*; **P**, *Picea abies*. (From Pigott, 1975; by courtesy of The Royal Society.)

shrubs as well as consuming the graminoids and forbs of the forest floor. Rooting by wild pigs (*Sus scrofa*) kills many existing plants, but often improves conditions for the subsequent establishment of tree seedlings by exposing mineral soil. Solitary boars or family groups destroy the herbaceous vegetation, remove the litter and tear up rhizomes or stolons of *Aegopodium podagraria*, *Asarum europaeum*, *Lamiastrum galeobdolon* and *Stellaria holostea*. Pigott (1975) noted that the patches disturbed in this way were similar in size to those occupied by groups of tree seedlings, and that the pigs might destroy the surface roots of trees and temporarily eliminate root competition as in trenched-plot experiments.

Tree seedlings develop beneath the tree canopy as well as where there are gaps. Measurements made in August indicated that the irradiance beneath the general canopy was greater than that known to be necessary (*c.* 200–300 kJ m^{-2} day^{-1}) for the growth of the seedlings of small-leaved lime, wych elm and hornbeam. A very deeply shaded site beneath an even-aged stand of lime gave values of less than 200 kJ m^{-2} day^{-1}. Seedlings of lime were completely absent from this and other very heavily shaded areas though scattered patches of wood sorrel and asarabacca (*Asarum europaeum*) occurred. Groups of young tree seedlings in the gaps often contain more than one species; if such groups later consist of only one species this probably results from competition. Groups growing beneath shade, however, usually consist of a single species. This may be due to different requirements for germination, but arises partly from the uneven distribution of the seed parents and the large variations in the amount of fruit produced in particular years by individual trees. Limited

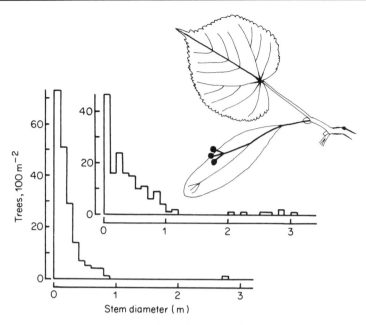

Figure 5.13 Frequency of trees of small-leaved lime (*Tilia cordata*) in successive girth-classes in two plots, each of 0.25 ha, recorded from the Białowieźa Forest in 1973. Note the large bracteole to which the flower buds are attached. (Histograms redrawn from Pigott, 1975; by courtesy of The Royal Society.)

seed availability may well lead to an area being colonized by a single species though the conditions within it are suitable for the seedlings of several.

Certain herbs can largely prevent the establishment of particular species of tree seedlings. Common stinging nettle (*Urtica dioica*) frequently establishes and grows vigorously in gaps, where the soluble phosphate in the top 5 cm of soil of the area tested was higher than beneath the canopy. Hornbeam seedlings were absent from all the examined regions where stinging nettle was growing, while lime seedlings were very sparse where nettle grew vigorously. Well-established wych elm seedlings were found in all the nettle clumps, growing at very low irradiance. This is interesting as *Ulmus glabra* seedlings of English origin tend, if anything, to be less tolerant of shade than those of *Tilia cordata*. The relative susceptibilities of various species of tree seedlings to fungal disease and predation by invertebrates may be an important factor in their survival under the moist shaded conditions of nettle clumps.

Tilia cordata is more abundant in the central region, which includes the National Park, than elsewhere in the Białowieźa forest. In 1928

Paczoski attributed this to the sensitivity of this tree to exploitation, the central area having been the least heavily managed. Much of the forest away from the National Park had the same herbaceous vegetation, differing only in the much smaller proportion of *T. cordata* and a corresponding increase in *Quercus robur* and *Acer platanoides*. The gap in the regeneration of *Tilia cordata* extends backwards from about 1923 to before 1870, a period when the area was maintained as a hunting reserve for the Czars. During this period the number of carnivores was kept low and many deer and European bison were kept for hunting. If this helped cause the failure of regeneration in small-leaved lime it did not prevent that of other trees.

Vigorous regeneration of lime since 1923 has caused the proportion of young trees in the population to become very high, so the forest still has features derived from selective thinning. Its previous condition represented a **plagioclimax** in which the proportion of limes was kept at an unusually low level by some factor connected with the management at that time.

In the oak–lime–hornbeam forest *Tilia cordata* appears to be the potential dominant. The mature trees are long-lived and very tall, and the species is very shade-tolerant when young. Hornbeam is shorter-lived and forms a lower canopy; it is likely to become displaced, at least for a time, from its present predominant role. *Quercus robur* is found most commonly in the National Park area; it is much less frequent in the whole forest than formerly having, like *Pinus sylvestris*, been heavily exploited for its timber earlier in the century. *Picea abies*, which is favoured by high numbers of game animals, now controlled, has diminished in numbers since 1950 and is also subject to the influence of acid rain.

There are some 990 species of vascular plant in Białowieża, which is a very complex forest with several distinct types of tree cover, and which now includes areas of abandoned farmland undergoing secondary succession. It is a forest of extraordinary interest in which long-term studies of forest dynamics, phenology, and variation within individual species have already yielded a wealth of data. Much has been learnt about gap formation, including that involving *Picea abies*; the mechanisms involved often include the washing out of shallow root systems which are frequently split from the substrate by expanding ice in early spring when thaw water freezes in the soil (Falinsky, 1986).

Although the Białowieża area has been allowed to develop naturally since the Second World War and contains some of the least disturbed mixed forest in the European lowlands, it is still responding to the effects of previous management and to environmental perturbations such as those caused by fire, wind and alterations in the level of the ground water table.

Development of beech and yew woods on chalk soils

Management is even more important in chalk grassland which soon reverts to scrub in the absence of mowing or heavy grazing by cattle, sheep or rabbits. Between the two World Wars Dr A. S. Watt described two main sequences in an eventual succession to beechwood on the chalk soils of the Chiltern escarpments; the sequence on the South Downs is similar. The 'hawthorn–mercury' sere was held to develop on the gentler, more sheltered slopes of the Middle Chalk where the soil is moist, quite deep and can support luxuriant growth. The 'juniper–sanicle' sere arose on the firmer and shallower soil over the harder strata of the Upper Chalk. This soil has a lower humus content in the surface layers, a lower water-holding capacity, and sparser leaf litter under the mature beechwood. A major difference between the two seres was that beech colonized juniper scrub directly whereas seral ashwood often, and on the South Downs usually, succeeded hawthorn scrub before beech entered the succession. In other sequences seral juniper scrub could give rise to yew wood directly, while hawthorn scrub passed through a yew–ash stage first.

When beech directly colonizes juniper (*Juniperus communis*) scrub on the exposed slopes of the Chilterns it rapidly kills the pioneer bushes, which are intolerant of shade, and forms closed beechwood. The transition from hawthorn (*Crataegus monogyna*) scrub is more gradual. Any juniper present is soon shaded out, while hawthorn gradually increases in abundance. Bramble, hazel and elder tend to remain as woodland species appear in the field layer. While *Sanicula europaea* is more constant on the shallow soils of the juniper sere, which will eventually give rise to a 'sanicle' beechwood, and *Mercurialis perennis* on the deeper soils of the hawthorn sere where a 'mercury' beechwood will develop, there is no profound difference between the field layers of the two seres. *Bromus ramosus* is more frequent in immature sanicle woods, and enchanter's nightshade (*Circaea lutetiana*) slightly more so in immature mercury woods.

In ash scrub on the deeper soils hemicryptophytes with basal rosettes, such as *Fragaria vesca* and *Viola hirta*, are later displaced by mercury with its tall leafy shoots. Seral ashwood contains many field layer species characteristic of the mature mercury beechwood which will later develop in its place.

The habit of the juniper bushes shows two extremes. One is erect and averages 1.8–2.4 m high (though an exceptional 6 m has been recorded), the other low and spreading. Hawthorn is the most widespread dominant in English scrub, flourishing on many soil types and occurring in the succession to oakwood as well as to beechwood. Dogwood (*Cornus sanguinea*), privet (*Ligustrum vulgare*), blackthorn (*Prunus spinosa*), buckthorn (*Rhamnus catharticus*) and wayfaring tree (*Viburnum lantana*) are very constant in chalk scrub, which has by far the most species of any British scrub type. Brambles and various species of *Rosa* are often pioneers. Spindle (*Euonymus europaeus*) and elder (*Sambucus nigra*) are also characteristic; field maple and hazel flourish particularly on deeper soils towards the bottoms of slopes, while *Clematis vitalba* is the characteristic and often luxuriant woody climber.

Whitebeam (*Sorbus aria*) is one of the commonest trees in chalk scrub and frequently persists in the canopy of the mature beechwood, as can the less abundant gean (*Prunus avium*). Pedunculate oak (*Quercus robur*) is unable to form trees of more than 4 m on the shallow chalk soils, though its seedlings are widespread. Common ash and beech have a high constancy in chalk scrub, though ash is much commoner on the South Downs than in the Chilterns. Yew (*Taxus baccata*) colonizes chalk scrub very readily. Individual trees frequently live to a great age, excluding competition from other trees and shrubs by means of the very deep shade which they cast. When Watt (1926) surveyed 10 yew woods on the chalk of the western South Downs it was the dead (relict) trees and bushes of scrub species such as whitebeam, hawthorn and juniper that convinced him that the shade-tolerant yew had grown through the other shrubs to form closed yew woods. More modern research confirms this view, but there is no evidence for Watt's suggestion that yew is a stage in the development of beechwood; indeed it is now known that many of the beechwoods of the South Downs originated as plantations.

Tittensor (1980) has shown that these yew-dominated woods on the Hampshire–Sussex border originated from abandoned sheep down or rabbit warrens during the last two centuries. When rabbit warrens were disbanded and the numbers of sheep reduced in this region during the Napoleonic wars, areas of previously heavily grazed ground retained bare patches long enough for more juniper, other shrubs and then yew to establish before coarse grasses formed a thick cover. The transformation of open sheep down to yew wood took from one-half to three quarters of a century. Kingley Vale, 'the finest yew wood in Europe', had changed from downland to juniper scrub by 1830 and then to yew wood by around 1870. Though this wood is considerably less than 200 years old the small grove in the valley bottom had some very old yews including one, dead in the 1950s, whose rings showed it to be 500 to 550 years old. Elsewhere it is possible that some of the seed parents were yews used as markers along parish boundaries.

These magnificent yew woods may in time present a conservation problem. Studies of girth-classes and ring counts at Kingley Vale suggest that the yew woods are virtually single-generation stands which formerly migrated across the countryside by edge regeneration. As yet there is little regeneration under the mature canopy or even in gaps. Though individual yews often have long lives these yew woods seem essentially seral and may ultimately be replaced by other species. In most places edge regeneration via juniper scrub and juniper–yew is now prevented by intense cultivation to the very limits of the yew woods.

Development of coniferous forest on glacial moraine

Successions from bare moraine to mature conifer forest afford striking illustrations of primary seres in which changes in the vegetation are accompanied by major alterations in soil conditions. A sequence of this type occurs in areas left bare by retreating ice in Glacier Bay, south-east Alaska. Burrows (1990) discusses ecological studies of this area which have continued for more than a century. Climax forest develops in about 250 years. Ring counts show the oldest trees on the last morainic ridge to be about 200 years old; the maximum age of trees on the moraines decreases progressively as the glacier is approached. The oldest trees in the area are over 600 years old and grow in places beyond the furthest reach of glacial activity. A simplified outline of the main stages of succession is shown in Table 5.1; on well-drained slopes the climax vegetation is coniferous forest dominated by *Picea sitchensis* with some *Tsuga heterophylla* and *T. mertensiana*. Where the ground is flat, or has only a gentle slope, water-filled *Sphagnum* comes to dominate the moss mat and the forest floor becomes soggy. Occasional scattered individuals of lodgepole pine (*Pinus contorta*) are the only trees which can tolerate the poor aeration of the resulting muskeg.

Marble is present in the area and the soil parent material has a pH of 8.0–8.4. Figure 5.14 shows the rates at which the pH of this material falls when left bare or covered by different types of vegetation. Sitka alder has a strongly acidifying effect, lowering the pH of the surface soil from approximately 8.0 to 5.0 within 30–50 years. The nodules on the alder roots contain actinomycetes of the genus *Frankia* which fix atmospheric nitrogen; the resulting increase in soil nitrogen is probably crucial to the initial establishment of the spruce, and soil nitrogen values decline after the elimination of the alder. The gradual accumulation of soil carbon is important in creating a good crumb structure, which assists soil aeration and the movement of soil water. Productivity increases greatly as spruce invades but with the formation of podzolic (Bh) soil horizons much nitrogen and phosphorus becomes effectively immobilized (Bormann and Sidle, 1990). Sustained growth of late-succession *Picea* in this area may depend on release of immobilized nutrients after soil disturbance by windthrow and other processes (section 9.6). Over very long periods nutrient losses, which encourage the development of muskeg, may occur in boreal forests such as those at Glacier Bay. Phosphorus and potassium may be permanently lost while nitrogen can be

Table 5.1 Outline of succession from bare moraine to mature forest and muskeg in the Glacier Bay region, south-east Alaska. Succession in this region, as elsewhere, varies in both its speed and end-point. Current estimates are that, starting with bare moraine, it takes approximately 250 years for mature forest to develop. In another 1250 years this changes to *Sphagnum*-dominated muskeg bog in suitable sites. Nitrogen fixation, a key process in this succession, is carried out by a variety of microorganisms, commencing with the Cyanophyceae present in the dark liverwort–lichen crust which initially covers the ground. The process is continued by *Rhizobium* in the root nodules of legumes, *Frankia* in the nodules of *Dryas*, *Alnus* and *Sheperdia*, and by free-living bacteria in the soil

Sphagnum-dominated MUSKEG on level and slightly sloping sites.	BOREAL FOREST
←----	Initially almost pure Sitka spruce; later with increasing amounts of western and mountain hemlock
	↑
	Alder–spruce (*Picea sitchensis*) forest.
	↑
	Sitka alder (*Alnus sinuata* = *A. crispa*) thickets with scattered cottonwood (*Populus trichocarpa*).
	↑
	Willow Stage (*Salix barclayi, S. sitchensis, S. alexensis*). These are at first prostrate but become erect and form dense scrub.
	↑
	Pioneer Stage. Moraine colonized by *Rhacomitrium canescens, R. lanuginosum, Epilobium latifolium, Equisetum variegatum, Dryas drummondii* and *Salix arctica*.

locked up in deep accumulations of peat, from which nutrients are not recycled.

Old field successions: monolayer and multilayer trees

Many of the principles governing forest succession are illustrated by the 'old field' sequences of the **secondary succession** which occurred as the American west opened up in the nineteenth century, and many eastern farms were abandoned. Natural regeneration was swift, even though the farmers had destroyed almost all the pre-colonial mixed conifer–hardwood forest. The resulting successions demonstrate the importance of the geometrical arrangements of tree leaves in suiting various trees to different light and water regimes, and thus to different stages in the sequence. Trees (Horn, 1971, 1975) may be considered to

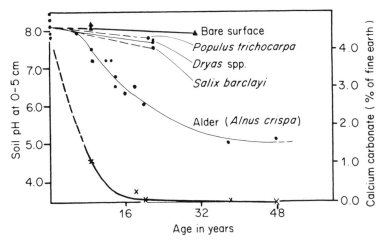

Figure 5.14 Rate of change of reaction in the 0–5 cm horizon relative to type of vegetation cover. Change in calcium carbonate content (×) under *Alnus crispa* is shown by the lowermost curve. (Redrawn from Crocker and Major, 1955; courtesy of *Journal of Ecology*.)

have two main patterns of leaf arrangement, the multilayer and the monolayer. The optimal arrangement for intercepting and utilizing light depends on the amount of light reaching the tree, whose number of leaf layers can be estimated using a light meter technique. Many tree leaves in temperate areas can reach about 90% of their maximal rate of photosynthesis when they are receiving as little as 25% of full sunlight. When incident light is less then this, **monolayer** species, in which there is a single layer of leaves in a shell around the tree, are the most efficient, intercepting all the available light at its highest intensity. In crowded stands of trees such as the sugar maple and the American beech (*Fagus grandifolia*) there may be just one unit of leaf area for each unit of ground area beneath the tree, i.e. LAI is one. This is a very efficient arrangement in late successional species which produce dense foliage at the branch tips. Sugar maple and American beech both cast heavy shade and can come to dominate secondary forest.

Multilayer trees on the other hand produce leaves along those interior branches which receive enough light to be above compensation point, as well as at the tips of well-lit branches. While measurements made on individual leaves will give values for compensation point and light saturation, it is the situation in the whole canopy which determines the use which the plant can make of the available light. In multilayer species the rate of canopy photosynthesis may well not saturate but rise linearly with increasing PAR. Trees of this form, including bigtooth aspen (*Populus grandidentata*), grey birch (*Betula populifolia*) and eastern red cedar

(*Juniperus virginiana*), are best able to utilize the abundant light available at the start of an old field succession when they can grow more rapidly than the monolayer species. In this early seral stage the pioneer multilayer forms may be expected to dominate the hot, dry and sunny canopy, while monolayer shrubs and herbs are likely to be the successful forms in the understorey which is cooler, moister and more shady. The late successional monolayer species have seedlings which can develop in such an understorey, eventually ousting the pioneer multilayer species beneath which they establish. Seedlings of the multilayer forms are at a disadvantage in such an understorey and are usually shaded out.

Horn (1971) describes the theoretical basis of the ways in which leaf size and distribution influence photosynthetic efficiency and drought resistance. The average length of the **umbra** of a leaf (the shadow within which the sun is hidden) is effectively proportional to the diameter of the largest circle which can be inscribed within the leaf. The distance at which the shadow of a circle vanishes is about 50–70 times its diameter with the sun at its zenith on a clear day, whereas under uniformly cloudy conditions light comes from a solid 180°, and the shadow of a circle disappears at a distance equal to its diameter. If no leaves are to be in total shade, the minimum distance between the layers of multilayer trees in sunny climates is thus much greater than in cloudy ones, assuming leaf size to be constant. Conversely leaf size can be much greater for a given interlayer distance in a cloudy climate than a sunny one. Light intercepted by the shoot but not reflected, transmitted, or used in photosynthesis, must be dissipated as heat. In a monolayer this heat load is concentrated in a single layer of leaves, but in a multilayer it is spread over several layers so this form is more resistant to drought. Further, the small size of leaves of multilayer plants increases the efficiency with which accumulated heat is carried away by convection currents. Horn argues that as leaves with a high heat load must transpire at a greater rate than those with a low one, water loss per unit of photosynthesis would be greater for a monolayer than a multilayer tree.

On these general theories we can expect – when soils are reasonably moist and light is the limiting factor – multilayer trees to invade open fields, for the seedlings of monolayer trees to develop beneath them and for the climax monolayer trees to cast so dense a shade that even their own seedlings would have great difficulty in establishing. This may well lead to a senile forest in which trees can establish only in gaps left by the decline and death of individuals which have lived for hundreds of years. In xeric successions, however, drought will prevent monolayer trees from competing effectively with multilayer species at late stages in the sequence.

In most forests the distribution of trees is influenced by so wide a variety of abiotic and biotic factors that diversity is maintained. There are also multilayer trees, including *Sequoia sempervirens*, *Sequoiadendron giganteum*, a fire-dependent species, and *Pinus strobus*, that are extremely persistent in the later stages of the succession as well as being very efficient pioneers. Thus, although we can undoubtedly gain valuable insights from generalizing theories such as those of Horn, it is important to consider the behaviour of all the individual species in attempts to understand the mechanisms of forest succession.

The changes which occur in the composition of communities associated with decomposition contrast with the autotrophic successions described in this chapter. A number of natural 'degradative successions' occur which end, not with a climax community, but a completely decomposed food material. The nature of these successions is described in section 8.3.

5.5 Succession on Krakatau: the natural experiments

The Krakatau group lies over an orogenic hot spot midway between Java and Sumatra. In the explosion of 1883 which started the present successional sequence the main island collapsed leaving Rakata, a major fragment with a peak some 800 m high, and the smaller and lower islands of Rakata Kecil (= Panjang) and Sertung. Sterilization of the islands, which were deeply covered in volcanic ash, appears to have been complete, but in the intervening century Rakata has been covered by a lush tropical rainforest, albeit one of low complexity in which the trees are restricted to a relatively small number of early successional species.

Although this great natural experiment has become more complicated with time, the succession on Rakata seems to have proceeded without interruption since 1883, apart from a fire on the western side of the island in 1919. The vegetational history of the first hundred years is summarized by Whittaker *et al.* (1989), who demonstrate the questions these islands pose with regard to succession, zonation, dispersal, gap dynamics, island biogeography and, most importantly, conservation of tropical rainforest.

Successional pathways and zonation

A few sporadic blades of grass were present on Rakata by September 1884, while in 1886 the pioneers of a new vegetational cover had developed from coast to summit. The colonists included six species of blue-green algae; this gelatinous layer of nitrogen-fixing Cyanophyceae facilitated the establishment of ferns which at this time formed the bulk of the vascular cover of the interior. Most of the few species of angiosperms were restricted to the beach.

By 1897 the interior had developed into dense savanna-type grassland dominated by *Saccharum spontaneum* and *Imperata cylindrica* with small groups of trees. Ferns now dominated only the summit and the higher regions, and even here the balance was swinging in favour of the flowering plants.

On Rakata the canopy of the coastal forests is typically dominated by *Terminalia catappa* and these communities are fittingly called 'Terminalia forests', though their seaward margins are in places composed of the thick irregular trunks of *Barringtonia* (another sea-dispersed species), that are undercut by the sea. Immediately inland of the coastal fringe *T. catappa* was in 1983 the dominant species. Understorey trees were 9–11 m high with emergents of *T. catappa* and *Erythrina orientalis* rising to 13–15 m. *Casuarina equisetifolia* is characteristic of disturbed or unstable habitats and a pioneer in coastal communities. In the 1900s it was a major component of the pioneer coastal forest on Rakata, where it competed with *Terminalia*. By 1979 this tree was restricted to one or two places on the south coast and to the precipitous north face of the island where normal forest cover could not establish. All these coastal communities are undergoing dynamic change but their species have tended to persist.

Further inland the predominantly coastal trees are joined by others such as *Ficus fulva*, *F. septica* and *Macaranga tanarius*, grading into a continuum that passes into the inland forests (Figure 5.15). Some features of the development of the inland forest were deterministic. The diminution of heliophytic ferns once the canopy began to close was inevitable, as was the steady build-up of the orchid population as suitable niches became available for exploitation by its minute windborne seeds. How far were stochastic events involved? If one or two tree species had been accidentally introduced earlier, or if some completely different species had migrated from the mainland, would things have progressed very differently and would the end-point, which has yet to come, have been different?

Neonauclea calycina forms over 50% of the canopy in the forest, dominating much of the interior. The patches of *Ficus pubinervis* forest found inland owe their mature appearance to its open nature and large individual trees. It contains a little *Neonauclea calycina* and is believed largely to have developed from *Casuarina* forest via a *Macaranga–Ficus* stage.

On Sertung and Rakata Kecil, where the inland forests are now dominated by *Timonius compressiocaulis* and *Dysoxylum gaudichaudianum*, it appears that the former species modified the environment and acted as a 'nursery crop' for the later successional *Dysoxylum*. The patchy distribution of *Dysoxylum* as a forest dominant is probably the result of

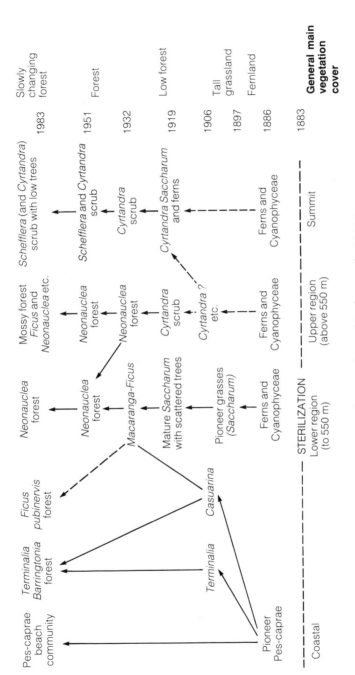

Figure 5.15 Principal successional pathways on Rakata. (After Whittaker *et al.*, 1989.)

continuing local disturbance within each island by ash falls from Anak Krakatau.

Vegetation and environmental gradients

There are clear gradients in the tree, shrub and bryophyte communities with increasing altitude on Rakata. The epiphytes of the uplands are much more luxuriant and differ in type from those of the lowlands. The occasional trees near the summit – *Ficus ribes, Villebrunea rubescens* and *Saurauia nudiflora* – are squat and merge with the shrub and ground layers, the latter being heavily shaded and species-poor. The sudden rise of multiple-trunked *Schefflera polybotrya* as a shrub layer dominant is a striking feature. Though the mossy forest has a greater depth of humus, there is no significant difference in the mineral nutrient content of the lowland and upland soils. Environmental changes, however, are very marked. The uppermost 100 m of Rakata is frequently shrouded in cloud and the high humidity, rain, wind and low temperatures cause a transition to montane or submontane mossy forest at 550–600 m. In larger land masses this vegetational transition occurs at much greater altitudes, 2000 m in the interior of Papua New Guinea, for example.

Species immigration and extinction

Experience on Rakata, where surveys have been conducted at irregular intervals and with various degrees of efficiency since 1883, shows that the floral recolonization of such an island requires a more complex interpretation than a balance between colonization and extinction rates as suggested by MacArthur and Wilson (1967) in the theory of **island biogeography**. Species turnover is not random and various groups of species are affected differently.

Recolonization is a successional process in which habitat development and dispersal – by sea, wind, birds, bats and occasionally humans – are key issues. **Community dynamics** influence rates of immigration, colonization and extinction. These processes vary in time and between different groups; for example, the proportion of vascular plant species contributed by the Pteridophyta fell from 42 to 10% between 1886 and 1908. Spores of many shade species of forest ferns must have blown in from a very early stage but establishment of these plants was difficult until tree canopies had developed. Fruiting trees whose seeds were originally introduced by animals attract more visitors and a positive feedback develops in which more animal-dispersed plants attract further bat and bird visitors. The number of species on Rakata has continued to increase over the last 50 years.

The example of Rakata demonstrates just how difficult it would be to create even a simulacrum of the species-rich ancient tropical rainforests that are being destroyed so rapidly today. Our knowledge of the complex interrelationships between the multitudes of organisms involved is simply inadequate, though the discussion of the *Ficus* species and their pollinating chalcids given in section 3.4 illustrates the type of phenomenon involved.

After a century the vegetation on the islands is still at a comparatively early stage. Even on Rakata the tree species of the interior are few in number and characteristic of relatively early stages in a full succession. The Krakatau Group was visited very soon after the initial explosion – a report being made by Verbeek in 1885 – and interest has continued ever since. The 1989 expedition (Whittaker *et al.*, 1990) found that significant changes had occurred even since 1983. It also established six large permanent plots whose development will be followed far into the future.

5.6 Origins of British woodlands and their floras

During early Tertiary times, which began 70 million years ago, the northern forests in what is now the Arctic Circle contained species of *Abies*, *Betula*, *Pinus* and *Populus*. Adjoining forests to the south possessed evergreens such *Abies*, *Chamaecyparis*, *Picea*, *Pseudotsuga*, *Sequoia* and *Tsuga* as well as modern angiosperm genera including *Acer*, *Aesculus*, *Castanea*, *Fagus*, *Fraxinus*, *Juglans*, *Quercus*, *Tilia* and *Ulmus*. At that time the breakup of the ancient supercontinent *Pangaea*, now explained in terms of tectonic plate theory, was at a comparatively early stage. A land bridge connected North America and Asia, while the straits separating Greenland, Iceland and Northern Europe were relatively narrow and many tree genera were widely dispersed right across the northern hemisphere.

Figure 5.16 shows the southward movement of the temperate Tertiary forest which was initiated by a drop in temperature in the Late Tertiary and culminated in the Pleistocene Ice Age which began some 2.5 million years ago. The Pleistocene is the first part of the Quaternary which is completed by Holocene, or Recent, time. Separate western and eastern forests developed in North America because the upthrust of the Rockies created the treeless rain shadow of the Great Plains. In Eurasia the western and eastern forests were separated by a central area whose low rainfall resulted from its wide separation from the sea. These four major forest regions developed differently; in North America the western forests became dominated by conifers while angiosperm trees remained abundant in the east.

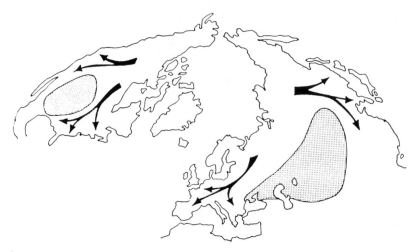

Figure 5.16   Map showing the southward movement of the temperate Tertiary forest to give the four distinct forest regions found in the northern hemisphere today. Note that in Europe the Alps and Pyrenees formed major barriers to north-south migration. (From Gibbs and Wainhouse, 1986; courtesy of *Forestry*.)

The natural distribution of the tulip tree (*Liriodendron*), which in the early Tertiary included stands in Iceland and Europe, is now restricted to south-eastern USA (*L. tulipifera*) and eastern China (*L. chinense*). *Sequoia* is now found only along the Pacific coast of North America. At the end of the Pleistocene glaciations, however, at least one modern species of many tree genera existed in each of the four major forest regions. These species have continued to evolve with their associated fauna, flora and microbial populations including pests and pathogens, a fact of considerable significance if members of the last two groups are transferred from one major region to another (Gibbs and Wainhouse, 1986, see also section 6.5).

Because of their east–west alignment the Alps and Pyrenees prevented the easy migration of trees and other plants to and from refuges lying to the south of them during the successive glaciations of the Pleistocene Ice Age, whose end was marked by the brief, relatively warm period known in Europe as the Allerød interstadial. In North America the major mountain ranges run north–south so that there, as in eastern Asia, the major elements of the rich Arcto–Tertiary forests were able to retreat south along continental migration routes, whereas in western Eurasia these complex forests were destroyed by the ice. Today we can only fully appreciate the former magnificence of the Arcto–Tertiary forests by visiting the mixed mesophytic stands of the Southern Appalachians or eastern Asia, areas far richer in tree and understorey species than any other temperate forests.

Interglacial sequences

At the start of each interglacial period the ice sheets wane in response to a gradual rise in temperature which reaches its peak at the **climatic optimum** and then declines. Annual precipitation increases as the temperature rises, often remaining at a high level for most of the interglacial period. Heavy leaching under cold conditions, in which chemical weathering proceeds slowly, causes the amounts of available mineral nutrients in the soil to dwindle, particularly in upland areas, as the interglacial draws towards its end. Between glaciations plants which survived in refuges in the south, or on elevated areas which escaped the ice, have recolonized areas devastated by glacial incursions.

The trees which recolonized the country during the interglacials came from areas far to the south and east. As one glaciation succeeded another the indigenous tree flora became more impoverished. Examples of losses include Norway maple (*Acer platanoides*) and Norway spruce which both grew in Britain during the last complete interglacial, the Ipswichian, and silver fir (*Abies alba*) present in the previous Hoxnian interglacial. *Pterocarya* was present in the Hoxnian, and occurred with *Carya* in the early Pleistocene.

Evidence of these previous floras comes from fossils found in sedimentary deposits, especially **microfossils** in the form of pollen grains. The exine (outer coat) of these grains is extremely resistant to decay, particularly in acid media, and much of our knowledge of Quaternary floras (Godwin, 1975; Pennington, 1974) is derived from the pollen complement of cores extracted from peat bogs, lakes and estuarine muds, often supplemented by **macrofossils** including wood, bark, leaves, budscales, fruits and seeds. The cores are laboriously studied layer by layer and reveal changes in the pattern of pollen deposition, and hence of the previous plant population, with time. Pollen analysis, **palynology**, presents many problems of interpretation: anemophilous plants produce far more pollen than those which are entomophilous and some pollen grains preserve better than others. Lime (*Tilia*) is extensively pollinated by insects and its pollen is not so widely dispersed by wind as that of other forest trees: as a result palynologists originally underestimated its importance relative to that of oak. Wetland species growing close to the deposit are over-represented, while the reverse is true of plants at a distance from it. Discontinuities in the deposition of the sedimentary material cause the record to be incomplete, and care must be taken that the fossils are contemporaneous with the deposits in which they are found, rather than derived from those of another age.

Pollen diagrams for the last three complete British interglacials, the **Cromerian**, **Hoxnian** and **Ipswichian**, provide evidence of the plants, and consequently the climates, which characterized them. In each zone

I, the **pre-temperate zone**, boreal trees (*Betula* and *Pinus*) are present together with light-demanding herbs and shrubs, especially members of the Ericaceae. Zone II, the **early temperate zone**, is dominated by trees of the mixed oak forest (MOF) – *Quercus, Ulmus, Fraxinus* and *Corylus* – and the humus type improves to mull. During the period in which zone III, the **late temperate**, deposits are laid down, forest trees such as *Carpinus, Abies, Picea* and perhaps *Tsuga* increase at the expense of the MOF. In the **post-temperate zone** of climatic deterioration (zone IV) the dominant tree genera are boreal (*Betula, Pinus* and *Picea*) and mor is the characteristic humus type. The forest thins and non-tree pollen types are frequent, especially those associated with damp heathland. The next glaciation approaches and the vegetation is reduced to treeless tundra.

Various points of ecological interest arise from our knowledge of previous interglacials. Several exotic trees, including the commercially important conifers *Picea abies, Sequoia*, and *Tsuga*, formerly indigenous in previous interglacials under climates like those of today, can be expected to become fully naturalized when planted in appropriate areas. Impoverishment of the woodland flora has other consequences; in former times the presence of a greater number of competing species led to the more rigorous zonation of what tend now to be rather broad habitat types.

Flandrian events

The ecological events of the present interglacial (Table 5.2) form a sequence which so far has many similarities to those of the previous three. Early in this century a careful analysis of the stratigraphy of Scandinavian peat bogs and lakes enabled Blytt and Sernander to interpret it as consisting of five post-glacial climatic periods. After the subarctic **Pre-Boreal** came the warm dry **Boreal** period, which was succeeded by the warm, wet oceanic climate of the **Atlantic** period. This gave way to the **Sub-Boreal** which was considered to have experienced a warm, dry continental climate, in contrast to the cool, wet oceanic climate of the **Sub-Atlantic** that still continues. Von Post began the pollen-zone scheme in southern Sweden and the Blytt and Sernander scheme, with which it was soon correlated, proved so convenient that it has been retained as a chronological system even though the original climatic interpretation given above has not proved entirely accurate. In the following account of the development of the vegetation the pollen zones are indicated by Roman numerals.

The Lower (I) and Upper (III) Dryas periods represent the end of the Devensian (= Weichselian) glacial in which the country was covered by

Table 5.2 Correlation table for the end of the last (Devensian) glacial period and the present interglacial. (Based on Godwin 1975, by courtesy of Cambridge University Press.) Radiocarbon years correspond well with those determined by tree-ring chronology (dendrochronology) up to 2000 years ago. They then begin to diverge with the age as determined by ring counts, being up to 900 years greater in 3000 BC than given by radiocarbon dating. The difference is thought to have resulted from changes in the concentration of ^{14}C in the atmosphere. BP ('Before Present') radiocarbon datings take the present as AD 1950.

DATE		VEGETATION			ARCHAEOLOGY		CLIMATE		
Radio-carbon years	Pollen Zones	BRITISH ISLES Ireland		N. Scotland	Forest cover	Cultures		Blytt & Sernander periods	
B.P. 0	VIII	Alder-birch-oak	ALDER-BIRCH-OAK (BEECH)	Lightly wooded heath	Clearing of forest by man	Norman	Warm spell	SUB-ATLANTIC	FLANDRIAN
2000						Anglo-Saxon			
						Romano-British			
	VIIb	Alder-oak	ALDER-OAK-LIME (Eln decline)			Iron Age (Plough)	Rapid deterioration		
4000						Bronze Age (Hoe)	Colder and drier	SUB-BOREAL	
						Neolithic (Flint axes)			
6000	VIIa	Alder-oak-elm-pine	ALDER-OAK-ELM-LIME	Pine-birch-alder			Climatic optimum		
				Pine-birch		Mesolithic (Hunter)	Warm and wet	ATLANTIC	
8000	VIb c	Hazel-pine	PINE-HAZEL (M.O.F.)	Birch-hazel			Warmer and drier	BOREAL	
	VIb a					Proto-maglemosian			
	V	Hazel-birch	HAZEL-BIRCH-PINE	Juniper-				PRE-BOREAL	
10 000	IV	Birch	BIRCH-(PINE)	Empetrum					
	III	Salix herbacea			Grass-sedge		Cold	UPPER DRYAS	LATE DEVENSIAN
	II	Birch			and	Upper Palaeolithic	Milder	ALLERØD	
12 000					open vegetation			LOWER DRYAS	
	I	Salix herbacea					Cold		

herbaceous vegetation. During the Allerød interstadial (II), a brief period of warming, and again in the Pre-Boreal (IV) at the start of the Flandrian, the climate improved sufficiently for birches (*Betula pendula* and *B. pubescens*) to form the first woodlands. At this time only a few of the major broadleaved genera associated with temperate climates such as hornbeam, sweet chestnut, beech, oak, lime and elm survived in southern Europe. In Britain, birch and pine, which now occur relatively early in the succession to climax woodland, remained dominant until the end of the Boreal though the amount of hazel increased greatly. The junction between pollen zones V and VI is based on the general replacement of birch by pine, but it was in zone VI that the mixed oak forest (MOF) began with the first major establishment of oak and elm as forest trees in Britain. The MOF began in the south, presumably on the better soils.

Pollen analysis of 'moorlog' – freshwater peats dredged from the sea floor – indicates that the southern part of the North Sea basin was above sea level during zone IV, the Pre-Boreal. During the Boreal there was a rapid and sustained restoration of water from the world's icesheets to the oceans. This led to the separation of Britain from continental Europe by the English Channel in about 5500 BC. The Irish Sea was either widened or separated Ireland from Britain at the same time. As the sea spread the climate became more oceanic and in the Atlantic period (zone VIIa in Britain) heavy rainfall accompanied the trend to greater warmth. The English Channel effectively ended the northward migration of plants to Britain; since its formation almost all woodland species which have established here have been introduced by man. However, the Atlantic saw the territorial expansion of the more thermophilous species which had already arrived and MOF became the climax vegetation of much of lowland Britain. The length of the **climatic optimum** of the present interglacial is in dispute but it certainly includes the whole of the Atlantic period.

The work of Birks *et al.* (1975) on some 140 pollen profiles from all over the British Isles has revealed the extent of the regional variation of the primeval forest just before the onset of the elm decline at the beginning of the Sub-Boreal. The picture of the British pre-historic forests – Wildwoods – which has been revealed is of considerable complexity. Rackham (1986, p. 70) recognizes five provinces:

1. the Pine Province of the eastern Scottish Highlands, with outliers on English and Irish mountains;
2. the Birch Province of the western Scottish Highlands;
3. the Oak–hazel Province of southern Scotland, Highland England, most of Wales, and parts of Ireland;
4. the Hazel–Elm Province of most of Ireland and probably of south-west Wales and
5. the Lime Province of Lowland England.

Alder (*Alnus glutinosa*) was present on damp land almost throughout the area, having reached Ireland just before the Boreal rise in sea level when lime had got no further than the Lowland Zone of England. The five regional types of 'dry-land' tree assemblages were each a mosaic of local forest variants. Traces of some of them still persist; the patchwork of pinewoods and oakwoods still present around Loch Maree dates from as long ago as 7000 BC. Though a few ancient Scottish Highland pinewoods still persist today, the Scots pine as a native tree is extinct south of Scotland and in Ireland. Perhaps the most significant contrast, one still partially reflected in the semi-natural woodlands that yet remain, is between the Highland area dominated by oak and hazel and the Lowland Province in which lime (*Tilia cordata* with some *T. platyphyllos*) was almost certainly the most important tree, being associated with oak, hazel, and various others including ash in Somerset and pine in the Breckland of East Anglia. This lowland mosaic was the most complex of the five main types of Wildwood found in Britain. Native lime has become much rarer in the last 5000 years though it is more common than had been supposed.

Influence of man on Flandrian forests

Until 5000 radio-carbon years ago the development of the forests of the Flandrian Interglacial followed a pattern very similar to that of the preceding warm periods. Since then the influence of man has become overwhelming, particularly in Great Britain where forest clearance was so extensive that even after the widespread plantings of the present century only 9% of the country is covered by woodland. The first major intervention by man may have been the lopping of elm branches for cattle fodder causing a decrease in elm pollen displayed in pollen profiles as the elm decline. An alternative explanation is that this may have been caused by early epidemics of Dutch elm disease. Evidence of the clearance of forest areas for the shifting cultivation of Neolithic man is provided by the pollen and charcoal records of the Landnam clearances (Iversen, 1941), though these clearances soon became long term (Fowler, 1983).

The beginning of the increasingly cool and dry Sub-Boreal period approximately coincided with the arrival of the Neolithic peoples in Britain, using flint axes with which they practised woodmanship. The earliest of the Neolithic trackways, such as the Sweet Track (Coles, 1989) laid across the Somerset levels, date from about 4000 BC and the wood used in its construction appears from some of the thickened curved ends to have been coppice poles from mixed coppice on a rather long rotation. The Bronze Age and Iron Age peoples introduced more efficient axes, the hoe and the plough. Woodland was felled and burnt to provide grazing as well as arable crops. Even so, when the Roman occupation ended in about 300 AD it is thought that out of a total of some 23 million hectares in Great Britain, the amount of land used for grazing was no more than one million hectares, with rather less used for cultivation. Since then man has been far more destructive of the woodland cover. In Britain only fragments of the original forest cover remain; even in North America and Continental Europe permanent clearance has been extensive.

The Saxons were able to cultivate the heavy clays on which much of the mixed oak forest had stood and at the time of the Norman Conquest (Cousens, 1974) four million hectares were in cultivation. At this time swine still fed in quantity on acorns and beech mast; the emphasis placed on the right of **pannage** for these domestic pigs suggests that the climate was more favourable for oak and beech than it is now, though it became even better in the thirteenth century, the Golden Age of British agriculture.

The ancient forests of England were large tracts of land bearing trees and undergrowth in many places but frequently with considerable areas of pasture and, as in many other countries, subject to special laws and belonging to the Crown. In England such 'Forests' were all royal, similar tracts controlled by nobles were called chases (e.g. Cannock Chase in the West Midlands), and at one time about one-fifth of England was covered by forest and chases though much of this area was not wooded. In the Middle Ages the word came to apply to the 'waste of the Forest' which was the area under Forest jurisdiction that was not farmland, private woodland or built up. Here farm animals were grazed, timber obtained from trees, wood from pollards, and poles from coppices. Venison was also produced, a function later taken over by enclosed parks in which deer still flourish. Now, however, the primary difference between forests and woodlands would be taken by many as one of scale; certainly most man-made forests are huge plantations with large areas of evenage and often monocultures at that. The importance of woodland as a resource was well understood in medieval times; the way in which man has influenced the woodland landscapes of Britain is described by Rackham (1976, 1986) who shows how woodland archaeology co-ordinates the evidence provided by documents, the timbers of old buildings, and the remnants of ancient trees, woodlands, ditches and boundary banks.

5.7 The ecology of coppicing

Coppicing ceased in most British woodlands after the Second World War, but we have descriptions of commercial **coppice cycles** in Hertfordshire by Salisbury (1924) and in Gamlingay Wood, Cambridgeshire, by Adamson (1912). In Scotland oak coppice reached its peak of profitability in 1800, but its value declined rapidly after 1815 as Hobson (1988) points out in her account of Methven Wood. In recent years the practice has gradually returned to favour, largely on a limited scale and for conservation reasons so that, as with charcoal burning, a practical knowledge of the craft is being re-established. Within a particular wood, different stages of a coppicing cycle provide a wide variety of structural and climatic conditions, so that the wood as a whole supports a **high diversity of species**, both plant and animal, and for this reason the ecology of coppicing is widely studied (Buckley, 1992). Old coppiced woodlands, such as Bradfield Woods in which coppicing has never been interrupted, may contain more than 300 species of herbs, while certain Kent woods are partially dependent on coppice management for their rich insect and bird faunas.

Coppicing involves the felling of the underwood in autumn and winter and its most obvious effect is on shade (Mitchell, 1992). As this varies at different times of year the effect on a particular species is related to both its tolerance of shade and its phenology (sections 2.4 and 3.5). Rackham (1975) estimates that coppicing '(a) increases summer light at least 20-fold, (b) increases spring light roughly twofold, and (c) extends the period of spring light by about three weeks (because the remaining shade is produced by standard trees which come into leaf later then small coppice)' in a typical English site. These increases in the amount of light reaching the ground layer after cutting the coppice enable many established herbs to grow more vigorously, while other species enter and become established. The exuberant flowering of vernal species such as primrose and wood anemone is an obvious feature during the first 2 or 3 years after cutting, for instance in oakwoods on clays and loams, while later in the year violets, red campion, yellow archangel, foxglove and rosebay willow-herb may flourish, the last being further favoured on burnt areas. As the coppice canopy closes the increased shading and crowding are tolerated by most of these species, although flowering may not be nearly so prolific. Others, however, may die out.

In Ham Street NNR, Kent, Ford and Newbould (1977) recorded the greatest number of species of herbs (30) in areas of *Castanea sativa* coppice which had been cut 5 years previously, but thereafter lack of space prevented further colonization and the total number of species declined towards the end of the 12- to 17-year cycle. Components of the **'shade flora'** such as bracken, bramble and bluebell persisted throughout the cycle, showing maximum above-ground production when the canopy was removed, except for bluebell which was most productive after canopy closure in year 7. Bluebell is a special case, being a sun plant which photosynthesizes vigorously before the flushing of the tree canopy (section

2.4). Wood sage (*Teucrium scorodonia*), on the other hand, was abundant in areas examined 2 and 5 years after felling, but absent from later stages. Figure 5.17 shows fresh growth in a recently coppiced *Castanea sativa* woodland in West Sussex.

Temperature is an important factor in coppice ecology. The productivity of most coppice trees, e.g. hazel and oak, is largely determined by site features such as soil pH and base content. Sweet chestnut, however, grows well even on markedly acid soils – in one instance raising the pH of its soil from 4.4 to 4.8 actually depressed growth – but is a thermophilic species whose production rate is largely controlled by summer temperature. A number of rare butterflies are dependent on the high temperatures and open nature of recently cut coppice (see also section 1.2) which in summer may be 5°C warmer than in mature coppice. Pearl bordered fritillary (*Boloria euphrosyne*), high brown fritillary (*Argynnis adippe*), and heath fritillary (*Mellicta athalia*), all occur in recently coppiced areas and die out towards the end of the cycle, though in Blean Woods, Kent, heath fritillary has been shown to survive longer in poor than in vigorous coppice. These insects are relatively immobile so continuity of management is essential. A freshly cut panel 300 m from an existing colony is likely to be colonized, but 600 m is probably too distant. Coppice panels need to be close and connected by wide corridors to act as flight paths.

Clearings with insects in turn attract certain birds such as grasshopper warblers, nightjars and tree pipits, while other species may be favoured by cover for nesting or feeding, or by the structural diversity of a particular stage of the cycle, for example scrub-like structure tends to support the largest number of breeding birds. In Ham Street NNR, the density of nightingales was found to be correlated with the presence of oak standards, and also with the age of the hornbeam coppice, which was most favoured when 5–8 years old, but quite unsuitable when older than 15 years, presumably because of the virtual absence of a ground flora. Fuller *et al.* (1989), who mapped the distribution of breeding birds within 30 ha of mixed coppice for a period of 5 years, found that bird species differed greatly in the stages of growth at which they were most abundant and in their habitat amplitude. Whitethroats come in early in a coppice rotation, chiffchaffs are characteristic of the middle stages, while the robin avoids the early sequence but becomes dominant after canopy closure. Coppice woodlands of various ages also support high densities of summer visitors, but though very many British birds flourish in this habitat no species is entirely dependent on it.

Felling the underwood also influences the bryophytes which are, in general, very good indicators of small-scale mosaic patterns involving soil pH, levels of mineral nutrients, humidity and illumination. After coppicing the relative humidity (RH) of the air is decreased, while ir-

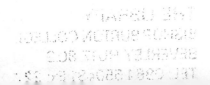

Figure 5.17 Coppiced sweet chestnut (*Castanea sativa*) near Chithurst, West Sussex. These trees had been cut to the ground a year previously, leaving a stool from which fresh shoots have arisen. The previous coppicing cycle lasted 15 years. (Photograph by John R. Packham.)

radiation and the temperature range are increased. This has a differential effect on the growth of those bryophytes which are either able to invade or are present already. Gimingham and Birse (1957) found that the sequence – dendroid forms (e.g. *Thamnium alopecurum* and *Mnium undulatum*) and thalloid mats (e.g. *Pellia epiphylla*): rough mats (e.g. *Eurhynchium striatum*): smooth mats (e.g. *Hypnum cupressiforme*): short turfs (e.g. *Ceratodon purpureus*) and small cushions (e.g. *Orthotrichum anomalum*)

– occurred along a gradient in which light intensity increased and atmospheric RH decreased. This sequence helps us to interpret the striking contrasts in growth-form distribution (section 2.2) which may be seen when tracing bryophyte communities along a stream which runs through both felled and unfelled regions of a wood.

The liverwort *Pellia epiphylla* is favoured, like the moss *Mnium hornum*, by surface acidity, whereas *Pellia endiviifolia* is calcicole. *Hookeria lucens*, a moss of heavily shaded moist places, such as the deeply incised Seckley ravine, Wyre Forest, is particularly susceptible to environmental change; exposure to direct sunlight kills it. The rugged bases of old ash stools in Hayley Wood bear species such as *Lejeuna cavifolia, Porella platyphylla, Homalia trichomanoides* and *Neckera complanata*, for which Cambridgeshire is otherwise too dry.

In a mature compartment moisture is mainly lost through the transpiration of standard trees and large coppice. When the large coppice is felled the water 'saved' is probably lost by increased transpiration from the ground vegetation, new coppice and the standards, all of which are more exposed to the sun than before and around which the wind can now eddy freely. Standard trees and ground vegetation in the experimentally coppiced areas of Hayley Wood often show more severe symptoms of drought damage than those in the uncoppiced wood. Small seedlings, especially those of rushes, are particularly vulnerable to drought in the first year after felling. Successions of ground layer species which developed in this wood since felling of coppice plots began again in 1964 are described in detail by Rackham (1975). They have been largely unpredictable, varying from year to year – partly because of variations in the weather – and even within plots. Enormous quantities of rushes (*Juncus* spp.) have come up in some years. Rushes do not grow in the surrounding uncoppiced woodland; the seeds involved may have been produced by plants which grew after the last coppicing 50–90 years ago, been buried, and then disturbed by the felling operations which exposed them to the light necessary for germination.

A low level of grazing by large herbivores can, however, play an important role in the conservation of semi-natural woodlands; indeed in the British uplands Mitchell and Kirby (1990) showed that it increased diversity in vegetation structure and species composition.

Hayley Wood is famous for its large populations of oxlips (*Primula elatior*); flowering in this species often increases seven-fold in the second spring after coppicing, but here its leaves and flowers are frequently eaten by fallow deer which also graze shoots developing on coppiced ash. The ash are now pollarded so that the young shoots are out of reach, while the deer have been excluded from the best oxlip area by a high fence similar to those used in Sweden to prevent elk wandering across major roads.

Muntjac and roe deer, both of which have a relatively simple gut, mainly graze on new coppice shoots with a high nitrogen content. Roe deer, which lack the ciliated protozoa associated with the ability to digest grasses found in cervids, can completely eliminate sensitive woodland plants. Fallow deer have a more complex gut and frequently consume more fibrous materials. Deer populations, like those of pheasants, are favoured by coppicing, tending to be highest just before canopy closure when their densities are so high that selective grazing effectively removes many plant species of conservation value. Ratcliffe (1992) concludes that deer numbers cannot be sufficiently reduced by shooting to protect the field layer; exclusion by fencing is the only effective deterrent.

In the Wyre Forest (Salisbury, 1925), where *Quercus petraea* was the most important tree before the Forestry Commission became a major landowner in 1926, most of the woods were coppiced on a compartment system at approximately 16–18 year intervals. Each area was clear-cut except for the few seeding trees. In some places there were two ages of coppice, with the general matrix being on a shorter rotation than the more sparsely scattered oaks which were cut at longer intervals. The larger coppice stems were used for pit props, while some of the branches were rent and used to make oak baskets. An interesting succession developed on the old hearths used for the production of charcoal from the smaller cord-wood (2.5–10 cm in diameter). Recent hearths were blackened areas devoid of vegetation, but these were soon colonized, being especially suitable for the development of seedling trees so that old hearths were frequently covered by a thicket of shrubs and trees. Hearth sites were favourable to plant growth in several respects, being initially devoid of competition and having a surface soil enriched with carbonates, nitrates and phosphates, as well as a pH above that of the surrounding area.

Marchantia polymorpha, *Funaria hygrometrica* and *Ceratodon purpureus* often occurred in great profusion during the early colonization of a hearth. Even in older phases when many woodland herbs, shrubs and trees had invaded, the more calcicole character of the hearth vegetation remained evident and species such as wood sanicle (*Sanicula europaea*) were often present, though absent from the surrounding vegetation. Charcoal burning in Wyre Forest was resumed very successfully in 1990 and if a new market, probably as a fuel for barbecues, can be created it should create both interest and diversity (Figure 5.18). Whether the coppice crafts formerly practised here (Figure 5.19) will be revived is another matter.

Paris quadrifolia, most orchids and a few other species, seem not to increase after coppicing (Rackham, 1975), while in *Mercurialis perennis* (Salisbury, 1924) the dry weight per unit area actually declines. The

Figure 5.18   Charcoal burning in the Wyre Forest. (Photograph by the late Mr Whitcombe.)

coppice cycle, when uncomplicated by grazing, benefits many herb species in turn and its effect in restricting the dominance of dog's mercury, so common in well-drained neglected woodland, also encourages variety.

A major consideration when establishing or re-establishing coppice in areas which have not been subject to such a regime for a very long time is the ability of stumps to coppice after felling. Some 70% of maiden oaks 100 years old do so, but this figure drops to about 40% in 160-year-old maidens (Evans, 1992). Beech does not usually respond well to coppicing in the UK, but does so in the Vosges, eastern France, where the humidity regime is such that stools tend not to dry out. Watkins (1990) provides a valuable guide to coppice management and conservation, noting that a key issue when deciding whether to resume coppicing is the period since the stools were last cut. Many plants of the glade margins can survive in the seed bank for 40 years, but after 70 years the losses are very great. Brown (1981) concluded that it was essential to retain remnants of the original **shade flora** within old coppice woods because the species involved were not likely to survive in the seed bank. Paradoxically the glade margin species were more successful in surviving as seeds. None of the herbs characteristic of the denser regions of deciduous temperate woodlands, including *Anemone*

Figure 5.19 Coppice craft: broom manufacture in the Wyre Forest before the Second World War. (Photograph by the late Mr Whitcombe.)

nemorosa, Hyacinthoides non-scripta (Figure 5.20), *Ranunculus ficaria* (shade-evaders), *Mercurialis perennis and Oxalis acetosella* (shade-tolerators), appear to persist for any length of time in this way. They are particularly vulnerable when former coppices are converted to plantations, failing to survive rotations of 50 years or more, though Kirby (1988) has shown that survival is greater under some tree species, such as oak, than others (beech and Norway spruce).

In Holland a resumption in 1977 of coppice management abandoned some decades earlier did not succeed in recreating the former attractive herb layer rich in geophytes, which had been largely suppressed by the shade-tolerant evergreens *Hedera helix* and *Lamiastrum galeobdolon*. After felling *Clematis vitalba* and a number of nitrophilous species grew so rapidly that the coppice shoots of the calcareous mixed oakwood near Schin-op-Geul, South Limberg, were partially overwhelmed (de Kroon, 1986). Pollarding would be a better approach in an area where there appears to have been a marked increase in soil nitrogen possibly due to atmospheric pollution.

In summary it can be said that many woodland management practices lead to a simplification of woodland ecosystems. Coppicing is often of value in conservation in circumstances where it actively encourages variety. It is, however, labour-intensive and should only be used in

ecologically appropriate situations, preferably where it has been possible
to establish a market for coppice products. When woodlands support
rare animal species as well as a highly diverse flora it is important that
management regimes should be adapted to encompass a range of habitat
requirements.

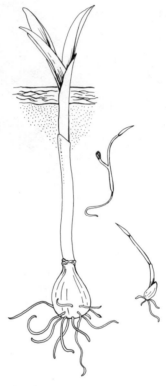

Figure 5.20 Bluebell (*Hyacinthoides non-scripta*), an important member of the
shade flora of coppiced woodlands. This shade-evading bulbous geophyte has a
pointed shoot which can penetrate thick litter (see section 4.5) and expands
rapidly on reaching the light. The black seeds germinate at the surface; as the
bulbs develop they form contractile roots which draw them downwards. The
rarity of bluebell on the chalk scarps of SE England appears (Blackman and
Rutter, 1954) to be due to the rapid drying out of the thin surface layer of soil
and the physical barrier of the underlying chalk (which prevents the bulbs being
drawn down into a deeper and moister zone), rather than to the high pH and
calcium content of the soil.

6
The herbivore subsystem and the exploitation of living autotrophs

6.1 The role of heterotrophs in woodland ecosystems

The tissues and exudates of trees and other green plants, rich in energy and nutrients, provide potential food for heterotrophs, ranging in size from microorganisms to birds and mammals. These may exploit the autotrophs either directly or, as with carnivores and fungivores, indirectly, via other heterotrophs. Many of the species composing this plant-dependent web affect green plants adversely, for example, through feeding or trampling, but the balance is redressed by pollinators and dispersers of fruits and seeds, and, of wider significance, by nutrient cycling (section 9.6).

The **grazing chain** (**herbivore subsystem**) starts with living tissues of autotrophs, the first link of primary consumers including exudate-feeders, herbivorous animals and parasitic plants. Their counterparts in the **decomposer chain** are detritivores and decomposer microbes. Certain carnivores act as links between the two chains, and eventually the dead remains of autotrophs and of heterotrophs, together with faeces, are mineralized.

This chapter considers various aspects of the exploitation of living autotrophs, including instances where the same organism proceeds to feed on tissues which it has killed. Decomposition of dead remains is discussed in Chapter 8, with particular reference to soil organisms.

6.2 Epiphytic microorganisms

The surfaces of living plants provide a substratum for autotrophic algae and lichens, and also support a varied heterotrophic microflora of bacteria, yeasts and filamentous fungi. Species found on or in seeds and buds provide an inoculum for the **phylloplane** (leaf surface; Dickinson and Preece, 1976), but the **rhizoplane** (root surface; Wood, 1989) community is mainly derived from the soil. Some of these epiphytes subsist on dead tissues, such as sloughed cells which accompany even the earliest stages of root growth, while others feed on plant exudates and animal products, including the sugary excreta, or honeydew, of aphids.

Root exudates are mainly produced from the elongating region a few centimetres from the tip, and also from lateral roots, root hairs and senescing or damaged tissues. Compounds identified from these exudates include sugars, amino acids, enzymes, growth factors and cyst-nematode hatching factors. Exudates are believed to stimulate the germination of fungal propagules, young roots being particularly susceptible to colonization by saprophytes and pathogens. Selected bacterial species also build up on the young root surface, which provides numerous crevices as microhabitats. Some species invade and disrupt the epidermal and cortical cells, leading to sloughing of debris (Russell, 1977).

Rhizoplane microorganisms may compete with roots for essential elements, but others, such as ectotrophic mycorrhizas, *Rhizobium*, and *Frankia* found in the actinorhizal plants such as *Alnus* spp. (Schwintzer and Tjepkema, 1990), are beneficial to higher plants (section 9.6). Nitrogen-fixing actinomycetes on tropical leaves provide another example of a symbiotic relationship. In contrast, certain microbes penetrate the surface defences of leaves, by means of enzymes such as pectinases, and exploit living tissues. As well as pathogenic bacteria and fungi, there are many weak parasites, some of which cause no visible disease symptoms, while others can invade only already damaged or senescing tissues. Senescence involves a lowering of defences against attack, as well as a greater release of exudates, so that species diversity normally increases with ageing. The successions of organisms which have been recorded during the lives of leaves presumably reflect the changing availability of different nutrients. *Aureobasidium pullulans* is a hyphomycete which has been reported from buds and leaves of a large number of coniferous and broadleaved species. This fungus grows on and in sycamore leaves for the first 2 months after the buds open, usually without necrosis, and then survives as resting chlamydospores (Pugh, 1974). Among the species colonizing mature sycamore leaves another hyphomycete, *Epicoccum nigrum*, is active until leaf-fall but invades internal tissues only during senescence.

Senescence is probably hastened by the activities of many of these epiphytes, particularly in the case of pathogens, and the fact that decomposition starts in the seedling stage illustrates the difficulty of separating grazing and decomposer chains.

6.3   Forest pathogens

Tree disorders can be caused by abiotic factors such as frost, lightning, wind, unsuitable light levels, variable soil water content, drought and mineral deficiencies as well as by pathogens, of which the most serious are fungi. Stressed trees succumb more easily to disease than those which are vigorous and fast-growing. Thus the bark of older beech trees stressed by drought, nutritional imbalance, root disorders, etc., can become susceptible to attack by *Nectria coccinea* even in the absence of the felted beech coccus (*Cryptococcus fagisuga*) which initiates beech bark disease in younger trees (Evans, 1984). Conversely, trees well suited to their physical and chemical environments withstand the attacks of disease organisms better than those which are not.

Diseases frequently influence the productivity of forests and alter the species composition of mixed woodland communities. Particular species are seldom completely eliminated by a disease, but their frequency and importance may be greatly reduced. In the early years of this century, for example, the American chestnut (*Castanea dentata*) was reduced from one of the most important upper-canopy species in the Appalachian forest to the status of an understorey shrub by a fungal pathogen. Diseases of the major trees and shrubs are particularly important in that these are the dominant plants of woodlands, moderating the habitats of other organisms in major respects. A full account of any microbial disease should include details of the pathogen and its host (or hosts), the symptoms, development and distribution of the disease, the damage which it causes and methods of control. An up-to-date view of the occurrence and treatment of tree diseases in Britain can be obtained by consulting the current Forestry Commission account of forestry practice (Hibberd, 1991). This is a major subject in its own right; we are concerned with its importance in forest ecosystems and the main types of disease caused, though section 6.4 treats the ecology of Dutch elm disease in detail. Nursery diseases influence trees at the seedling stage where 'damping off', which is often caused by *Pythium* spp. and *Rhizoctonia solani*, is an important cause of loss. The other main groups of fungal diseases attack the leaves and young shoots or stems (causing diebacks), or the roots.

Fungal diebacks

Stem diseases of the bark and cambium such as that caused by the ascomycete *Endothia parasitica* in *Castanea dentata* often cause wilting and dieback of the whole crown, or of branches, if the shoot concerned is completely girdled. *Endothia parasitica*, the Asiatic Blight fungus, was noticed on trees in New York Zoological Park in 1904, having apparently entered America on Asian chestnut trees imported from Japan. Within two or three decades the American chestnut was destroyed as a commercial crop over the whole of its natural range in the eastern USA. The fungus is spread by sticky orange spores which ooze from the pycnidia and are carried long distances by insects and birds, including migratory woodpeckers, and also spread by wind-borne ascospores shot into the air from perithecia. Cases where vigorous recovery shoots arose immediately below the lesions, or where the cankers healed in instances where the fungus failed to girdle the shoot, indicated the acquisition of at least some field resistance to the fungus. American attempts to combat Chestnut Blight were largely concerned with the development of disease-resistant hybrids by crossing Asiatic species such as *Castanea mollissima* and *C. crenata* with the American chestnut. In 1938 *Endothia parasitica* was discovered on *C. sativa* in Italy where trees were often felled before the fungus reached the base of the trunk. The uninjured rootstocks coppiced freely; any sprouts which became infected were also cut back. When healthy coppice resulted the trees could be allowed to develop to high forest.

Fungal infection of trees often occurs through wounds made by insects, such as the two-lined chestnut borer (*Agrilus bilineatus*) with Chestnut Blight and felted beech coccus (*Cryptococcus fagisuga*) in the case of Beech bark disease, now causing damage in English pole-stage stands. *Nectria coccinea* which causes Beech bark disease is endemic on *Fagus sylvatica* in Europe. It is also spreading on American beech (*F. grandifolia*) following the accidental introduction of the minute sap-sucking felted beech coccus into Nova Scotia in about 1890.

The destruction of the American chestnut by *Endothia parasitica* is an example of a severe ecological imbalance between an exotic parasite with high reproductive ability combined with severe pathogenicity and a new host plant. Similar situations involving different trees and pathogens will doubtless arise in the future. In view of this it is significant that the natural remission of Chestnut Blight on *Castanea sativa* first noted in Italy in 1951 resulted from the dissemination within the fungal population of a cytoplasmically transmitted agent, considered to be a double-stranded RNA, that has since been successfully transferred to wild populations of *E. parasitica* in North America. In Italy the epidemic is now well into its decline and the disease now tends to kill twigs and branches rather than whole trees (Rackham, 1986). The origin of the hypovirulent agent is not yet known, but Elliston (1982) has suggested that it may have entered the *E. parasitica* population as a result of interaction with the related European species *E. radicalis*.

Peridermium Stem Rust, caused by *Peridermium pini*, differs from most other rusts in being monocyclic, i.e. spores from a rust on one pine are able to infect another pine directly. It is a highly specialized parasite which can live in the cortex and outer wood of the shoot for many years. In the Scots pine girdling and death of large branches or the trunk, together with a massive exudation of resin, is eventually likely to ensue. In Thetford Forest, East Anglia, the susceptibility of *Pinus sylvestris* to this disease has favoured the adoption of Corsican pine, which is seldom infected, for the second rotation (Gibbs *et al.*, 1991).

Brunchorstia dieback has long been an occasional cause of damage to pine, especially Corsican pine (*Pinus nigra*), in the northern and wetter parts of Britain. In years when May and June are wet young developing buds of Corsican pine are often infected by airborne spores of *Gremmeniella abietina*. If the following dormant season is cold the fungus kills the buds and the older shoots may die back progressively during the ensuing May and June. Spores of the asexual stage (*Brunchorstia destruens*) are formed on the dead needles and shoots. Although *Pinus sylvestris* is considerably less susceptible to *Brunchorstia* than *P. nigra*, it too suffers from dieback in the uplands of north and west Britain where this species is debilitated by a needle-cast fungus (*Lophodermium seditiosum*) and further damaged by pine shoot beetle (*Tomicus piniperda*).

During 1979 many Scots pine trees in the Scottish border counties died from attacks begun in the previous year. Other diebacks in 7- to 20-year-old plantations of Scots and Corsican pine in the English Midlands during 1979 had symptoms resembling those of *Brunchorstia* with buds failing to flush in spring. The fungus consistently isolated from the dying branches, however, was *Cenangium ferruginosum*, a common saprophyte on pines in the UK although weakly pathogenic in North America and parts of Europe. It had not previously been considered to act as a pathogen in Britain; in this instance severe spring weather may have disposed the pines to fungal attack.

British plantings of lodgepole pine (*Pinus contorta*) were in the late 1970s found to be suffering from a dieback similar to that caused on Corsican pine by *Gremmeniella abietina*. Infection of developing shoots in wet early summers was again involved. The cause, a previously undescribed fungus named as *Ramichloridium pini* in 1982, affects tissues up to 1 year old and can occasionally kill whole trees in the thicket or early pole stage. Usually, however, the death of a few scattered shoots has little influence on tree growth though certain provenances, such as Lulu Island, are especially vulnerable.

Root rots

Heterobasidion annosum (*Fomes annosus*) causes Fomes root rot which kills conifers and occasionally deciduous trees. It is the most important forest pathogen in Britain, and is especially prevalent in pines on alkaline soils. Economic losses result mainly from extensive rotting of stem wood, particularly in spruces, larches, western red cedar and western hemlock. The orange-brown fruit body of this bracket fungus, which causes both root and butt-rot, is perennial, has a white margin and is formed close to the ground. Cut stumps are easily infected by its basidiospores and the fungus spreads into the attached root system, passing by root to root contact into healthy trees. Modern methods of control are aimed at preventing infection of the cut stump by treatment with a disinfectant, usually urea. The difficulty is that a badly knocked stump or root will have fresh surfaces exposed so that *H. annosum* can enter and form a fresh focus of disease. In pines, but not in other genera, an effective alternative treatment is to inoculate stumps with spores of the saprophytic basidiomycete *Peniophora gigantea* which competes with the parasite, rotting the *Pinus* stumps and preventing successful colonization by *H. annosum*. This method of biological control, which destroys the substrate instead of just capping it off, was developed by Rishbeth (1963) who prepared and packeted suspensions containing the oidia; this control in now used by the Forestry Commission in whose woodlands *Peniophora gigantea* can often be seen looking like spilt candle wax on the dead stumps. The rapid wood colonizing and rotting ability of *P. gigantea* means that pine logs left lying in such woodlands may quickly suffer degrade and loss of value. In Thetford Forest, where many soils are on chalk, an attempt is being made to eradicate the disease from severely infected sites by destumping before replanting.

Heterobasidion annosum does not grow freely in soil, but a number of other root fungi such as *Fomes lignosus* and Honey Fungus (*Armillaria* spp.) spread through it by means of bootlace-like rhizomorphs (tough masses of entwined hyphae) provided the fungus has a source of nourishment. The hard brown or black rhizomorphs of *Armillaria* can extend more than 5 m from a fallen log and thus enable some species of this basidiomycete, whose fruiting bodies disappear with first frosts, to reach and destroy a wide range of herbaceous and woody hosts. Honey Fungus has a world-wide distribution and the tree genera in which it causes *Armillaria* root rot include *Pinus, Picea, Larix, Castanea, Fraxinus, Quercus, Salix* and *Taxus*. *Armillaria* often kills small trees outright but death may be ultimately caused by drought, waterlogging, windrock or other adverse factors which the tree could have withstood had its root system been complete. Honey Fungus was formerly regarded as a single species (*A. mellea*) but is now known to include several species which differ in their form and pathogenicity. Of the six species of *Armillaria* found in Europe (*A. obscura, A. mellea, A. borealis, A. cerastipes, A. bulbosa* and *A. tabescens*), the first two are strongly parasitic. The others are saprophytic or weakly parasitic.

The colonization by fungi of hardwood stumps is affected by the type of chemical treatment used to control regrowth. Treating the stumps with 2,4,5-T encourages the establishment of *Chondrostereum purpureum*, which infects the cut surfaces and rots the wood relatively slowly. This fungus is often replaced by *Armillaria* which colonizes the stumps by means of its rhizomorphs. Ammonium sulphamate, on the other hand (Rishbeth, 1976), is translocated in the stumps and generally favours species causing rapid decay, such as *Bjerkandera adusta* (*Polyporus adustus*) and *Coriolus* (*Polystictus*) *versicolor*, some of which compete well with *Armillaria*.

Non-fungal pathogens

The importance of an increasing number of non-fungal tree diseases is now being recognized. Elm phloem necrosis, for example, is caused by a mycoplasma and has resulted in severe losses of American elm (*Ulmus americana*) and winged elm (*U. alata*) in the USA. Early symptoms are similar to those of Dutch elm and other vascular diseases; trees often die in the year they are first infected. Viruses cause many tree diseases including elm scorch and oak ringspot. Wetwood of elm results from infection by the anaerobic bacterium *Erwinia nimipressuralis* which produces methane and can cause sufficient pressure to split the trunk releasing a watery 'slime flux'. Bacterial canker of poplar, the most damaging disease of poplars in the UK, is caused by *Xanthomonas populi* and renders the trunks of non-resistant varieties useless for veneer cutting.

Importance of forest pathogens

In the constant battle against forest pathogens there are always areas where particular species are causing unusually heavy damage, or where the entry of exotic parasites would cause severe problems. The present activities of *Phytophthora cinnamomi*, which is slowly working through large areas of Australian *Eucalyptus* forest, show it to be one of the most ecologically damaging pathogens, killing most native trees, understorey and ground-cover plants. This phycomycete has an almost world-wide distribution in temperate and tropical regions, is listed as attacking 900 species and grows on a wide range of conifers, broadleaved trees, shrubs and herbaceous plants. The zoospores of *P. cinnamomi* show remarkably active chemotactic movements towards the growing regions of avocado roots in response to a root exudate, and cause lesions within days.

In British forests Ink disease of sweet chestnut is the only *Phytophthora* disease of any importance. No fruit bodies or mycelia are produced by this very simple fungus: attacks only become evident when coppice stools or whole trees start to die. The inky blue-black colour common in dead *Castanea sativa* roots gives its name to the disease, though *Phytophthora* is not responsible for the pigmentation. The disease occurs after wet periods. If improved drainage fails to arrest it, the only effective procedure is replanting with species other than beech or Lawson cypress; both these species are very susceptible.

The genetic basis of a plant's resistance to a pathogen has important ecological consequences. Though vertical (**race-specific**) **resistance** is very effective, breakdown, when it occurs, is usually complete. Horizontal (**race non-specific**) **resistance** usually develops selectively after exposure of the host to disease and is commonly polygenic. This type of defence is at a lower level, but much more general and not liable to a complete and sudden breakdown. The employment of vertical resistance in new hybrids of short-lived crop plants is often an effective strategy. With trees horizontal resistance, already widespread in many natural populations, may ultimately be more beneficial.

6.4   Dutch elm disease

Dutch elm disease (Webber and Brasier, 1984) results from infection of the internal tissues of the trunks and branches of elms by *Ophiostoma* (= *Ceratocystis*) *ulmi*, an ascomycete spread by insects and sometimes by wind, which produces toxins which cause the wood parenchyma to exude gum leading to the formation of tyloses. These bladder-like intrusions pass through pits in the cell walls into the vessels, which become blocked causing the foliage to wilt. The disease, recently rampant in the UK and North America, has in Britain been rapidly dispersed by the bark beetles *Scolytus multistriatus* and *Scolytus scolytus* (*S. destructor*). Aggressive strains of *Ophiostoma ulmi* attack *Ulmus americana* and the English elm (*U. procera*) severely, wych elm (*U. glabra*) being a little more resistant. The foliage of diseased trees yellows and dies, often in midsummer, and trees showing such symptoms are often dead within a few weeks. The purified fungal toxins consist of a number of unusual glycopeptides capable of reducing water conduction in cuttings from elm seedlings.

The characteristics of *Ophiostoma ulmi* populations in Britain have been investigated by Brasier and Gibbs (1973, 1975) who showed that the present epidemic was begun by an aggressive 'fluffy' strain, rather than the 'waxy' strain which is slow-growing, non-aggressive, has few aerial hyphae and appears to be a residuum of the 1930s epidemic, which killed 10–20% of elms in England. The fast-growing aggressive strain was imported into Britain on bark-bearing logs of Rock elm (*U. thomasii*) from Toronto; its initial dispersal was probably accomplished by the North American bark beetle *Hylurgopinus rufipes* and by *Scolytus multistriatus* growing in galleries in the bark. In a typical infestation the beetles, carrying the spores of *Ophiostoma*, fly to the crowns of adjacent healthy elms. The fungus is inoculated into the vascular tissues of the tree, at the

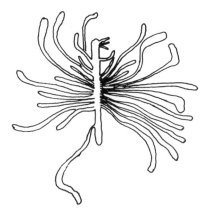

Figure 6.1 The 'signature' of *Scolytus* engraved on elm sapwood. Note central chamber and radiating larval galleries.

junction between the leaves and twigs and at the crotches of the twigs, during maturation feeding of the beetles when their gonads mature. Once the foliage has wilted the shoot tips curl over forming 'shepherds' crooks'. If an infected twig is cut through, black spots, marking areas containing tyloses and gum, can be seen in the youngest annual ring.

Fertilized female scolytids lay their eggs in galleries excavated under the bark of trunks and branches of trees affected by *O. ulmi*. The resulting larvae tunnel out further galleries (Figure 6.1) and pass through five instars before pupation. Even after the death of the tree, bark remains suitable for scolytid development for up to 2 years. *O. ulmi* may survive for many months as a saprophyte, its one sexual and two asexual phases sporulating luxuriantly in the galleries and in crevices in the bark. The beetles play a vital role in spreading *O. ulmi*; they carry spores on their antennae, mouthparts and legs but unless they create a feeding injury extending into the xylem infection will not normally occur. Ambrosia beetles form galleries inhabited by fungi in a similar way to the scolytids, but whereas they are dependent on the fungi, which they consume, the scolytids feed on components of the wood. Scolytids apparently do not eat *O. ulmi* and their relationship with the fungus seems to be coincidental.

Dutch elm disease is thought to have entered Europe from Asia during the First World War and was first described in The Netherlands in 1921. Epidemics of the disease might be prevented by controlling the vector, controlling the fungal pathogen, or replacing the existing elm population with hybrid trees bred to resist the disease. Crown spraying with insecticides to prevent maturation feeding and to kill the scolytid vectors cannot be recommended for general use. The nematode *Parasitaphelenchus oldhami* has little effect on either *S. multistriatus* or *S. scolytus* (Hunt and

Hague, 1974) though it parasitizes both. However, it might eventually be possible to find an effective means of biological control. The '**cordon sanitaire**' policy is to remove dead and diseased trees promptly, and to prevent the removal or importation of diseased trunks. In 1976 there were few parts of southern England where elm disease was not extensive, but in east Sussex less than 20% of the original elm population was dead or dying. A **cordon sanitaire** policy operated in this area but not in west Sussex where over 80% of the trees were dead by 1976. The relatively light infestation in East Anglia was probably related to the predominance of the more resistant smooth-leaved elm (*Ulmus carpinifolia*). As the fungus can spread via root grafts between trees whose trunks are 10 m apart, attempts to save healthy members of a partly diseased group may involve digging trenches and killing connecting roots. Protection of healthy trees by injection with systemic fungicides such as benomyl is only partially effective; it is expensive, has to be repeated every year and damages the trunk. A cure may be possible in the early stages with 'ceratotect' (Hibberd, 1991).

A pheromone-baited trap tree technique of controlling elm bark beetle vectors of Dutch elm disease gave quite promising results in the USA. In north-west England, however, the number of beetles attracted to the trap trees was not great. Colonization of these trees by the bark saprophyte *Phomopsis oblonga* following cacodylic acid treatment made them unsuitable to beetles for breeding, though many beetles landed on the trees. It seems likely that they penetrated the bark and detected the saprophyte (or compounds produced by it) by 'tasting' the phloem (O'Callaghan *et al.*, 1984).

Recent western European outbreaks of Dutch elm disease have been caused by two genetically distinct aggressive strains, one from North America and another, whose colonies are powdery and irregular in shape when grown on a culture plate, from central Europe or further east. The former passed from North America through the UK to France, Germany and Italy, while the latter caused outbreaks from Iran to Italy (Brasier, 1983). Both were involved in the epidemics which occurred in Ireland in 1977. In view of the considerable variation exhibited by *Ophiostoma ulmi*, Brasier suggests that new elms should be bred from stocks in the Himalayas or southwestern China, where pathogenicity is not a problem; although the origin of Dutch elm disease seems to lie in this region, it also has the greatest diversity of elms in the world.

While certain Dutch elm cultivars were more resistant than the British native elms they were attacked by the fluffy strain of *O. ulmi*: the search for resistant hybrid trees continues. The greater resistance of wych elm may be due to reduction in vector activity, either by the less favourable climate in the north and west of England where wych elm is preponderant, or by the presence of the fungus *Phomopsis oblonga* (Webber, 1981). Perhaps the best hope for the survival of elms in Britain lies in a genetic change occurring in the fungus, leading to the fluffy strain becoming less pathogenic. It is encouraging that some of the trees which develop from the stumps or suckers of felled trees survive. The Forestry Commission surveyed five 1 km² plots of hedgerow elms in the River Severn floodplain and found that all 764 trees over 6 m in height were dead or dying in 1976. Scott (1985) resurveyed these plots in 1984 and found 535 elms over 6 m, apparently free of disease.

Fungal diseases and insect pests are often of greater significance in plantations and regeneration forests having a very limited number of species than in more diverse natural systems. Even so, it seems likely that fungal epidemics occurred in the past and were responsible for widespread destruction of particular tree species. Indeed, *O. ulmi* may have caused the post-glacial elm decline (section 5.6), a view supported by Rackham (1986). While it is feasible that the parasitic activities of an exceptionally aggressive strain of *O. ulmi* could cause the extinction of the elm itself, the scolytid vectors would survive as they can develop in trees other than elms (Archibald and Stubbs, 1980). Relatively unaggressive strains such as the 'waxy' isolates of *O. ulmi* are, on the other hand, better adapted to continued parasitic life in that a large host tree population remains available to future generations of the fungus. Brasier (1983) considers that attenuation of pathogenicity within the present aggressive strain is more likely than its replacement by the present non-aggressive strain.

Dutch elm disease, together with *Scolytus multistriatus*, reached North America, where the new aggressive strain later evolved, on European elm burl logs imported to manufacture veneer. It is ironic that a similar importation into Britain should have caused the death of over 20 million elms, out of an original population of 30 million, between the beginning of the outbreak in the late 1960s and 1982. The effects in the central and southern regions were most marked, the southern population being reduced from 23 million to 3 million (Burdekin, 1983), posing severe problems with regard to falling trees and replanting. It also emphasizes the dangers of introducing more aggressive strains of existing pathogens into Britain, as well as such serious new diseases as American Oak Wilt (*Ceratocystis fagacearum*) and Chestnut Blight.

6.5 Exotic forest pests

Native heterotrophs, such as insects and fungi, living in balanced relationships with their host trees in a particular forest region may cause major damage if moved elsewhere. Gibbs and Wainhouse (1986) illustrate this point by describing the spread of forest pests and pathogens in the northern hemisphere between the four main forests derived from the relatively homogenous Arcto-Tertiary forests (Figure 5.16). Though the original communities were similar subsequent evolution of the pests, pathogens and host trees has created a potential for substantial damage to forest trees when pests and pathogens are moved from one major area to another usually by man.

It is difficult to predict whether or not a particular forest herbivore will develop into a major pest if introduced into a fresh area, though the damage potential of many, such as the great spruce beetle *Dendroctonus micans* (an American species first reported in Britain in 1982), is obvious. The large European elm bark beetle *Scolytus scolytus* has reached North America many times but failed to establish. Similarly the American bark beetle *Hylurgopinus rufipes* failed to establish when it reached Europe. Other introduced herbivores have had an initial explosive burst in population size and then declined. The brown tail moth (*Euproctis chrysorrhoea*), a pest on several tree species, entered North America from Europe around 1890 and spread through New England and into several Canadian maritime provinces in about 30 years. The irritant hairs of its caterpillar were a public health hazard and the pest also caused defoliation of several species of trees and shrubs. The range of the moth began to recede in the 1920s, for reasons not properly understood, and it is now restricted to a few coastal habitats in the USA.

The winter moth (*Operophtera brumata*) became a major pest of fruit trees and oak after being introduced into Nova Scotia. Its progress was abruptly halted by two parasites, the tachinid fly *Cyzenis albicans* and the ichneumonid *Agrypon flaveolatum* introduced from Europe. Neither is important in regulating population density of winter moth in Europe (sections 7.2 and 7.3) and this example shows how an environment suitable for an introduced pest may favour its parasites even more.

Balsam woolly aphid (*Adelges piceae*) appears to have been introduced to several locations in North America on nursery stock prior to 1900. It is now widespread throughout maritime Canada and the northeastern USA. The crawler stage is dispersed passively but spread to new areas is mainly by movement of infected nursery stock, Christmas trees and infested logs. In North America *A. piceae* causes great loss of timber and is a serious pest of *Abies* spp., especially *A. balsamea*, heavily attacked trees dying within 3 or 4 years. Salivary secretions injected while the adelgid is feeding on bark parenchyma result in the formation of abnormal xylem known as redwood or 'rotholz'. In contrast the insect is not regarded as a serious pest in Europe, where native firs possess a genetically based resistance that is maintained when the trees are grown outside their normal range.

Many of the most striking examples of epidemics resulting from the movement of forest pests and pathogens involved the movement of timber and planting stock into North America during periods when human immigration into that area, particularly from Europe, was high. The dangers which could result from further movements can be illustrated by the impact of pine wilt disease in Japan where by 1981 10 million native pines (*Pinus densiflora, P. thunbergii* and *P. luchuensis*) were dying annually with a timber volume loss of 2×10^6 m^3. The pathogen is the pinewood nematode (*Bursaphalenchus xylophilus*) with a probable origin in North America, where it is widespread but causes little damage to native pines, while the vector is the Japanese pine sawyer (*Manochamus alternatus*).

Although many of its coniferous plantations are now into their second rotation, commercial forestry in the UK is still relatively immature and hence is susceptible to exotic pests and diseases, or to unexpected changes of status as we shall see in the case of the pine beauty.

6.6   Woodland herbivores

Certain mammals (and less importantly, birds) can cause serious damage to woodland herbs, shrubs and trees, by eating foliage, severing roots, stripping bark, and trampling or even felling whole plants. They also play an important role in woodland regeneration, as feeders on fruits, seeds and seedlings.

Temperate forests lack specialized arboreal leaf-eating mammals, although the canopy-dwelling fat dormouse (*Glis glis*) includes tree leaves in its diet. Squirrels and various voles and mice, such as *Apodemus sylvaticus*, feed on buds in spring, after the mast crop has been depleted. Grey squirrels (*Sciurus carolinensis*), first introduced into Britain in 1876, often occur at much higher densities than in North America, where they are rarely pests. As a result, certain individuals are forced into suboptimal territories, which they mark by chipping off bark. Large-scale removal of bark between May and June can kill susceptible hardwoods such as *Fagus sylvatica* and *Acer pseudoplatanus* (Gurnell, 1987). In Colorado, the porcupine *Erithezon epixanthum* feeds almost entirely on the phloem of *Pinus edulis*, selecting trees whose branches have high sugar concentrations, associated with attacks by root fungi (Spencer, 1964).

Many kinds of deer and other large herbivores rely on woody browse during the winter and dry seasons, and some show marked preferences for particular species. If populations are allowed to increase excessively, as when protected from predation, the consequences may be marked. The eruption of mule deer (*Odocoileus hemionus hemionus*) on the Kaibab Plateau, Arizona, during the first quarter of this century, led to the virtual elimination of their preferred species of *Rubus* and *Salix*.

Deer also damage trees by fraying the bark with their antlers (Putman, 1988). In some species this occurs only when the velvet is being removed, and also at traditional rutting stands. Roe deer (*Capreolus capreolus*) bucks also cause widespread fraying during territorial disputes. Bark can be stripped off by deer and certain other mammals, including elephants, but the latter also push over whole trees to reach fruit, leaves and bark. The threatened re-introduction of beavers (*Castor fiber*) into Britain would at least provide the UK with an example of a tree-felling animal.

Turning now to invertebrates, Southwood (1973), in a discussion of the various chemical and physical hurdles facing potential insect feeders on angiosperms, points out that relatively few insect orders contain appreciable numbers of species which feed on or in foliage, the major woodland examples occurring within the Lepidoptera, Coleoptera, Diptera, Hymenoptera and Hemiptera. Even these may have to adapt their feeding strategies to the phenology of the plants, since much of the greenery of the forest is only potential food, protected for most of the year by the defences of the plants (Crawley, 1983). Herbivory below ground has been somewhat ignored by ecologists even though its effects can be just as great as its foliar-feeding counterpart (Brown and Gange, 1990). Root-feeding insects have to cope with a food source which although abundant may be of exceptionally low quality. Even though developmental times may be long as a result of this, root-feeding can seriously reduce the vegetative growth of plants, causing perhaps 50% loss of total plant biomass in the field situation.

Foliage feeders may do no more than raise a pimple on a leaf surface, or they may cause reduced growth or even death of whole trees. Partial defoliation of trees may alter light penetration and hence the performance of the shrubs and herbs below, while wide-scale mortality of mature trees, followed by natural regeneration or replanting, will have far-reaching effects on the ecology of the community. Whatever the scale of the damage, in all cases energy and nutrients are moved from the tree to the animal, and so to diverse food webs.

Galls are abnormal growths, each of a characteristic form which results from the unique interaction between a particular species of plant and the gall-causer. The latter may be a microbe, mite or insect, the gall providing food and protection for the developing organisms. Although in some instances, such as Knopper galls on acorns (Figure 6.2) and certain adelgid galls on conifers, the plant is adversely affected, it is assumed that the relationship is usually harmless, even when for example the burden may reach several hundred spangle galls per oak leaf, covering 80% of the surface. Darlington (1974) has suggested that galls benefit not only the animal but also the plant, by localizing the damage rather than allowing a free range. They certainly benefit the communities of parasites and inquilines ('lodgers') within the galls, and also supplement the diet of birds such as tits and gamebirds.

Sycamore aphids (*Drepanosiphum platanoidis*) are an example of a more

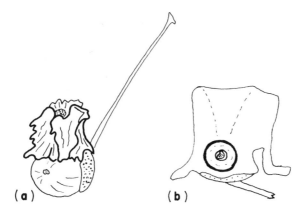

(a)   (b)

Figure 6.2 Knopper gall (*Andricus quercus-calicis*): (a) on undistorted acorn; and (b) in section, showing larval cell above degenerate acorn. In late summer these galls are bright green and sticky; they later turn brown and woody. They were first recorded in Britain in 1961, and are a potential threat to seed collection stands, since acorns are usually destroyed. The sexual generation occupies 1 mm swellings on male catkins of *Quercus cerris*.

mobile burden, whose reproduction is favoured by the higher concentration of soluble nitrogen in the phloem sap of actively growing or senescing leaves (Dixon, 1970). During midsummer, when gravid females are virtually absent, aggregations tend to occur in the cooler lower canopy, under leaves sheltered from the wind and not likely to be brushed by other leaves. Their sugary excreta, or honeydew, support sooty moulds, which may interfere with photosynthesis, and also channel nutrients into the forest floor. The siphoning of energy through aphid populations, as they feed selectively for nitrogen, is partially compensated by the production of leaves with a higher chlorophyll content, but reduced growth rates of leaves, stems or roots, recorded in sycamore and lime, are probably associated with the injection of growth substances in the aphids' saliva. Various adelgids (which, like aphids, are true bugs, i.e. Homoptera) cause distortion of conifer needles, sometimes leading to defoliation, but spraying of plantations is usually not justified economically; it is a different matter in seed orchards and for nursery stock, while large-scale planting of the silver fir (*Abies alba*) in Britain has been thwarted by dieback caused by *Adelges nordmannianae*.

As the leaves of dominant species in temperate woodlands age they not only become tougher but also tend to have increasing concentrations of resins, tannins or other compounds which are thought to defend them against various herbivores. A majority, but not all, species of forest Lepidoptera considered to be pests feed on new foliage in the spring. Winter-moth caterpillars (*Operophtera brumata*) reared on oak leaves picked

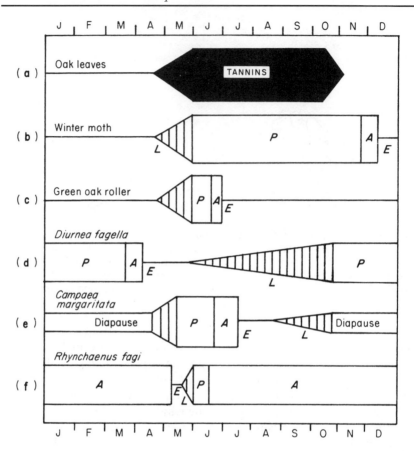

Figure 6.3 Phenology of foliage-feeding insects on oak and beech. (a) Tannin content of leaves of *Quercus robur*. (b)–(e) Phenology of four species of Lepidoptera on oak (from Varley, 1967). (f) *Rhynchaenus fagi*; beech leaves are mined by the larvae and skeletonized by the adult weevils (data from Nielsen, 1978, *Natura Jutlandica*, **20** 259–72.) E, egg; P. pupa; A, adult; L, larval growth periods, shaded.

at the end of May were found to develop into smaller (and therefore less fertile) adults than those which fed on younger leaves, with lower tannin levels and so more available nitrogen (Feeny, 1970; Mattson, 1980). Many Lepidoptera and Coleoptera with larvae feeding on oak or beech leaves typically develop early in the season. Among species which feed on older leaves, most, including leaf-miners, develop slowly, over several months, in one season, while some complete their development in the following year (Figure 6.3). Oaks and other broadleaved trees in Britain are sometimes defoliated by caterpillars such as those of *Operophtera* and the green oak roller (*Tortrix viridana*) especially when flushing and egg hatch coincide. Damage was widespread in England and Wales in 1979, 1980 and 1990. In Chaddesley Woods, Worcestershire, at the end of May 1980, after a month of high temperatures and virtually no rain, numerous oaks, hazels and even common ash were completely defoliated, although individual oaks which had only just begun to flush escaped attack. Falling frass was distinctly audible as it hit the bone-dry litter, and trunks were festooned with silken skeins produced by caterpillars descending to seek food or to pupate. Yet by mid-July these ravages were largely masked by lammas growth. Outbreaks of this kind are relatively minor compared with the plagues of gypsy moth caterpillars (*Porthetria dispar* = *Lymantria dispar*) in New England, which originated from small numbers introduced from Europe in the 1860s.

Yew (*Taxus baccata*) provides a British example of a tree with alkaloids in the foliage and seeds which are poisonous to man and livestock, while β-ecdyson, a moulting hormone commonly found in gymnosperms, presumably disrupts development in herbivorous insects. Alkaloids, tannins and other 'defensive compounds' are at their greatest diversity in the tropics, a feature perhaps reflecting the wealth of potential herbivores (Rosenthal and Janzen, 1979). In general such plants are protected from severe defoliation, although specialists can feed on them, for example the proboscis monkey on tannin-rich foliage in Malaysian mangrove swamps. Plants growing on tropical white-sand soils are particularly rich in defensive compounds, for example greenheart (*Ocotea*) and mora (*Dimorphandra*), whose timbers are extremely resistant to decay. The foliage of plants from New Jersey pine barrens has also been described as being 'unharvestable' by herbivores, and as yielding 'a medicine man's warehouse' (Janzen, 1974). Reviews of defence strategies, both constitutive and induced, include Mattson (1980), Schowalter *et al.* (1986), Speight and Wainhouse (1989) and Karban and Myers (1989). Edwards (1989) challenges some of the assumptions commonly made in this area, especially relating to plant evolution. He points out that we know very little about the influence of grazing on plant fitness, in the genetic sense, and advocates restricting the term defence to situations where there is strong evidence to suggest that herbivory was a major selective factor (e.g. bruchids and legumes, section 3.4).

Further north, a native lepidopteran, the eastern spruce budworm (*Choristoneura fumiferana*), normally occurs at low densities. Periodically, approximately every 35 years, it erupts to produce deadly conifer-defoliating epidemics which last 5–10 years; in Quebec these are known to have occurred back to 1704. Only 1–2% of species of forest Lepidoptera reach outbreak densities or show **cyclic dynamics** (Myers, 1988). Many of the 18 species in North America and Europe that show outbreak densities also show a clear pattern of periodic fluctuations, with peaks every 8–10 years. There is no clear evidence of cycles in tropical

Lepidoptera. For a given geographical area, the phases of increase can vary in the timing of their initiation, rates of increase, peak density and amount of damage, yet still the populations decline more or less in synchrony. Population cycles characteristically occur towards the elevational and latitudinal edge of the species distribution; for example, larch budmoth (*Zeiraphera diniana*) cycles are best demonstrated in populations above 1700 m (Baltensweiler, 1984). The few experimental manipulations that have been attempted demonstrate that it is difficult to perturb the basic dynamics of the population fluctuations. Many studies of forest insects have examined mortality closely but this has not led to a general understanding of the driving mechanism behind the cycles. The four main hypotheses for cycles are (a) variation in insect quality, (b) climatic release, (c) variation in plant quality, and (d) differences in disease susceptibility. Myers (1988) speculates that the 8–10 year population cycles of forest Lepidoptera are caused by prolonged effects of disease or microparasites reducing the vigour and fecundity of populations for several generations after the beginning of the population decline. These species of forest Lepidoptera with outbreak population dynamics appear to have evolved in such a way that they have little long-term detrimental effect on the forests. This can be contrasted with the situation with bark beetles that kill trees and therefore have a much larger impact on forest dynamics.

In the western USA, the bark of *Pinus contorta* is attacked by females of the mountain pine beetle (*Dendroctonus ponderosae*), thus liberating terpenes and an aggregating pheromone which attracts other egg-laying females (Amman, 1977). If only small numbers are involved, eggs or larvae may be controlled by resin. Large-scale invasions girdle and kill trees, possibly through the action of blue-stain fungi, introduced by the beetles, which maintain moisture conditions favourable for larval development. Major infestations occur at 20- to 40-year intervals, devastating the largest trees over vast areas and thus altering community structure.

The generally low population densities of most forest insect pest species may be contrasted with the situation in agriculture, where many species are persistent pests. These endemic populations of forest insects were likened by Bevan (1974) to 'a porpoise close behind us'. One example is the defoliation of Scots pine in the UK by pine looper moth caterpillars (*Bupalus piniaria*). This is a native species, but although periodic epidemics had been recorded in continental Europe for over 100 years, large-scale damage was not seen in the UK until 1953. Patterns of population fluctuation have been analysed by Barbour (1988, 1990) who showed that although a few UK populations show cyclic fluctuations similar to the highly synchronized ones in Germany, in general there was poor synchrony between different forests. Synchrony may be critically dependent on the amplitude of the cycles, and so may be almost absent in low-amplitude fluctuations such as those of the British *B. piniaria* populations. Aerial spraying with insecticides, which reduces populations to non-outbreak levels (Figure 7.7b) has been used on eight occasions in Britain, up to 1991.

Pine looper caterpillars attack needles after buds have been formed, so that defoliated trees can flush again in the following year, and therefore recover from attack, but this ability may be upset by drought or secondary pests. After the 1953 Cannock attack, 50 000 defoliated *Pinus sylvestris* were further exploited, as breeding sites, by pine shoot beetles (*Tomicus piniperda*; Figure 6.4) which had previously existed at a low endemic level in an apparently healthy forest. Usually this species frequents freshly-felled logs for breeding, attacks on living trees being limited to shoot-pruning due to maturation feeding. Breeding in healthy trees is normally resisted by resin, but not in the case of *Bupalus*-weakened trees. Even after removal of timber, however, replanting has its pest problems in the form of *Hylobius* (Figure 6.4) and *Hylastes* beetles which breed in the remaining stumps and then attack the stems and roots of seedlings.

Another indigenous species, the pine beauty (*Panolis flammea*), known to be endemic throughout Britain and to have caused serious damage to Scots pine in Central Europe, has only recently (1976) become a pest of British forests, but has been described as the most important entomo-logical event in the history of the Forestry Commission, necessitating aerial spraying on a scale unprecedented in Britain. Outbreaks in Scotland are associated with lodgepole pine rather than Scots pine, particularly lodgepole pine growing in deep unflushed peat. *Panolis* larvae are affected by spatial and temporal variability in the quality of the foliage of their plant hosts. Larvae reared at the top of the pine crown survive and grow better than those reared at the base; young larvae are unable to feed on the previous years' foliage but older larvae feed on both old and current years' foliage (Watt, 1990). Larval growth and survival (in the absence of natural enemies) tends to be greater on Scots pine than on lodgepole pine; larvae grow and survive least well on lodgepole pine growing in deep peat. However, the mortality in cohorts exposed to natural enemies is greatest on lodgepole pine. Currently the possibility is being explored of control by nuclear polyhedrosis viruses (NPV), including genetically manipulated material (Payne, 1988).

Woodwasps (Siricidae) are of little importance to conifers in Britain, unlike the situation following their introduction into Australia and New Zealand, where *Sirex noctilio* kills trees, especially *Pinus radiata*. Heavily trimmed or lopped trees are most likely to be killed; translocation, tran-spiration and phloem respiration rates are altered in these 'stressed' trees. Subsequent changes in bark permeability allow a greater loss of water vapour and volatile monoterpenes, thus attracting *Sirex* females. These introduce their eggs in mucus, as well as spores of the basidiomycete *Amylostereum areolatum*, beneath the bark. Fungus and mucus both cause

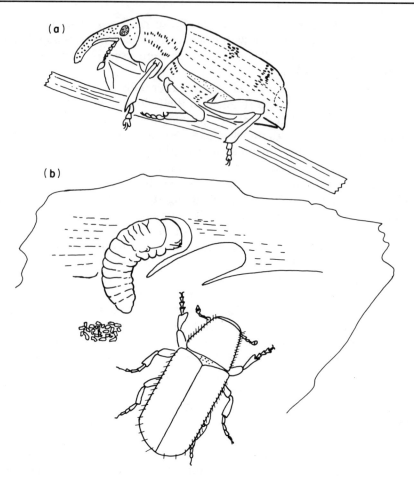

Figure 6.4 Beetle pests from Cannock Forest, Staffs, June 1980. (a) *Hylobius abietis* (large pine weevil). Larvae develop in stumps and logs; adults damage bark of young pines. (b) *Tomicus piniperda* (pine-shoot beetle.) Larva and adult under bark of a pine log, a potential source of infestation of weakened standing trees.

further inhibition of translocation, so that more woodwasps are attracted. The fungus-softened wood provides food for tunnelling larvae, but this fungal development is also the cause of death of the tree (Madden, 1977).

6.7 Influence of defoliation on trees

It might seem obvious that insect defoliation causes damage to trees (Gradwell, 1974; Schowalter *et al.*, 1986; Speight and Wainhouse, 1989) but insects can also be regarded as regulators of primary production in natural ecosystems (Mattson and Addy, 1975). Grazing affects flux rates by increasing light penetration through the canopy, reducing competition for abiotic resources, altering plant species composition, increasing the rate of nutrient leaching from foliage, and increasing the rate of fall of nutrient-rich litter into the decomposition subsystem. It also enhances the abiotic flux by stimulating the redistribution of nutrients within plants from boles and branches to components such as leaves and buds with high turnover rates. This promotes consistent and optimal output of plant production over the long term.

Deciduous trees produce new leaves throughout the summer whether attacked by insects or not. They appear either continuously in an indeterminate fashion, as in ash, or in distinct determinate flushes, as is usual in oaks. The second flush of young yellow leaves often stands out against the dark green of older leaves and usually appears on oak at the end of June. Less frequently there is a third flush at the beginning of August, giving rise to the English description **lammas growth**, often used to describe any conspicuous growth flush after the first. These flushes are normal in oak but are often ascribed to the need to replace leaves which have been destroyed by insects (section 9.3). Certainly the removal of leaf material stimulates new leaf flushes which may then occur earlier than the end of June. Especially if defoliation is severe, they also occur with little of the stem extension that usually accompanies second and third flushes.

These defoliation effects were conspicuous in a simulation on 196 2-year-old oak seedlings in a plot at Wolverhampton (Hilton et al., 1987) which was designed to determine the effect of defoliation on production and growth. The oaks were given one of four defoliation treatments in which all, two-thirds, one-third or none of the leaf surface present in mid-June was removed by hand cutting. The position of the terminal bud in spring 1981 was marked and the diameter just below this mark was recorded each year as a measure of growth increment and production. After 3 years, half the trees were removed and weighed to establish a correlation between biomass and diameter, leaving room for the remainder to grow on without defoliation treatment while their recovery was observed.

Relative growth rates were compared on the basis of area of cross-section calculated from the diameters, shown in Figure 6.5. This shows the markedly reduced growth rates of two-thirds and entirely defoliated trees, but only slightly reduced rates for one-third defoliated trees. Figure 6.5 also shows how the growth rates of the trees recover when defoliation ceases, even though the biomass of heavily defoliated trees is much reduced. The other feature of note is that the variation between years is as great as the variation between treatments. Although the reduction

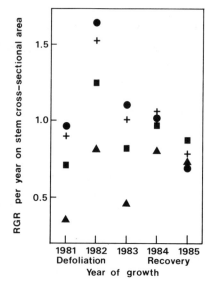

Figure 6.5   Mean relative growth rates (based on cross-sectional area of stem) for oak saplings subjected to artificial defoliation during the years 1981–1983 by removal of a proportion of the leaf area present in mid-June.

● – control, not defoliated   + – one third defoliated
■ – two thirds defoliated   ▲ – totally defoliated

(From Hilton Packham and Willis, 1987; courtesy of *New Phytologist*.)

in growth increments of defoliated trees may be small, it is important to realize that they are placed at a competitive disadvantage in a natural population, especially if they are saplings growing under shaded conditions.

Defoliation may also affect other aspects of plant growth, including fecundity. Wong *et al.* (1990) showed that heavily defoliated crowns of the tropical tree *Quararibea asterolepis* produced significantly fewer flowers and fruits than lightly defoliated crowns. Flowering also occurred later and was less synchronized for heavily defoliated trees. These differences in reproduction did not persist beyond the outbreak year.

6.8   The importance of woodland carnivores

Carnivores, from tiny invertebrates to large birds and mammals, are often said to be of major importance in the regulation of certain herbivore populations (section 7.4). Loss of the larger carnivores, including wolves, wolverines, lynx and bears, from most European forests has destroyed the predator–prey relationships which formerly kept deer numbers within bounds. The resulting increase in deer populations, particularly in the British Isles, has caused tree regeneration rates to fall sharply and communities of less common plants to require protection. Control of roe deer (*Capreolus capreolus*) in agricultural areas is often particularly difficult, the animals raiding arable crops from woodland refuges. Adequate culling by shooting is difficult to achieve and often unpopular with the public.

Problems of overbrowsing and bark stripping by red deer (*Cervus elephus*) occur in south-east Poland where overpopulation, underculling and increased density are associated with a change from clear felling to selective cutting which maintains both food and cover for the deer (Bobek *et al.*, 1984). Deciduous trees provide a higher quality of forage than do conifers and deer feeding is often very selective. In northern Sweden the now abundant elk (*Alces alces*) sometimes browse heavily on young *Pinus sylvestris* but not on *Picea abies*. Feeding preferences thus influence tree regeneration. In a coastal montane *Abies anabalis* forest of the Cascade range, Washington State, USA, Hanley (1984) found that wapiti (*Cervus elephus nelsoni*) feed on graminoids and forbs while black-tailed deer (*Odocoileus hemionus*) consume woody browse and forbs in more xeric habitats.

Even in forests with major carnivore populations which prey on cervids, as in North America, the behaviour patterns of the carnivores have often been so modified by contact with man that herbivore populations are now subject to more major oscillations.

The traditional explanation of herbivore eruptions is that reduced predation (usually due to the intervention of man) leads to a massive increase in numbers. Lack of food then causes a crash to a lower, food-limited equilibrium. Caughley (1970) showed through his refutation of the story of the Kaibab deer how such ideas can become dogma when, in fact, few well-documented cases follow this pattern. Sinclair (1989) suggests that there is an incorrect general perception amongst North American wildlife ecologists that large mammal herbivores are regulated by predators. That populations increase when predators are removed certainly indicates that predators are limiting their prey populations, but it does not necessarily imply regulation. The effects of predator removal can be used to advantage as a recent 6-year study in southern England by the Game Conservancy has demonstrated. The numbers of surviving game birds became much higher when predators were controlled by gamekeepers, a scientific confirmation of traditional practice.

Tawny owl (*Strix aluco*), a K-species, capturing its prey the bank vole (*Clethrionomys glareolus*), an r-species. (Drawing by P. R. Hobson.)

7
Community change and stability

7.1 Population stability: selection and strategy

The continuity of natural woodlands may, for long periods, be one of their most reassuring features: from year to year the changes in the trees, shrubs, and woodland herbs, and in the seasonal patterns of activity within the woodland community often seem slight. Yet things are never quite the same, there is always a change of balance and from time to time sudden catastrophic changes result from the action of great storms, devastating fires, or epidemics of disease which can kill many millions of trees within a comparatively short time. As the environment changes, perhaps owing to management policy, one species will increase and another decrease. More than once in European prehistory the vegetation has altered as the climate changed from cold and dry to mild and wet, as at the beginning of the Atlantic period (c. 5500 BC) when oak, elm and lime were replacing the pine in Britain. Such climatic changes often take a very long time; with plants it is usually reproduction that is first affected. As oaks live for three centuries or more, and redwoods for three millennia, the resultant changes in the populations of trees are not easily noticed, though effects on the relatively short-lived herbs, particularly those near the limits of their geographical range, may be more conspicuous.

Compared with other ecosystems, however, mature woodlands provide essentially stable habitats, saturated in the sense that all the niches they contain are usually occupied. Here all the organisms present compete for survival with others that are also well suited to life in climax forest. Nevertheless, there are some places within the woodland mosaic where **open habitats** exist. Recently burnt or coppiced sites, and those where trees have been windthrown, are for a short period open to colonization by a wide variety of life forms, though within a year or two they will usually be occupied by a closed community. In the woodland as a whole, however, such transitory sites are virtually always available to 'opportunist' or 'fugitive' species such as *Chamaenerion angustifolium* (fireweed, rosebay willow herb) and various bryophytes, of which *Funaria hygrometrica* is prominent in the first and second seasons after burning.

In the spruce–fir forest of the Rocky Mountains fireweed grows along streamsides, on land slides and in clearings, but after forest fires whole mountainsides become covered with fireweed whose seeds are rapidly dispersed by wind. After a few years the trees grow again and the fireweed gradually diminishes. In Britain this plant has spread widely in the present century and is now, at least in some areas, a troublesome weed. The annual seed output per plant averages 76 000, but *C. angustifolium* also propagates rapidly by adventitious shoots which arise from the long spreading roots. A comparable insect opportunist is the plague caterpillar, *Tiracola plagiata*, which exploits ephemeral clearings in southeast Asian rainforest (Conway, 1976). The strong flying moths can lay more than 1000 eggs which develop into adults in 30–40 days, so that populations of several million can build up to strip a wide range of plants, either in temporary habitats or in agricultural crops.

There is wide variation in the weights of individual propagules of plants within the various habitat groups (Salisbury, 1942), but mean propagule weights increase in the sequence: open habitat – semi-closed or closed non-shady habitat – herbs of scrub and woodland margins – shade species – shrubs – trees. Of the herbaceous groups the shade species found in woodlands have the lowest seed output; their increased seed weight on the other hand enables the seedlings of many of them to reach a relatively large size before becoming entirely dependent on their own photosynthetic production. In shaded habitats some species circumvent the problem of producing seedlings large enough to survive independently by reproducing vegetatively. *Lamiastrum galeobdolon* (yellow archangel), for example, reproduces by seed when growing in warm, well-lit habitats and by long stolons in cool shaded places.

Population growth and the r–K continuum

In and around woodlands competition and selection have operated differently on various plants and animals; these differences often result from the nature of the particular habitat. Southwood (1976, 1977) has likened an organisms's habitat to a templet (literally a 'mould') against which evolution has fashioned for the organism a specific **ecological 'strategy'** which maximizes its chances of survival. To understand these strategies, and their relevance to competition and stability, we must consider the nature of population growth (see also section 3.1 for C–S–R strategy).

Under particular environmental conditions, when there is no restraint due to population size, an animal population with overlapping generations will increase geometrically, i.e. at an exponential rate. Its **growth rate**, representing the difference between birth and death rates, is constant

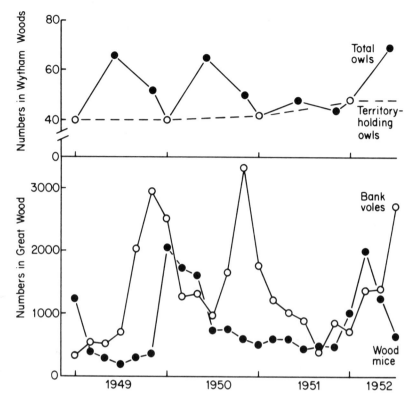

Figure 7.1   Population fluctuations in animals with different longevities in the same locality: tawny owls (*Strix aluco*) in Wytham Woods (525 ha), near Oxford, and mark-and-recapture estimates of bank voles (*Clethrionomys glareolus*) and woodmice (*Apodemus sylvaticus*) in Great Wood, Wytham (95 ha). (Data from Southern, 1970; redrawn from Southwood, 1976.)

and independent of population density; this is the intrinsic or maximum per capita rate of increase for that species under that particular set of conditions (r_m). Such unrestrained growth cannot be maintained for long, even in the laboratory. One way in which the growth rate can be diminished is by progressively reducing the actual rate of increase per individual, **r**, to zero as the **population density**, **N**, approaches the **carrying capacity**, **K**, which is the maximum number of that species sustainable by the habitat (Whittaker, 1975).

Regular oscillations are characteristic of some natural populations, such as the small mammals of northern latitudes, e.g. lemmings. Generally, however, one finds either erratic fluctuations, as in many pest species, or relatively stable populations (Figures 7.1 and 7.2). The components of the contrasting strategies which are presumably responsible for these differences can be considered in relation to heterogeneity in the basic habitat dimensions of space, time and physical conditions. For example, adaptations of organisms can be related to an **adversity axis**, as in Raunkiaer's classification of life forms of plants (section 2.1), while the **r–K continuum** of MacArthur and Wilson (1967), still much used by ecologists, is one of many relating species distribution to the durational stability of the habitat. Stability ranges from ephemeral to relatively permanent habitats, and in all cases the important relationship is between the generation time (t) of a species and the duration (H) for which a habitat remains suitable for breeding. The following summary describes some of the characteristics associated with the spectrum of strategies from r to K.

Extreme **K-species** (i.e. those which display K-strategies, evolved through K-selection) are adapted to living in basically stable, 'permanent' habitats, in which H/t is large, where they tend to maintain their relatively constant populations at or near the carrying capacity (K). Examples include certain long-lived species of vertebrates and trees. In contrast, **r-strategists**, such as the classic pest and weed species, are continually colonizing unpredictable or ephemeral habitats, for which H/t is small. Considerable investment in reproduction enables these species to exploit favourable conditions by rapid population growth (hence 'r'). This is followed by a marked decline (which may be partially related to interspecific competition for food or space) as individuals die, form dormant propagules, or migrate. Since the next generation must normally arise elsewhere, overshooting the carrying capacity in one place is irrelevant, as long as migration occurs. This is the 'boom and bust' strategy of the opportunist.

K-species, on the other hand, experience low mortality and recruitment rates, and have evolved high interspecific competitive ability, requisite for success in such 'desirable' habitats; considerable investment is made in defence, for example by plants against herbivores. Although fecundity is relatively low, much energy may be expended on each offspring, for instance by parental care. Some of the more obvious characteristics at the extremes of the r–K continuum are shown in Table 7.1. As climax forest is one of the most stable of ecosystems a preponderance of K-strategists is to be expected, with r-strategists occupying spatial (woodland rides and streamsides) and temporal (caused by fires and clear felling) gaps in the mature system.

Mature (but not senescent) individuals of large K-species such as bears, oaks and redwoods are seldom killed in nature by the attacks of other species, while r-species largely avoid predators by their higher rates of increase and mobility (a 'hide and seek' strategy), rather than by chemical or physical defences. Most species lie between these two extremes and

Table 7.1 Features of extreme r- and K-strategists (modified from Southwood, 1977)

r-species	K-species
Opportunistic, exploiting temporary habitats (*H/t* small)	Equilibrium species, of stable habitats (*H/t* large)
Small sized and short-lived (relative to other members of the same taxon)	Large and long-lived
High fecundity	Low fecundity or iteroparous* (repeated breeding) with 'masting'
Rapid development (generation time, *t*, short)	Slow development (*t* long)
High capacity for increase (r_m)	r_m low
Low investment in 'defence'	High investment in 'defence'
Time efficient	Food and space resource efficient
Population density very variable – 'boom and bust'	Population density relatively constant from generation to generation $\simeq$ K
High rate of dispersal (e.g. migration)	Low rate of dispersal

* Semelparous species, in contrast, have only one period, usually short, of sexual reproduction.

here natural enemies may play a role in establishing a stable equilibrium well below what would otherwise be the carrying capacity. As Conway (1976) points out, biological control involving the use of predators, parasites or pathogens is likely to be most effective against intermediate pests.

High rates of increase enable r-species to recover quickly from unfavourable periods: their short response times cause the population to track environmental fluctuations much more closely than K-species, which have long response times for return to an equilibrium following disturbance. (Contrast the fluctuations of bank voles and wood mice with the relative stablity of tawny owls, whose main diet consists of these rodents; Figure 7.1). This often presents a problem for conservation as populations of K-strategists are not well fitted to recover from population densities substantially below their equilibrium level, and small populations of them are far more likely to die out than are those of r-selected species.

7.2   Population fluctuation: key-factor analysis

It is often taken as axiomatic that there exists a 'balance of nature', at least in ecosystems with minimal human interference. It is in fact unlikely that such complex systems can be studied in their entirety to demonstrate such a balance; indeed Connell and Sousa (1983) suggest there is little or no reason to suppose that natural populations actually possess a steady-state equilibrium, let alone that it is stable. In practice studies are usually restricted to, at most, a few species.

The numbers of individuals at a given stage in the life cycle usually fluctuate in successive generations, the relative change (best demonstrated by logarithmic plots) varying in different species (Figure 7.2). However, due to the complexities presented by different mean population levels and by spatial and temporal scales of sampling, what can actually be said about differences in species population variabilities is strictly limited (McArdle et al., 1990). The major causes of intergeneration fluctuation in species with discrete, non-overlapping generations, can be determined by **key-factor analysis** (Varley et al., 1973). Despite criticism, most workers still use this method (Sinclair, 1989). Basically it involves following the progress through time of a single cohort (the progeny of a population born at approximately the same time) and so constructing an age-specific life table, to show the numbers surviving at a particular time. As Gilbert (1989) suggests, this approach assumes the very existence of key factors, when, however, the influences on the dynamics may interact in complex ways. The decrease in numbers during successive time intervals is a measure of mortality, including not only death but also other processes such as emigration which result in a decrease. If abundance is expressed in logarithms, the difference represents the killing power or k-value of the process or factors responsible for the measured mortality:

$$k\text{-value} = \log_{10}N_s - \log_{10}N_{s+1}$$

where N_s and N_{s+1} are population densities on successive sampling occasions.

The total generation mortality, K, equals the sum of the individual, sequential k-values for that generation. The contribution of each to changes in K can be assessed from a number of generations in order to determine which factor(s) (represented by k_n) is the main **cause** of population change, the '**key factor**'.

The results of Varley and Gradwell's classic work on winter moth (*Operophtera brumata*) at Wytham Wood, near Oxford, can be used to illustrate key-factor analysis. Sampling techniques and results are summarized in Table 7.2 and Figure 7.3. Visual examination shows that

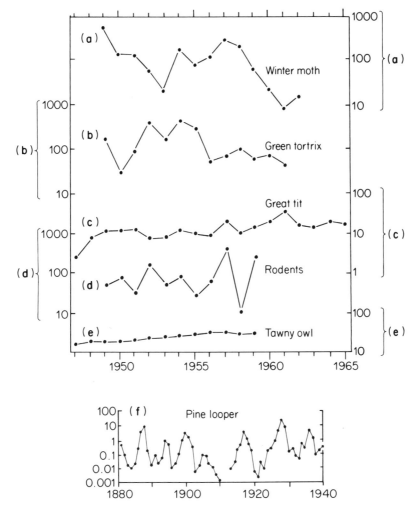

Figure 7.2 Examples of population fluctuations. (a)–(e) at Wytham, Oxford: (a) *Operophtera brumata*, larvae per m²; (b) *Tortrix viridana*, larvae per m²; (c) *Parus major*, breeding pairs per 10 ha; (d) *Apodemus sylvaticus* plus *Clethrionomys glareolus* per 5 ha in June; (e) *Strix aluco*, pairs per 525 ha (= Wytham Estate). (f) At Letzlingen, Germany; pupae of *Bupalus piniaria* per m². ((a) and (b) from Varley, 1970; (c) from Perrins, 1980; (d) and (e) after Southern, 1970; (f) Varley, Gradwell and Hassell, 1973.)

Table 7.2   Life table for winter moth at Wytham Wood, 1955–56 (from Varley *et al.*, 1973). Actual samples yielded the values (numbers per m²) shown in bold type, from which the remainder are derived. Sampling was confined to five oak trees (*Quercus robur*), of total canopy area 282m². The wingless females were trapped in November and December as they climbed up the trunks to lay their average complement of 150 eggs each. After feeding in April and May, caterpillars were collected as they descended on threads to pupate in the soil; they were examined to estimate numbers due to be killed by particular parasites, such as the tachinid fly *Cyzenis albicans*, whose eggs are ingested by caterpillars. The number of healthy pupae (15.0) is double the number of adult females emerging during the following winter, assuming a 1:1 sex ratio. Pupal parasitism was estimated by trapping the emerging adults of the wasp *Cratichneumon culex*. The remaining difference between healthy larvae (83.0) and healthy pupae was taken to have been caused by predation, acting before pupal parasitism

	No. killed	No. live	Log no. live	k-value
Females climbing trees, 1955		**4.39**		
Maximum oviposition		658.0	2.82	
(= no. of females × 150)				
Fully grown larvae	551.6	**96.4**	1.98	$0.84 = k_1$
Attacked by *Cyzenis*	**6.2**	90.2	1.95	$0.03 = k_2$
Attacked by other parasites	**2.6**	87.6	1.94	$0.01 = k_3$
Infected by protozoan	**4.6**	83.0	1.92	$0.02 = k_4$
Pupae killed by predators	54.6	28.4	1.45	$0.47 = k_5$
Pupae killed by parasites	**13.4**	15.0	1.18	$0.27 = k_6$
				$1.64 = K$
Females climbing trees, 1956		**7.5**		

changes in K are most closely correlated with changes in k_1, which is the key factor for this population. In certain cases it may not be so easy to distinguish the key factor by visual correlation. By plotting candidate k-values on the correct generation mortality, and determining the one with the greatest slope (max b = 1) the key factor(s) may be readily determined (Podoler and Rogers, 1975).

The key factor for winter moth at Wytham, k_1, corresponds to the reduction in numbers, described as **winter disappearance**, between the calculated initial population of eggs and the mature larvae which descend from the tree to pupate. Observations in the canopy showed that neither egg mortality nor feeding by birds on caterpillars contributed much to k_1. However, examination of individual trees revealed that winter disappearance was least on trees which flushed early allowing the first instar larvae access to nutritious foliage of low tannin content (Figure 6.3). Most larvae on late-flushing trees failed to find food and emigrated on silken threads, giving at least some a chance of surviving elsewhere. The

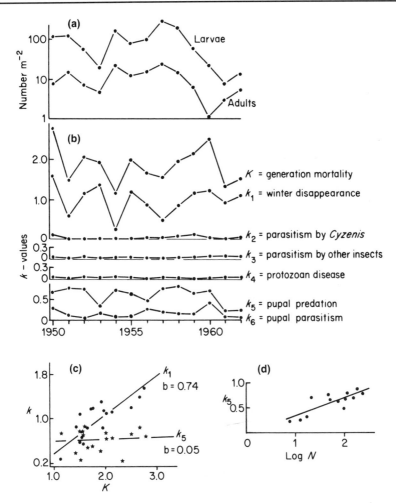

Figure 7.3 Population dynamics of winter moth (*Operophtera brumata*) in Wytham Wood, Oxford. (a) Generation curves. (b) Graphical key-factor analysis. (c) Key-factor analysis by regression coefficients. (d) Density-dependent action of pupal predation (k_5). ((a), (b) and (d) from Varley *et al.*, 1973; (c) from Podoler and Rogers, 1975.)

degree of synchronization between egg hatching and bud burst is largely determined by the effects of spring temperatures on egg development, the time of flushing of individual trees being less dependent on the external environment. A similar relationship has been detected in *Tortrix viridana* infestations on oaks in the English Lake District.

The effects of parasites which attack winter moth caterpillars ($k_2 - k_4$) showed very little variation from year to year (Figure 7.3b), unlike pupal predation (k_5) which often varied in the opposite direction to k_1, suggesting a possible compensatory or regulatory role.

7.3   Regulation

Regulation, in the sense of restoring a population towards its characteristic equilibrium level after disturbance, implies negative feedback, involving **density dependence** (Solomon, 1976). A density-dependent mortality process (e.g. disease, competition) or factor (e.g. a specific pathogen) has **proportionately** more adverse effect on a high density population than on the same species when less abundant (Figure 7.4). Although regulation depends on density dependence, density-dependent processes are only regulatory if they are of sufficient magnitude and act at the right time to offset the effects of disturbance.

Varley and Gradwell tested for density-dependent relationships in their winter moth life-table data by plotting k-values against the logarithms of the population densities on which they acted. Only k_5, **pupal predation**, showed a positive regression which was statistically significant, indicating density dependence (Figure 7.3d). The relative contributions of various beetles and small mammals to this predation are incompletely known. Somewhat surprisingly, larval parasitism does not appear to vary in a density-dependent fashion; although percentage parasitism by *Cyzenis* tends to be greater on trees with higher numbers of caterpillars, this is less likely in years when winter moths are generally abundant. The small effect of *Cyzenis* is largely a result of the strongly density-dependent mortality experienced by both parasite and host as pupae in the soil. Winter moths introduced into Nova Scotia and British Columbia lacked the regulatory influence typical of Wytham populations and increased to very high levels. Subsequent release of *Cyzenis* and *Agrypon flaveolatum* resulted in effective biological control (section 6.5). As winter moth populations have declined, the pattern increasingly resembles that observed in Britain, where parasitism plays a minor role, and predation on pupae a major regulating role.

7.4   Dynamics of specific woodland animals

Conifer caterpillars

Population monitoring of Pine Looper in commercial pine plantations involves winter sampling of litter and soil for pupae. If numbers exceed 30 m^{-2}, foliage is sampled for eggs in the following summer, before deciding whether control by insecticide is necessary. Barbour (1985,

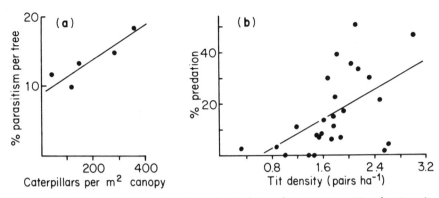

Figure 7.4 Examples of density-dependent relationships. (a) Parasitism by *Cyzenis albicans* of winter moth caterpillars on five oak trees, Wytham, 1958 (from Hassell, 1980). (b) Predation by weasels on nests of titmice (*Parus* spp.) in nest-boxes, Wytham, 1947–1972. (From Dunn, 1977.)

1988) summarizes the results of pupal counts from 47 British sites between 1954 and 1978. Most of these populations, as for example at Thetford in East Anglia, were relatively stable, with no evidence of subsequent damaging outbreaks of caterpillars (section 6.6). A comparable situation occurred in the forest in The Netherlands where Klomp (1966) studied changes in abundance of various stages of *Bupalus* over a 14-year period. Mortality of early larval instars, which averaged *c.* 65–70%, varied widely from year to year and was identified as the key factor. Regulation might have been by bird predation on larger caterpillars, which was thought to be density-dependent, or by parasites, which caused a marked reduction in numbers between pupation and adult emergence. The relationship between larval density and the fecundity of the ensuing females (Figure 7.5) apparently involves bodily contact between caterpillars, leading to fewer instars and so smaller pupae.

By contrast, certain British sites resemble the German forest whose population data are shown in Figure 7.2, with oscillations producing outbreaks at intervals of 6–7 years (Figure 7.6). Soils in outbreak areas are typically poor and sandy and of intermediate dryness, while annual rainfall is usually 500–700 mm. In German forest lore outbreaks are held to occur one or two years after vintage wine years, which are characterized by warm, dry summers. A relationship between drought years and subsequent outbreaks is borne out by events at Tentsmuir, but not by Cannock data. It has been suggested that higher levels of available nitrogen in the needles of drought-stressed trees might reduce the mortality of early instars, so initiating an outbreak. The subsequent cyclical oscillations involve deviations from the local equilibrium density being compensated after a time lag of exactly one generation, implying a strong

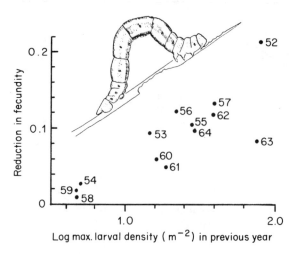

Figure 7.5 Influence of larval density on fecundity in *Bupalus piniaria* at 'de Hogue Veluwe', in the Netherlands. (Modified from Dempster, 1975; after Klomp, 1966).

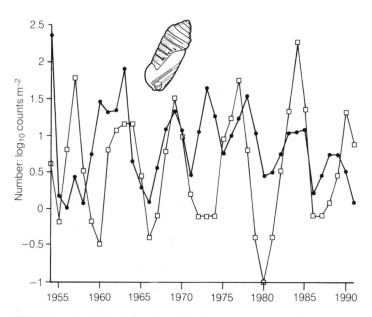

Figure 7.6 Pupal counts of *Bupalus piniaria*, representing the highest count from any compartment at Cannock, Staffs, (●) and Tentsmuir, Fife (□). (Data by courtesy of the Forestry Commission.)

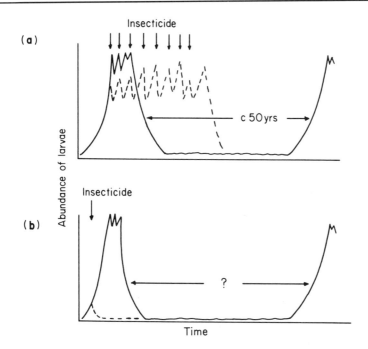

Figure 7.7 Fluctuations of conifer pests treated with insecticide (–––) and untreated (——). (a) Spruce budworm in New Brunswick; (b) pine looper in Britain. In contrast to results obtained with spruce budworm, high populations of pine looper can be rapidly reduced to low, endemic levels by a single application of insecticide. (From Way and Bevan, 1977.) (Reproduced by permission of the Council of the Linnean Society of London.)

delayed density-dependent influence. The percentage of pupae parasitized varies throughout each cycle, peaking 1–2 years after pupal maxima, but the magnitude of such parasitism is insufficient to drive the cycles. A more likely candidate is the host-specific ichneumonid parasitoid *Dusona oxyacanthae*, which kills prepupae of *Bupalus*. The abundance of pupae of host and parasitoid at Cannock fluctuated out of phase by one generation, while data from six sites showing marked cycling revealed a highly significant negative correlation between parasitoid abundance and the subsequent trend in *Bupalus* populations.

Various components of the complex of pest and natural enemies can now be used to predict the likelihood of an outbreak following 2–3 years of increases in looper populations: a significant increase in mean pupal weight, pupal parasitism <25%, or <2 m^{-2} pupal cocoons of *Dusona*. The aim of control measures should be to time the application of insecticide so that larval populations are reduced by 80–90% before caterpillars have become large enough to cause damage (Figure 7.7b). In order to encourage integration with natural control, the use of insecticides should be kept to a minimum. Pre-emptive spraying at an early stage of the cycle is likely to encourage pest resurgence, through the short-circuiting of delayed density-dependent controls. Extreme fluctuations may be avoided by deliberately leaving a reservoir of looper and associated parasitoids in unsprayed enclaves within the overall treated area, and also by increasing the diversity of natural enemies by mixed plantings of pine and hardwoods. Unlike DDT and organophosphates, the insecticide diflubenzuron, which upsets moulting, has no effect on adult insects, including parasitoids, so that the few loopers left after spraying are heavily parasitized.

Host–parasitoid models for *Bupalus* and *Dusona* have revealed the importance of the value of the maximum per-capita rate of increase for *Bupalus* in determining population fluctuations, ranging from stable equilibria to unstable cycling. Simulation modelling has been used to explore the likely effects of different management strategies, including the use of different provenances of *Pinus contorta*, on Pine Beauty, *Panolis flammea* (Watt and Leather, 1987).

The economic importance of outbreaks of spruce budworm in North America led in the 1960s to this species being among the first subjects of attempts to model insect population dynamics. During the following decade there was a proliferation of extremely detailed, and complex, simulation models for various pest species. One for *Choristoneura fumiferana*, described by Peterman *et al.* (1979), incorporated data for 30 years, and involved interactions between the pest and its enemies, the weather and management practices. This model helped to explain why control measures against outbreaks of this species are liable to be locked into a fate like that of Sisyphus, who was punished in Hades by repeatedly having to push a huge stone uphill, only for it to roll down again (Figure 7.7a). Latterly there has been a return to simpler models, incorporating ecological concepts such as density dependence, which appear to provide more powerful insights into the pest–forest system, at least for certain species, so leading to improved predictions. Various examples will be found in Berryman (1988) and Watt *et al.* (1990).

Aphids

Key-factor analysis has not been applied to aphids, because of the complexities associated with overlapping generations, polymorphism and migration. However, by looking for relationships between the number of aphids present at the beginning and end of each year and between years the analysis of the population dynamics of certain host-specific deciduous tree-dwelling species has been simplified (Dixon, 1990). The seven species studied so far show similar population dynamics: overcompensated density dependence within years (producing instability), and strong density dependence between years (introducing a degree of stability). The strong density-dependent factor operating between years tends to dampen the disturbing effect of the overcompensating density-dependent factor operating within years. Unlike other insects feeding on trees predators and parasites do not play an overwhelming role in the population dynamics of at least four of the species that have been studied in detail. Changes in aphid quality caused by intraspecific competition appear to be the major factor regulating their numbers.

Sycamore aphid (*Drepanosiphum platanoidis*) overwinters as eggs laid on sycamore twigs (Dixon, 1977, 1979). Larvae which emerge from these eggs before the buds of a particular tree have expanded in spring are especially vulnerable to predation and to being dislodged by rain. However, if egg-hatch and bud-burst are synchronized, the larvae are able to feed on young leaves with a high amino-nitrogen content; these larvae develop into adults which are potentially larger, and so of a higher fecundity, although this may be diminished as a result of over-crowding. There are two further parthenogenetic generations before males and egg-laying females gradually appear among the late-summer population, so that some eggs can be produced even if there is an unusually early leaf-fall. An inverse relationship has been observed between the abundance of aphids colonizing the leaves in spring and the numbers of egg-laying females in the following autumn. This was once thought to be a consequence of the reproductive rate in early autumn being influenced by aphid-induced changes in the quality of the host plant. During the autumn, various nutrients are withdrawn into the trunk from the leaves of trees (section 9.6). Senescing leaves of infested sycamores retain more nitrogen, implying that less is available to aphids in the phloem (Figure 7.8a). It has since been shown that autumnal fecundity is determined largely by the size of the second-generation adults, which is itself related to the degree of crowding earlier in the summer (Figure 7.8b). The timing of the autumn increase depends on the duration of the period of reproductive diapause during June and July, when virtually no larvae are produced, this aestivation being prolonged the more the aphids are crowded. At this time, when there may be a hundred or more adults per leaf, mutual disturbance results in reduced rates of feeding on leaves with a low nitrogen content. These density-dependent consequences of crowding are considered by Dixon to be capable of regulating populations of sycamore aphids, and in fact they tend to overcompensate for the disturbing effects of weather, especially autumn winds. The action of parasites and predators, such as anthocorid bugs, is not considered to be regulatory; by killing large numbers of young aphids in those autumns when the fecundity of crowded populations is already low, they may merely add to the overcompensation. Similar mechanisms seem to apply to the lime aphid (*Eucallipterus tiliae*), where high densities lead also to increased migratory activity, while in this species and in the green spruce aphid (*Elatobium*

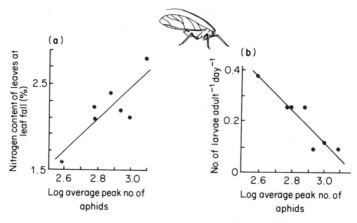

Figure 7.8 Aspects of the population dynamics of sycamore aphids (*Drepanosiphum platanoidis*). (a) Relation between nitrogen content of leaves at leaf fall and spring–summer population density of aphids. (b) Fecundity between 15 July–15 August in relation to aphid abundance earlier in the year. Average peak number = 1/2 [(peak no. on buds + peak at budburst)/2 + summer peak]. (From Dixon, 1979.)

abietinum) there is stronger evidence that aphid-induced changes in food quality are important in the population dynamics of aphids. In other parts of the world various arboreal aphids, including *E. tiliae* and the walnut aphid (*Chromaphis juglandicola*), have been controlled by parasites, showing that the role of natural enemies cannot be totally discounted when considering changes in aphid abundance.

Great tits

Among woodland vertebrates, long-term studies at Wytham have provided a wealth of data about populations of great tits (Perrins, 1980), small rodents and tawny owls (Southern, 1970; Flowerdew, 1987), and to a lesser extent weasels (King, 1980, 1989) (Figure 7.2).

Great tits (*Parus major*) feed their young for about two weeks in the nest (built in a tree-hole or nesting box) and for a similar period after the young have fledged and left the nest. Spring temperatures affect the date when the first eggs are laid (Figure 7.9b), basically through the availability of food such as March flies (*Bibio* spp.), needed for the production of eggs; an average clutch of 8–10 eggs in as many days is

Figure 7.9 (a) Number of pairs of great tits (*Parus major*) in Marley Wood, Wytham, Oxford 1947–1991. (b) Great tit, mean laydate (1 = 1 April) in relation to warmth sum, the sum of maximum temperatures each day during March 1 –April 20. (c) Blue tit, clutch size and density, in Bean Wood, Wytham. (d) Great tit, clutch size in Marley Wood, with density of both blue and great tits in the same year. (From C. M. Perrins, personal communication.)

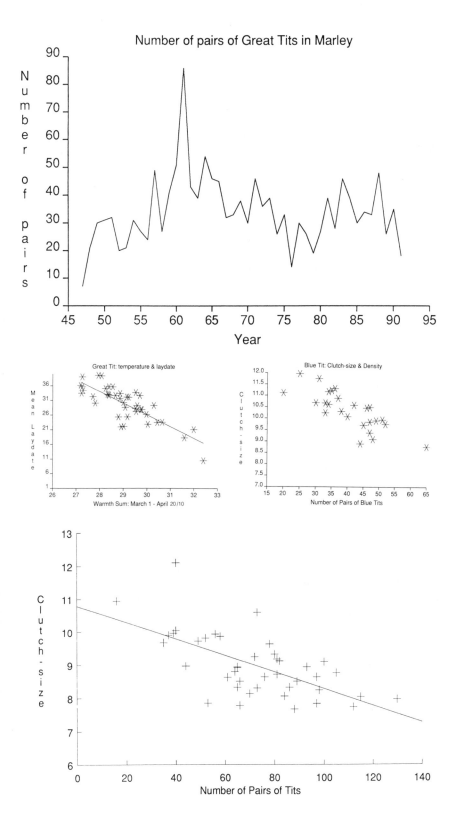

almost equivalent to doubling the female's body weight. Hatching dates for eggs of moths such as *Operophtera* and *Tortrix* are also influenced by spring temperatures: the ensuing caterpillars provide the major source of food for the tit nestlings. Some correlation has been found between the time when caterpillars of early-summer species are most abundant (section 6.6), and the average date when nestlings are 11 days old and require the greatest number of food items (Perrins, 1991). This peak feeding rate may involve the parents in as many as a thousand visits to the nest per day, yet it has been estimated that tits cull only 1–2% of the total caterpillar population. Young from early broods tend to be heavier than those reared later, and heavy fledglings are much more likely to survive and breed than are lighter birds. It is apparently advantageous for the tits to breed as early in the summer as the food supply for the females permits, thus producing the greatest number of offspring to survive the summer, when the young have to fend for themselves on small insects high in the canopy.

Between 1947 and 1968 in Marley Wood at Wytham the average number of fledglings per pair was six. Only about half the number of breeding adults died between one year and the next, so that if the breeding population was to be kept stable a marked reduction in the numbers of young before the following breeding season seemed inevitable. 'Mortality outside the breeding season' (i.e. between fledging and the setting up of breeding territories, which is usually in January) had been identified as the key factor for great tits in Marley Wood. This has now been separated into late summer/autumn mortality (the key factor) and recruitment into the breeding population (density-dependent) (McCleery and Perrins, 1985). There was a positive correlation between survival of juveniles and the size of the mast crop of *Fagus sylvatica*, but this cannot be a direct causal relationship, since mast becomes available only after the survival rate has been determined; possibly mast crops are correlated with insect abundance during the summer. In contrast to this variable survival of young until the autumn, winter mortality is fairly constant from year to year, although during the severe winter of 1962–63 the total population was reduced by 75%, and the young by 90%.

Disappearance need not necessarily imply death: it has been shown that some of the surplus of young birds could be accounted for by their emigrating to less suitable breeding sites such as hedgerows, having been unsuccessful in establishing woodland territories. Territorial behaviour does not set an upper limit to the number of breeding birds in a given area of woodland, but appears to have a restraining effect on settlement, although further settlement does not seem to be completely prevented. The size of individual territories seems to be related to the pressure exerted by intruders and consequently is inversely correlated with nesting density. Smaller territories generally provide less food for the female, while the time available for gathering food may be reduced by territorial disputes, the result being fewer eggs. Although great tit (and blue tit) clutch size is significantly negatively correlated with density, great tit clutch size is significantly more closely correlated with the combined blue and great tit density (Figure 7.9c, d). Attacks on nests by weasels (*Mustela nivalis*) can lead to death of laying or incubating females, or of eggs or fledglings. The closer together the nest boxes at Wytham, the higher the proportion of nests attacked (Figure 7.4b). Although predation acts in relation to the spacing out of nests, it does not have a regulatory effect on the population as a whole since hatching mortality and fledging mortality do not show density dependence. An additional complication is that reproductive success of great tits is negatively density-dependent on the density of blue tits (that use the same nest boxes) as well as other great tits (Minot, 1978). Once on the wing the young birds are still subject to predation: as many as 35% may be taken by sparrow-hawks.

The flexibility of the territorial system in great tits permitted their breeding numbers at Wytham to fluctuate considerably over a 30-year period (Figure 7.9a), as did numbers of woodmice and bank voles, the major food of strongly territorial tawny owls (Figure 7.10).

Rodents, weasels and tawny owls

Woodmice (*Apodemus sylvaticus*) are strictly nocturnal, but bank voles (*Clethrionomys glareolus*) may also be active during the day. At Wytham woodmice were found to subsist mainly on seeds, especially of oak and sycamore during the autumn. Voles ingested large amounts of green leaves from trees and shrubs, as well as seeds with soft testas, and also bark and leaf litter. In both species the diet was supplemented in May by an influx of defoliating caterpillars. The breeding season normally extends from May to October, with four or five litters. Since populations are at their lowest level in early summer, survival of young is probably poor at this time, but numbers typically pick up from midsummer, reaching a peak in autumn or early winter (Figures 7.1 and 7.10). During the breeding season, there is a gradual change in age structure and associated social behaviour, the older generation being replaced by younger and less antagonistic adults. **Overwintering survival** was broadly correlated with the size of the acorn crop of the previous autumn; a similar relationship between bank voles and ash seeds was observed in Lathkilldale (Flowerdew and Gardner, 1978). Woodmice continued to breed into or even through the winter in 1956, 1958, 1962 and 1964 at Wytham: all were bumper years for acorns. Winter recruitment and improved survival resulted in the anomalous situation of

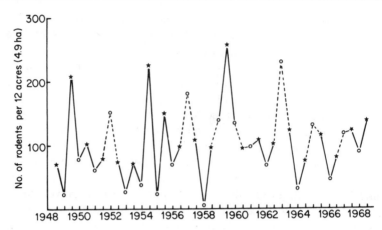

Figure 7.10 Abundance of woodmice plus bank voles in Wytham Wood, estimated by live-trapping in December (★) and June (○). Interrupted lines indicate reversal of usual winter increase and summer decline. (From Southern, 1970.)

the subsequent summer population being larger than that of the previous winter (Figure 7.10). In years when acorns were scarce, however, survival during the winter was poor, presumably reflecting competition for food.

Fluctuations in availability of acorns were thus primarily responsible for the considerable variation observed in the spring density of woodmice. However, this variation was found to be largely counteracted by a density-dependent postponement of the increase in numbers in late summer, this being later in years when spring numbers were high. The duration of this period of increase varied by as much as 4 months, and since numbers approximately double each month once it begins, this postponement had a considerable regulatory effect. Despite this regulation, numbers fluctuated markedly around a mean value of c. 20 ha⁻¹, and these small mammals were probably near the limit of their food supply on average one year out of two at Wytham.

It was concluded that weasels had no observable effect on rodent density or survival in Marley Wood (King, 1980), in contrast to the control exerted on great tits. Apart from the influence of nest boxes on predation rates on tits, the major differences include predator/prey ratios, with many more rodents than tits per weasel, and the fact that the replacement of killed individuals occurs much more readily among the rodents, with their several litters, whereas lost clutches of eggs are rarely replaced.

Weasels are opportunistic feeders, who are more likely to exploit birds' nests when rodent numbers are low. They expend considerable energy in hunting, especially for rodents, a food source for which they are most likely to be in competition with tawny owls (*Strix aluco*). Although they can exploit rodents, including young, in their tunnels (a resource denied to owls), they may occasionally be attacked by owls. Weasels usually breed only once in a lifetime which is normally less than one year, in contrast to long-lived K-selected owls, which employ a sit-and-wait hunting strategy. Like weasels, owls appear to have little effect on the population density of rodents at Wytham, but they themselves are influenced by the abundance of rodents.

Between 1947 and 1959, the breeding success of tawny owls at Wytham was found to be closely linked to the abundance of woodmice and bank voles, especially between about mid-March, when up to four eggs are usually laid, and mid-May, when the chicks can be left alone in the nest. During the summer owls transfer their attention to other foods, such as moles, beetles and worms. In years when the combined density of the two species of rodent, assessed in June, was below the average of 20 ha^{-1}, owls failed to breed, or at best produced very few eggs (Figure 7.11a). Failure to breed was identified as the key factor at Wytham during 1947–59. The number of pairs which failed to breed was rarely less than 30% of the total, and in 1958, when rodents were particularly scarce at Wytham, no owls bred at all. Hatching failure largely reflected desertion of eggs by females leaving the nest to supplement food brought by their mates, while failure to fledge the young was often also a result of food shortages; in both instances wet weather had a deleterious effect on hunting, the rodents being less audible in the sodden forest floor. The number of young fledged per year was usually *c.* 20, except in 1951, 1955 and 1958, when rodent numbers were very low. The mean expectation of life is about 5 years, some adults occupying the same territory for more than 7 years. Consequently, there are usually far more fledglings than are needed to compensate for the deaths of territory-holding adults.

Owlets which failed to find a vacant territory either died or emigrated, thus contributing to 'overwintering loss'. This loss varied from year to year in a density-dependent fashion, a greater proportion of owlets disappearing when their numbers were high, and with immigration in years when insufficient offspring were produced at Wytham (Figure 7.11b). In contrast to the situation in great tits, recruitment to the adult population is limited by a remarkably rigid territorial system, which has persisted at Wytham with little change over many years. With *c.* 30 breeding pairs, the average size of territory is *c.* 16 ha, being greater in regions of denser cover. The chief benefit of this system is apparently that breeding success improves as the residents gain hunting experience within their territory. However, even experience cannot guarantee success every year.

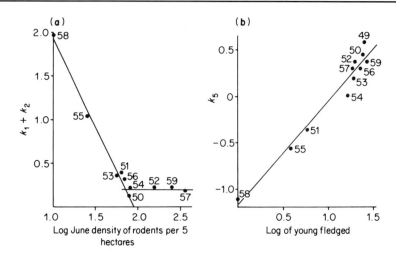

Figure 7.11   Tawny owls (*Strix aluco*) on Wytham Estate. (a) Influence of rodent density on population losses through failure to breed (k_1) and to achieve maximum clutch size (k_2). (b) The density-dependent nature of overwinter disappearance (k_5). (From Southern, 1970.)

Summary

From this review it can be seen that the breeding success of certain woodland animals depends on the sychronization of their life-cycles with food availability. Lean years may have far-reaching consequences, as when the acorn crop failed at Wytham in 1957: the following spring rodent numbers were at their lowest for the decade and tawny owls failed to breed. There is a variety of influences which can act as key factors, but even for the same species these may differ between sites and between years.

  Although the significance of key factors is now generally recognized, there is less agreement about the occurrence of regulation in general, or of density dependence in particular cases, such as over-wintering loss in tawny owls (Dempster, 1975). It is particularly difficult to assess the role of predators in regulation, as this requires long-term studies of changes in the proportion of prey killed as prey density alters. Winter moth populations were regulated at Wytham by predation on pupae, but there is insufficient evidence to decide whether predation by birds on pine looper caterpillars is density-dependent. Among sycamore aphids and the various vertebrates studied at Wytham, regulation seems not to depend on natural enemies, with the exception, possibly artificial, of the box-nesting tits. In these species, and possibly also in pine looper, the degree of crowding, itself often dependent on food availability, determines the extent of the subsequent input to the population, sometimes through behaviour-mediated changes in fecundity and timing of increases. Such populations are said to be largely self-regulated. Tawny owls, as top carnivores with virtually no enemies, exercise density-dependent control over membership of their exclusive territories.

Among pest species, greater understanding of the factors responsible for outbreaks should enable us increasingly to adopt integrated pest management strategies, thus reducing our dependence on insecticides.

7.5   Changes in community structure

In recent years various natural events have considerably changed the structure of British woodlands. A series of major gales, especially those of 1987 and 1990, have led to the loss of many trees, while the drought of 1976 hastened the end of many over-mature birch and beech. Additionally there has been an enormous loss of elm; the likely effects of Dutch elm disease have been reviewed by Archibald and Stubbs (1980). Of the 150 species of invertebrate associated with elm in Britain, only 38 are restricted to elm, including a gall mite and the white-letter hairstreak butterfly, which might become locally extinct. There is some evidence of regional shortages of nesting places for owls, and this would also be expected to apply to other birds which habitually used elms for nesting or roosting. Insects associated with bark beetle burrows have naturally increased, but the influence on decomposers is less clear. In shaded and damp situations they will be favoured by the increased availability of dead wood. In sites exposed to extreme insolation and desiccation, however, standing or fallen trees are virtually immune to attack, being likened by Elton (1966) to kiln-dried timber. Truly woodland communities, as opposed to those of hedgerows and copses, should show little change, except where susceptible elms are locally abundant, as on scarp slopes of Jurassic limestone in the Cotswolds where short-term reversion to scrub may be expected.

If bark beetles are to breed in them, elms must be above a certain size, so it might be possible to perpetuate the species (and perhaps some of the community) by cutting back suckers before they become vulnerable. Similarly, mountain pine beetles tend to attack the largest diameter trees of *Pinus contorta* (lodgepole pine; see section 6.6), so that infestations decline when few large trees are left alive (Amman, 1977). In some regions of North America where lodgepole pine is seral, its survival and that of the beetles depends on fire, since otherwise the lodgepole pines are killed by beetles, and eventually succeeded by shade-tolerant tree species including Douglas fir, a large conifer with very thick bark. Extremely hot fires will kill even this species, while favouring seed release from the semi-serotinous (resin-sealed) cones of *P. contorta*. As the stand matures it becomes increasingly susceptible to beetle attack, producing the necessary tinder for further fires and the perpetuation by fire of *P. contorta*. This provides a clear example of the effects of biotic and abiotic factors on competitive interactions and hence on the structure of a woodland community.

7.6   Diversity and stability

The idea that the balance of relatively simple communities of plants and animals is more easily upset by destructive oscillations or invasions, than that of more diverse communities, has become enshrined in the literature as 'complexity begets stability.' However, this may be more appropriately expressed as 'stability allows complexity' (May, 1976). Epidemics of pests or diseases provide an example of instability; they are normally thought to be typical of artificial monocultures rather than of natural stands, and to be more prevalent in non-tropical woodlands, although even tropical rainforests are not as free from pest outbreaks as used to be supposed.

Tropical rainforest includes the most complex communities in existence. For example, the 17 km² of Barro Colorado Island, Panama, harbours over 700 species of trees, shrubs and lianas. Here the opportunity for diversification of feeding niches, provided not only by the great range of plants but also by the profusion of flowers and fruits, would be expected to result in great stability.

One theory which attempts a partial explanation of the diversity of tropical forests was proposed independently by Janzen and by Connell in 1970, and involves the participation of natural enemies (Hubbell, 1980). Seed predation by insects (many of which were assumed to be host-specific, e.g. bruchids in Costa Rica, see section 3.4) and by more catholic vertebrates, would be expected to kill virtually all seeds within the immediate vicinity of the parent tree. Similarly, any seedlings which manage to develop would be more likely to be killed the closer they are to parental sources of host-specific defoliators or pathogens (cf. oaks in the UK; Shaw, 1974). Enemies which respond to the density of seeds or seedlings would also cause disproportionately greater mortality close to adult trees. Consequently only those seeds which are dispersed beyond a certain minimal distance from the parent would have a significant chance of escaping discovery, germinating and surviving as seedlings (Figure 7.12a). This is the essence of the **escape hypothesis** of Howe and Smallwood (1982).

It was predicted that this minimal distance effect would result in a low density of **conspecifics** (individuals of the same species), with adults tending to be spaced out, corresponding to the generally held view that

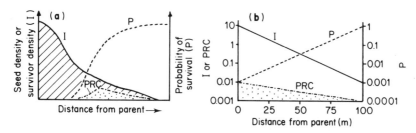

Figure 7.12 Graphical models of the effects of seed or seedling predation on tree spacing. (a) Janzen's model, which suggests that recruitment of new adults (as shown by the population recruitment curve, PRC) occurs only beyond a certain minimal distance. (b) Hubbell's re-scaled model, which assumes that a small fraction of a large number of seeds next to the parent tree escapes predation. (From Hubbell, 1980.)

tropical trees are thinly spread and fairly evenly spaced. So long as other species are able to colonize the minimal area denied to conspecific offspring, dominance by a single species should be prevented. However, Hubbell has shown that nearest neighbour distances between conspecific adults are insufficient for this spacing by predation to account for the observed numbers of co-existing species.

Hubbell stresses the high variability which characterizes the results of many of the experiments designed to test the escape hypothesis, and concludes that there may not necessarily be any clear relationship between seed predation and the spacing of trees. Within one site, the number of seeds produced by a species varies between individuals and between years, resulting in intraspecific variation of minimal distances and in the possibility of clumped patterns of dispersion. Surveys of two sites in Costa Rica showed that of the 114 tree species for which there were sufficient data all were either clumped or randomly dispersed, but none showed uniform spacing. The density distributions of young trees were equally revealing: some showed an exponential decrease away from the parent, others remained almost constant with distance, but very few showed a reduction close to the parent, as predicted by the escape hypothesis. Re-scaling of Janzen's **seed-shadow** model (Figure 7.12b) emphasizes the importance of the actual numbers of seeds surviving. Even if only a small proportion of the bulk of the seeds which fall near the parent manage to survive, they will contribute the largest numbers to the recruitment curve, so encouraging clumping.

A later review of studies of the spacing dynamics of tropical trees (Clark and Clark, 1984) concluded that most of those concerning seeds were consistent with the escape hypothesis. However, evidence based on seedlings was less convincing, since mortality factors were not always identified and might well have involved competition between seedlings, or with the parent, rather than herbivory. An exception is provided by the work of Auspurger (1983) who found that population recruitment in *Platypodium elegans* (Leguminosae) was strongly influenced, in a density-dependent fashion, by fungal damping-off of seedlings. The resultant dispersion merely sets the scene for the subsequent drama, where the choice of leading players is largely determined by the unpredictable spotlight provided when gaps form in the canopy.

Complex tropical communities, although intrinsically fragile, may remain stable over very long periods, but are ill-equipped to withstand perturbations imposed by man. Whatever the explanation of their diversity, we can ill afford to disrupt these unique communities with their myriads of species, many of which are undescribed or little known.

8
Decomposition and renewal

8.1 The vital key

Decomposition is the key which releases the mineral nutrients essential for the growth of green plants. Though forest trees receive some atmospheric inputs – notably of nitrogen compounds – via their foliage, the major pool of inorganic nutrients is in the soil (Figure 1.4). The levels of these nutrients are partly controlled by the chemical nature of the rock from which the mineral portion of the soil is derived, and by the weathering/leaching ratio which is itself dependent on climate. Weathering of surface rock is most rapid in warm climates; the loss of mineral ions from the soil solution by leaching is greatly increased when precipitation exceeds evapo-transpiration.

This chapter considers the more immediate problems associated with death and decay: of how decomposition releases nutrients locked within the dead bodies, shed parts and excrement of woodland organisms, to be mineralized and returned to the soil solution and the atmosphere so that life can go on and the forest be renewed.

8.2 Decomposition: resources and processes

Within the forest whole organisms or their component tissues and organs are continually dying or being killed, thus providing the resources of energy and nutrients which sustain the decomposers, either directly or via members of the grazing chain. Instead of the word detritus (which has several meanings, including rock fragments) **necromass** will be used as a general term for this dead material whatever its source: sloughed root cells or animal cuticles, dung and corpses, shed leaves, reproductive structures, branches and trunks, and, not least, standing dead matter. This chapter is largely concerned with decomposition of **litter** in the strict sense of shed plant remains.

Various agencies, acting simultaneously or in sequence on necromass, can bring about **chemical change** and **structural breakdown**, the twin facets of decomposition. Soluble components are leached out by rainwater, while degradation of chemical substrates is brought about mainly by catabolic activities of decomposers, especially fungi and bacteria, as well as by autolytic enzymes of the dead tissues and by fire. **Comminution** (fragmentation) can be caused by wind and rain, by freezing and thawing, or by movements of animals and plants; the feeding activities of many **detritivores** (used here in the broadest sense, to include decomposer animals which exist primarily on fungi or bacteria, as well as those which ingest necromass; see Swift *et al.* (1979) and Moore *et al.* (1988) for a more detailed terminology) are of importance in this respect. As time passes, the structural framework and chemical make-up are altered and the necromass components are dispersed into the air, water or soil, or among members of the decomposer communities (Figure 8.1). In this chapter emphasis is placed on how these organisms bring about decomposition, most of the quantitative aspects, including nutrient cycling, being discussed in Chapter 9.

That interacting agencies are involved in decomposition is suggested by the well-known fact that, even under similar conditions, the leaves of, for example, ash and elder decompose more readily than those of oaks and conifers. Various authors have claimed that different features of leaves have an overriding influence on decomposition rates; these include toughness and thickness, content of calcium, carbohydrates or nitrogen, the C/N ratio (Figure 8.2) and the initial lignin/nitrogen ratio. However, the demonstration of correlation between a single feature and rate of decomposition does not preclude the possibility of interaction (e.g. between softness and nitrogen content), while the ease of decomposition of leaves of certain species, especially relative to a particular group of decomposers, could be determined largely by features which differ from those of the majority. Nor does correlation imply causation, as Satchell and Lowe (1967) pointed out when discussing the order of removal of leaf discs by *Lumbricus terrestris* from a range of different tree and shrub species. Palatability was correlated with nitrogen content, and with the amount of soluble carbohydrate present (Figure 8.3a,b), but there was no evidence of how the worms made these distinctions: they were unable to differentiate between various concentrations of sucrose, and failed to discriminate between individual amino acids. On the other hand, the inverse relationship between palatability and polyphenol content (Figure 8.3c) suggested a possible role for distasteful substances, and in this case different polyphenols were distinguished by the worms. Tannins seemed to be implicated (Figure 8.3d), but the discrimination still occurred in their absence, as with fresh elm, alder or sycamore, and also, with most species, after weathering for 6 weeks. The fact that paper discs soaked in leaf extracts were selected in the same order as unweathered leaf discs implied that differences in palatability were caused

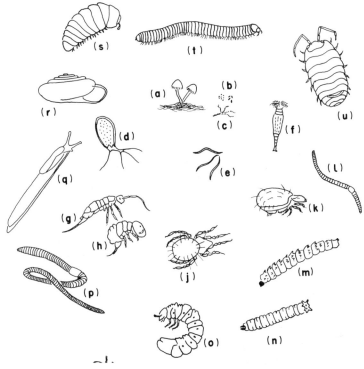

Figure 8.1 Representatives of the major groups of decomposers in litter and soil (not drawn to same scale). (a)–(c) Microflora: (a) fungi; (b) bacteria; (c) actinomycetes. (d)–(f) Microfauna: (d) shelled amoeba; (e) nematodes; (f) rotifer. (g)–(k) Microarthropods: (g), (h) Collembola (springtails); (j), (k) Cryptostigmata (oribatid mites). (l)–(u) Mesofauna and macrofauna: (l) Enchytraeid (potworm); (m), (n) Diptera larvae (bibionid and tipulid); (o) Scarabaeid beetle larva (white-grub); (p) lumbricid earthworm; (q), (r) molluscs; (s), (t) millipedes; (u) wood-louse. (For further details, see Wallwork, 1970.)

by unidentified compounds, possibly representing a balance between distasteful and tasty. Weathering appeared to involve microbial break-down of these compounds, rather than leaching. In this instance tasting is only the beginning of the process of decomposition, yet already it has involved chemical composition and the possible influence of microbes on the feeding of detritivores, features which commonly recur in studies of decomposition. Cooke (1983) showed that *L. terrestris* could differ-entiate between various species of fungi inoculated onto filter paper, and surmised that palatability might reflect the moisture content of the substrate, which in turn could indicate the availability of microbial food supplies.

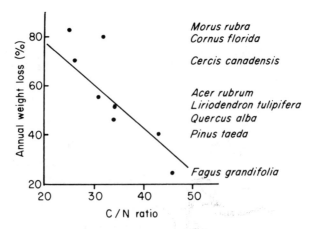

Figure 8.2   The relationship between decomposition rate and C/N ratio of leaf litter of eight tree species on mull sites in Tennessee. $y = 113.2 - 1.75x$ (P < 0.05). (From Reichle, 1971; © UNESCO 1971. Reproduced by permission of UNESCO.)

8.3   Degradative successions

As an individual unit of necromass progressively decays, there is a change in the availability of particular resources, such as energy or nutrients, or living space for feeding, shelter or oviposition. These changes are accompanied by qualitative and quantitative alterations in the species composition of the associated decomposer community. Ultimately the dead tissue loses its identity completely, its components being dispersed into the non-living environment and among the decomposers which are then dependent on further supplies of necromass. This 'anticlimax' is in marked contrast to the situation during successions of autotrophs, where plant biomass increases until the climax community is reached; the diversity of heterotrophs associated with these 'classical' successions also tends to increase. Because of their dependence on diminishing resources, community changes occurring during the decomposition of units of necromass will be referred to as **degradative successions**.

The complexity and opacity of most natural forms of necromass, together with problems of isolating microorganisms and deciding when they have been growing actively (rather than existing as spores or other resting structures) have made it difficult to determine which decomposers are active within particular tissues at a given time, quite apart from the role that they play in decomposition.

Cellophane

Isolation and observation are made easier by using a simplified bait such as cellophane sheet, which can be buried in litter or soil for periods

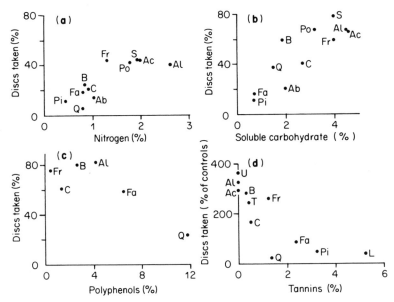

Figure 8.3 Palatability of leaf litter to *Lumbricus terrestris*. Leaf discs (1 cm diameter) or needles (2 cm lengths) were placed on the surface of large bins of garden soil, containing *L. terrestris*, kept in a cellar. Relative palatability was assessed in terms of the proportion of discs of each species removed by the worms over a period of time (up to 33 days), or, as in (d), as numbers removed compared to control discs of moist paper. Each of the four experiments was in a different year. Graphs show relation between palatability and chemical composition (% dry weight). (a) Nitrogen content. Nitrogen-rich species are generally more palatable, but alder and oak are less so than their nitrogen contents might suggest. (b) Soluble carbohydrate. (c) Total polyphenols (determined by the Folin procedure on 50% acetone extracts.) (d) Condensed plus hydrolysable tannins. Ab, *Abies grandis*; Ac, *Acer pseudoplatanus*; Al, *Alnus glutinosa*; B, *Betula pendula*; C, *Corylus avellana*; Fa, *Fagus sylvatica*; Fr, *Fraxinus excelsior*; L, *Larix decidua*; Pi, *Pinus sylvestris*; Q, *Quercus petraea*; S, *Sambucus nigra*; T, *Tilia europaea*; U, *Ulmus glabra*. (From data of Satchell and Lowe, 1967.)

ranging from days to months (Tribe, 1957). Colonizing organisms can be picked off and cultured, or stained *in situ*. **Cellulases** are necessary for chemical decomposition; these are formed mainly by certain species of fungi (e.g. *Botryotrichum piluliferum*) although the characteristic trails formed by the rasping radulae ('tongues') of molluscs act as a reminder that many slugs and snails also produce cellulases. Fungal hyphae ramify over and within the sheets, clear areas around their active tips indicating areas of extracellular decomposition. The resultant sugars are absorbed by the cellulolytic fungi, but they may also provide a potential food source for other organisms, such as sugar fungi (e.g. *Pythium oligandrum*), bacteria, protozoa, and nematodes, most of which lack cellulases (Figure 8.4). Alternatively these organisms may feed on healthy or moribund hyphae or spores, which are also extensively grazed by arthropods, especially mites and Collembola. These animals, as well as segmented worms, may also ingest the permeated cellophane, which consequently gradually becomes dispersed. Such decomposers in turn may be consumed by carnivores, including nematode-trapping fungi, while corpses and faeces are further worked over, even long after the cellophane has disappeared (Harding, 1967).

This fairly complex community constitutes a food web which is partially based on a very simple substrate. With the exception of the breakdown of cellulose, the changes responsible for succession occur within and among the decomposers themselves, rather than in the cellophane. Nevertheless, this technique provides a further example of necromass requiring some form of **conditioning** (in this case by cellulolytic species) before certain organisms can make use of it. As well as fungal and bacterial grazers, most of the major groups of detritivores include species which ingest litter, and their distribution in the forest floor and their feeding and survival in the laboratory all indicate a preference for conditioned litter, often at a particular stage of decomposition. Similarly, certain phthiracaroid mites ('hinged' oribatids; e.g. *Steganacarus*, Figure 8.7c) normally lay their eggs on conifer needles, cones or beech cupules only after a period of microbial colonization (Harding and Easton, 1984). A closer look must therefore be taken at patterns of development of microorganisms on natural debris, in an attempt to generalize about these successions and their causes, as well as the roles and interactions of the constituent species. It should be borne in mind that natural plant debris differs from cellophane, not only in its structural complexity, but also because it was once alive, and the cause of death, especially if parasitic microbes are involved, may predetermine the course of succession.

Conifer needles

The classic study of decomposition of needles of *Pinus sylvestris* by Kendrick (summarized by Millar, 1974) concentrated mainly on fungi, which, rather than bacteria, were found to be favoured by the acid conditions of a Cheshire podzol site. The absence of lumbricid earthworms reduced disturbance to a minimum, so that the remains gradually progressed down through the sub-horizons of a mor-humus profile. A variety of isolation and observation techniques showed that certain species were characteristically abundant, and by implication active, at different stages of decomposition. The phylloplane microflora (section 6.2) included the

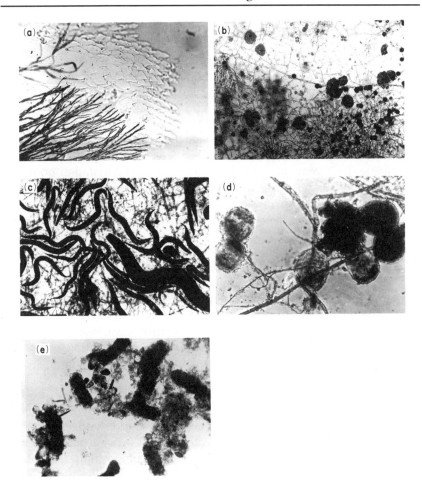

Figure 8.4   Colonization of cellophane buried in broad-leaved litter. (a) Fungal hyphae and evidence of rasping by radula of a mollusc (2 weeks). (b) Hyphae, bacteria and amoebae (3 weeks). (c) Nematode worms, moribund hyphae and bacteria (5 weeks). (d) Arthropod faecal pellets (diameter 45 μm), hyphae and bacteria (13 weeks). (e) Disintegrating pellets (100 × 35 μm), with spores, worm chaetae and shelled amoebae (21 weeks). (Photographs (d), (e) by Dr P. W. Murphy).

ubiquitous *Aureobasidium pullulans*, as well as other primary saprophytes and parasites with a more restricted host range. The weakly pathogenic Ascomycete *Lophodermium pinastri* remained quiescent on healthy needles until senescence set in. During the first 6 months after needle fall, in the loosely textured and rather dry litter layer (L = A$_{00}$, see Figure 4.1), cell contents were further decomposed by some of the phylloplane species. The litter-dwelling Ascomycete *Desmazierella acicola* then colonized the needles, being particularly active within the phloem during the 2 years or so in the F_1 (fermentation) layer, where the needles were more compacted and moist. Regions of internal attack were broken down further by **endophagous animals** (i.e. living within their food) such as the oribatid *Adoristes ovatus*, forming pellet-filled cavities in the mesophyll. Meanwhile, other microarthropods and enchytraeid worms grazed on hyphae or spores on the needle surface, including those of two species of Hyphomycete (*Sympodiella acicola* and *Helicoma monospora*) which replaced *Aureobasidium pullulans*. The needles were still superficially intact, but during the 7 years in the F_2 layer the major fragmentation occurred, due mainly to animals. Fragments which had not previously been attacked internally were colonized by Basidiomycetes, including *Marasmius androsaceus*, which can decompose cellulose and lignin, and by common soil fungi such as *Trichoderma* and *Penicillium*. The net result was the conversion of needles into an amorphous mixture of faeces, fungi and recalcitrant humus in the H layer. This was further processed by enchytraeids and chitinolytic fungi, the resulting ammonia enabling actinomycetes and bacteria to occupy more alkaline microsites.

The chemical and physical nature of the needles exerted a strongly selective influence on potential initial colonizers, so that most of the common primary saprophytes (see below) were excluded. Elsewhere, it has been shown that the relative competitive abilities of different species of fungi can be influenced by animals, such as Collembola, exerting differential grazing pressure, sometimes because certain fungi produce toxins. In turn, the particular species of pioneer may play a key role in the subsequent succession. For example, it was found that *Lophodermium pinastri* persisted until long after needle fall in Corsican pine, but with little evidence of other fungi or of noticeable decomposition. In contrast, needles infected with *Lophodermella* spp. were later colonized by fungi such as *Marasmius*, with appreciable depletion of cell walls. The relative importance of certain species can be altered by changing the pH or nitrogen status of the litter. Consequently site characteristics, as well as different techniques, may account for some of the discrepancies between accounts of succession on litter of a particular species.

Broad leaves

The numerous studies of microorganisms associated with broad leaves have been summarized by Hudson (1968), Jensen (1974) and Hayes (1979). Basically, Hudson recognizes three stages in these successions, comparable to the pine-needle situation. The first stage involves **colonization by phylloplane species**, some of which attack living tissues. The ascomycete fungus *Gnomonia errabunda* is specific to beech, acting as a parasite initially but persisting for several months on dead leaves, whereas other species may be less restricted but also less persistent.

Many saprophytes also occur in the phylloplane, being well placed to show appreciable activity during the second, or **senescent, stage**. **Common primary saprophytes**, with a wide host range, are particularly important here. They include the fungi *Aureobasidium pullulans*, *Epicoccum nigrum*, *Cladosporium herbarum*, *Alternaria tenuis* and *Botrytis cinerea*, and, in the tropics, genera such as *Nigrospora* (only *A. pullulans*, however, occurs on pine needles). Restricted primary saprophytes, with limited host ranges, may also be present at this stage. Some of these parasites and primary saprophytes may persist after leaf fall, but the third, **dead, stage** is largely characterized by **secondary saprophytes**, including Basidiomycetes and various soil-inhabiting Mucorales and Penicillia.

Degradative changes

Swift (1976) has pointed out that successions, in the strict sense of directional changes, occur only on individual units of organic matter, and that within a particular area of the forest floor community changes are cyclic, involving colonization and exploitation of a resource unit, followed by dispersal and recolonization of a new unit (Figure 8.5). Since studies of succession are normally based on samples of numerous units, the details of individual successions become blurred. The apparent importance of particular species or groups of microorganisms may be considerably coloured by the techniques used, and environmental differences between sites will also lead to varying results on the same type of resource (e.g. conifer needles). A survey of fungal studies carried out as part of IBP includes an examination of the concept of succession (Kjøller and Struwe, 1982). It is widely believed that different resources have their own characteristic communities and successions, and that individual sequences are not determined by chance alone.

Assuming that degradative successions do occur, can we explain why? Microclimatic changes could have differential effects on microbial species, as in the case of fungi which colonize freshly fallen pine needles but become active only in the damper F_1 layer. As a more general explanation, however, changes in species composition could result from changing availability of resources, brought about by abiotic and biotic agencies and with the possibility of competition for these resources. Living space within a unit alters as tissues are penetrated and opened up, providing microhabitats for motile and non-motile organisms, as shown by endophagous fauna. Chemical resources, in the sense of energy and nutrient supplies, may be inaccessible to a particular species because

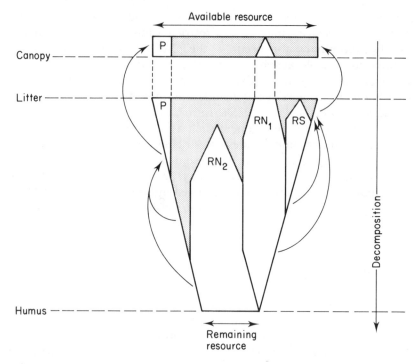

Figure 8.5 Diagrammatic representation of successional changes within the fungal community on a 'typical' resource unit, such as a pine needle or beech leaf. Non-stippled areas represent extent of fungal occupation of the available resource; arrows indicate dispersal of propagules to other resource units. P = parasites, some of which persist after litter fall. RN_1 = primary resource-non-specific saprophytes, some originating in the phylloplane. RS = resource-specific saprophytes. RN_2 = secondary resource-nonspecific saprophytes, which eventually replace the primary colonizers. This pattern may be disrupted if the unit is consumed by detritivores. (From Swift, 1976.)

they are masked physically or chemically (e.g. cellulose by lignin or tanned protein) or because the species lack the necessary enzymes, or as a result of some other difference in competitive ability compared with other species.

The observation made by Garrett (1951) that on herbivore dung the general succession passed from Phycomycetes to Ascomycetes to Basidiomycetes led to a functional explanation, based on the dominance of '**substrate groups**', **sugar fungi** being replaced by **cellulolytic species** and finally by **ligninolytic forms**, with secondary sugar fungi existing on the breakdown products of the last two groups. It is now realized that biochemical capabilities are not so strictly demarcated between the taxonomic groups, while sugar fungi are conspicuously absent from the initial stages of most successions. Cellulolytic ability seems more widespread than previously thought; there is, however, very little evidence of the role played by particular species, either individually or in combination, at particular stages of a given succession, for example little is known of the role of Basidiomycetes in needle decomposition. Enzyme complements presumably differ between species, but the availability of particular nutrients or the ability to tolerate polyphenols may be just as important as the possession of cellulases.

The order of disappearance of species from the community almost certainly reflects the results of competition for food resources, leaving the highly persistent Basidiomycetes, with their great diversity of enzymes, and the antibiotic-producing Penicillia. On the other hand, the order in which species appear may be largely a reflection of their characteristic distributions within the woodland: many of the initial colonizers are members of the phylloplane flora, well placed to exploit senescing leaves, while the aerial dispersal mechanisms of many primary saprophytes contrast with the passive, localized transfer of secondary saprophytes such as *Penicillium*.

Successions among detritivores

So far, discussion has centred on microbial successions. Anderson (1975) recorded changes in numbers and species composition of various groups of animals found among beech and sweet chestnut leaves enclosed in mesh bags and placed in the floor of a *Castanea sativa* woodland in Kent. Some of these changes were seasonal, but some reflected changes in suitability of the leaves for colonization. During the first autumn and winter, gut contents of enchytraeids and chironomid (midge) larvae consisted mainly of fungal remains, as did those of oribatid mites, most of which belonged to three mycophagous species. Even phthiracaroids, which are normally considered to be litter ingesters, contained mainly fungal material at this time. From April onwards, the numbers of many of the mycophagous species declined, the increasing species diversity resulting partially from an influx of litter feeders. Certain species, such as the phthiracaroids, changed their feeding habits, concentrating on the conditioned litter (Figure 8.8). This change coincided with a reduction in polyphenol levels of the leaves which could have influenced the behaviour of those species of microbes and animals which are intolerant of tannins and other related compounds. The increase in species diversity to a maximum in the second autumn was assumed to accompany sub-division of food resources, and possibly increasing microhabitat complexity. In other words, a diversity of substrates and living places became available within and upon the leaves, including those provided by faeces and corpses. Further examples of arthropod successions are discussed by Usher and Parr (1977) and by Wallwork (1983).

Wood

The ecology of wood decomposition, reviewed by Kaarik (1974) and Swift (1977a) provides further examples of successional changes, usually over periods of several years. In reviewing the factors which affect fungal community development in wood, Rayner and Boddy (1988) emphasize spatiotemporal changes and the roles of stress in pattern initiation, rather than seeking simple, single causes of change. Communities of animals associated with dying and dead wood are described by Elton (1966), faunal participation being of great importance in many types of wood decay.

Adjacent regions of an individual unit of necromass, such a pine needle, can undergo different pathways of decay, and this is even more likely with woody remains, because of their sheer bulk. Heterogeneity is further increased by situation: dead trees may stand for many years, being slowly attacked by a small number of species (as shown by the heartwood of many elms felled five or more years after death due to *Ophiostoma ulmi*) in contrast to the greater diversity of organisms in the more equable climate of the forest floor. Nevertheless, some general observations can be made about wood decay, taking the scheme of Swift (1977a) as a framework (Figure 8.6).

The **colonization stage** is characterized by the death of the relatively few living cells in wood, mainly in the inner bark and medullary rays. These cells 'defend' the adjacent dead wood which may be further protected by tough cell walls or by biologically active compounds (poisons or 'deterrents') in the heartwood. Breaching of these defences may be effected by certain organisms independently, but some are aided by wounding of tissues caused by wind, fire, frost, or other organisms (including man). The Basidiomycetes *Heterobasidion annosum* (*Fomes annosus*) and *Armillaria mellea*, are examples of fungi which can penetrate healthy tissues, causing root rot, and which can also colonize freshly cut stumps (section 6.3). Among insects, some termites (e.g. certain species of *Coptotermes*) habitually consume the wood of living trees, while bark beetles feed mainly on cell contents of tissues under the bark of living or recently felled trees. The introduction of spores of *Ophiostoma ulmi* during maturation feeding by elm bark beetles eventually leads to lowered resistance by the elms to oviposition and larval development (section 6.4).

Wounds on branches in the canopy may be colonized by bacteria, causing limited damage to the walls of ray parenchyma cells, as do soft-

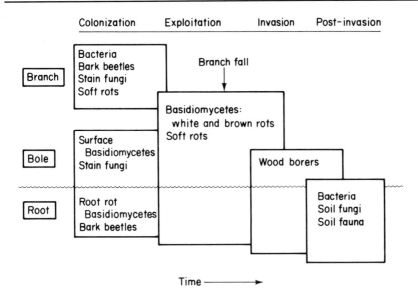

Figure 8.6 Patterns of succession among the dominant organisms associated with decomposition of 'typical' woody material in Britain. (From Swift, 1977a.)

rot fungi, while cell contents are attacked by blue-stain fungi (often associated with *Tomicus* outbreaks, see section 6.6). These primary attacks may facilitate invasion by wood-decaying Basidiomycetes, such as *Phellinus* spp., although more aggressive species, such as *Stereum* (*Chondrostereum*) *purpureum*, colonize freshly exposed surfaces and are inhibited by non-Basidiomycetes. Heart-rots (e.g. *Ganoderma applanatum*, responsible for the precarious state of many beech trees) often have their origins in wounds.

The **major decomposition stage** is dominated by the activities of Basidiomycetes and specialized woodboring insects which degrade cell walls. Some pioneer species of fungi may continue to attack dead wood, whereas others are replaced by saprophytes. Fungi causing white rot can attack both lignin and cellulose, either successively (e.g. *Heterobasidion annosum*), or simultaneously (e.g. *Coriolus* (*Polystictus*) *versicolor*). In contrast, brown rot is characterized by patchy attack on cell-wall polysaccharides, leaving the lignin framework intact. White or brown rot may be accompanied by various Ascomycetes and Fungi imperfecti which slowly attack cellulose but not lignin, forming chains of cavities in tracheids and fibres, characteristic of soft rot.

Undecayed dead wood is attacked by certain termites and beetles, some of the latter producing their own cellulases (as in *Anobium punctatum*, the wood-worm, and certain cerambycids or timber beetles) while others are dependent on endosymbiotic yeasts or bacteria. Wood-boring 'ambrosia beetles' transmit specific fungi which grow on comminuted wood in the faeces and provide their main food source. Other insects, including many termites, characteristically feed on wood after a certain amount of microbial decay. It was found that invasion by wood-borers into branches of various broadleaved species at the IBP site at Meathop, Cumbria, occurred about 2 years after branch fall, which was preceded by about 6 years of predominantly fungal decay (section 9.6).

Branches or stumps which are protected from desiccation by bark, moss or lichens may support a rich community of invertebrates, including many litter dwellers such as molluscs, woodlice, millipedes, mites and springtails, and beetle and fly larvae, e.g. wireworms, stag beetles and leatherjackets. Some of these, such as insect larvae and mites, may burrow into the heartwood which is reduced to friable **frass**. The activities of the general soil fauna and microflora (e.g. Mucorales) dominate the **terminal stage** of wood decay, when the remains are finally comminuted and incorporated into the soil.

To summarize, although certain species of fungi and animals can bring about wood decay on their own, the process normally involves the integrated physical and chemical action of a succession of organisms. The structural complexity of woody remains, together with the diversity of associated organisms, makes it difficult to explain why particular successions occur. However, the durability of these remains normally ensures an abundant supply of potential food resources, protected from climatic extremes, as well as refuges for hunters and hunted.

8.4   Divers detritivores

Several hundred species of microbes may be associated with decomposition of a particular type of litter in a single temperate woodland. A square metre of the same forest floor might well contain more than 1000 species of animals, ranging in size from Protozoa to earthworms. Apart from carnivores and root-feeding herbivores, the majority are detritivores. Of those feeding among plant remains, some are **microphytophagous**, ingesting predominantly fungal hyphae, spores or bacterial colonies, but most subsist on decomposing higher plant litter, including varying amounts of microflora in their diet (**macrophytophagous**) (including the panphytophages of Luxton, 1972). In culture, individual species often appear to display preference for particular microbes or leaves of particular species. However, within each feeding category, the gut contents of various species of animals taken from the forest floor usually appear remarkably similar to those of certain other species, which may even belong to different classes or phyla. This apparent **lack of food-niche differentiation** is in marked contrast to the specialization typical of herbivores.

Co-existence could be accounted for by surplus food, implying too much litter for too few detritivores. Fluctuations in space and time in amounts of litter do not necessarily coincide with variations in animal abundance, so that by chance certain populations may have plentiful resources. Alternatively, animal numbers could regularly be kept below the level at which interspecific competition for food occurs by the action of pathogens or predators, or, less dependably, by the influence of climatic factors. Very few quantitative data are available relating to such influences on abundance, let alone on competition. Intraspecific competition for other resources, such as oviposition sites, could also be intensified by climatic extremes; there is some evidence of higher mortality during the winter among mite eggs laid in beech cupules.

Assuming that variations do occur in the balance between supply and demand, what happens at those places and times with a shortage of food? **Interspecific competition** could be avoided by reducing the amount of overlap between various components of the niches of different species, especially when and where they feed, and what they feed on.

Separation in time occurs between species with markedly different phenologies, but overlap is particularly likely among those groups, such as many soil arthropods, which have developmental periods of a year or more. Time spent in moulting and other non-feeding phases such as diapause will reduce the number of potential competitors.

As regards **vertical distribution**, within the major groups of soil animals certain species are known to be characteristic of particular sub-horizons (Figure 8.7), although others are more wide-ranging. Among oribatid mites this distribution has been correlated with that of 'preferred' food, whether specific microbes or particular stages of litter decay (Luxton, 1972). Within each sub-horizon at a single site, however, there are likely to be individuals of different species which have similar gut contents. The possibility of further partitioning of space within sub-horizons has been explored by Anderson (1978b), who recognized two dozen microhabitats, ranging from intact leaves, twigs and roots to faeces and soil cavities, in gelatine-embedded sections from a range of humus forms under *Castanea sativa*. Species diversity was correlated with microhabitat diversity, but the precise requirements of individual species are insufficiently understood, partly because of failure to recognize and measure the components of living places which are crucial to particular animals, rather than to the observer. However, on the assumption that species differ in these requirements, they can be visualized as being distributed throughout a vertical and horizontal mosaic of microhabitats, and so kept largely separate. Wallwork (1983) suggests that gradients of

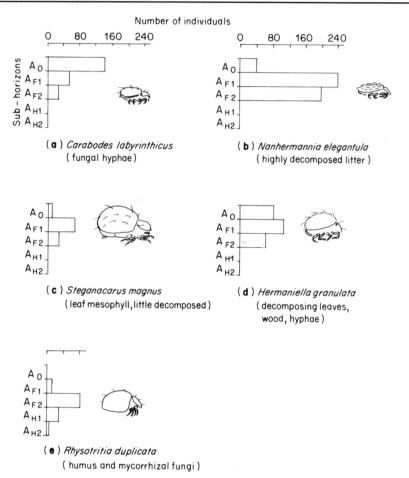

Figure 8.7 Vertical distribution of adult oribatid mites from a *Castanea sativa* site at Blean Woods, Kent. The histograms represent the number observed in a total of 780 soil sections obtained from monthly samples between September 1967–August 1968. The major components of the gut contents of each species are indicated in parentheses. (From Anderson, 1971.)

biochemical diversity (e.g. increasing between L and F layers) could provide a framework for the partitioning of resources.

The gut contents of each individual will indicate some of the food resources available within recently visited microhabitats. These resources may change with time, as shown by the phthiracaroids feeding on fungi and later on leaves (Figure 8.8a). On the other hand, some may be common to various microhabitats, so explaining similar contents in different species. At times of increased likelihood of competition, as when frost, drought or flooding lead to a compression of the normal

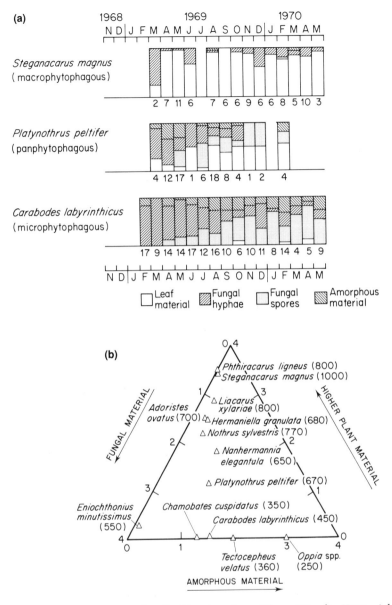

Figure 8.8 Gut contents of oribatid mites from Blean Woods, Kent. Adult mites were recovered over a 20-month period from mesh bags buried in litter of *Fagus sylvatica* and *Castanea sativa*. Contents occurred as 1–3 discrete boluses, which were dissected and scored on a 0–4 scale, indicating the relative amounts of each of four major food materials. (a) Temporal variation in feeding by three species of oribatid from *Castanea* litter, showing the average composition of a bolus at each sampling occasion. The number of boluses examined is shown under the histograms. (b) Cumulative mean proportions of the three major food items in those species which were most abundant in the two types of litter during 20 months. Specialized feeders occur near the apices of the triangle, with more catholic species nearer the centre. Values in parentheses are body lengths (μm). (From Anderson, 1975.)

vertical distribution, feeding niches may become narrower, each species having its own feeding 'refuge', similar to the average components shown in Figure 8.8b.

Finally, energy and nutrients could be shared between different species which ingest the same sort of food (and therefore have similar gut contents) by variation in **digestive abilities**. Unfortunately the techniques used to demonstrate the presence of certain enzymes, such as cellulases, often give equivocal results, and even if present, these enzymes may not be effective on natural litter. Nevertheless, possible digestive strategies may be envisaged.

Digestive strategies

Microphytophagous species (including the mycophagous majority among microarthropods) feed predominantly on the microflora, whose protoplasm is exposed to digestive enzymes, either after cutting or chewing through hyphae or possibly by chemical penetration of, for example, spore walls (Moore *et al.*, 1988). **Chitinases**, which have been demonstrated in microphytophagous oribatids (Luxton, 1972), presumably attack hyphal walls, while the possession of **trehalase** enables these mites to utilize the fungal storage compound, trehalose. It is, then, assumed that animals which 'specialize' on fungi or bacteria have the necessary enzymes to obtain their energy and nutrients from this diet.

For litter-ingesting species, the major components remaining in litter are cellulose and lignin, most of the more readily assimilable compounds having been 'mopped up' by herbivores and primary saprophytes. Molluscs, certain wood-boring beetles and at least one species of termite can decompose cellulose by means of their own **cellulases**, but these may be supplemented by microbial enzymes. In all other instances, decomposition of cellulose and lignin seems to be brought about by the **microflora**. A clearly **symbiotic relationship** is sometimes involved, such as with many termites and tipulid larvae (leatherjackets), in which certain species of microorganisms occur, often in a particular region of the gut.

Termites are known to derive most of their energy from the digestion of polysaccharides, especially cellulose and hemicellulose (Lee and Wood, 1971). Symbiotic flagellates (or bacteria, in certain species) decompose cellulose anaerobically to acetic acid, which the termites absorb. The mechanism of lignin decomposition is obscure, since in fungi this is an aerobic process; species of the Basidiomycete *Termitomyces*, which grow on the fungus combs of one subfamily of termites, possibly attack lignin and expose cellulose to the gut symbionts, as well as being consumed themselves. Despite their symbiotic complements, many species of termites tend to feed on wood and other litter at a particular stage of decay. In some instances it is known that microbial conditioning involves the breakdown of repellents or poisons; in others the microflora may provide vitamins or even form the bulk of the diet.

Symbiotic bacteria have been found in the larvae of *Tipula maxima* which apparently digest cellulose, for example in beech leaves, while wood-boring beetles either have endosymbionts or feed on ambrosia fungi. In other groups the microbial connection is more tenuous, although there is some evidence of characteristic gut floras in certain Collembola, oribatids, bibionid larvae (Diptera) and woodlice, perhaps involving the selection of certain ingested species.

The fact that litter feeders often show poor survival or cease feeding altogether after sterilization of their food suggests that they are somehow dependent on living microorganisms. These could form the major part of the assimilated diet, so that the habit of microbial browsing is merely extended into the substance of the litter. Some litter feeders can be reared on a purely fungal diet. The necessity for live microbes could imply that certain species of, for example, litter-dwelling bacteria multiply in the gut, among the comminuted litter, and are continually cropped, but only if further supplies are ingested; this seems to occur in at least one species of woodlouse (Reyes and Tiedje, 1976), and in the pill millipede, *Glomeris marginata* (Anderson and Bignell, 1980). Certain macrophytophagous oribatid mites, such as phthiracaroids, lack chitinase and trehalase, but could presumably subsist on microbial protoplasm and the breakdown products (exometabolites) released by microbial activity in the litter. Cellulolytic activity, demonstrated in the intestines of certain woodlice, millipedes and, possibly, phthiracaroids, is again lost when the food litter is sterilized. Continuing microbial action may also be of importance in **coprophagy**; for example, certain millipedes resemble rabbits in needing to ingest their own faeces, in order to obtain sufficient energy and nutrients for normal growth. Feeding on the faeces of other species, with the possibility of differential digestion, provides a further way of sharing resources; this successive reworking of material, which does not occur in herbivore communities, may be one of the factors accounting for the great diversity of the decomposition subsystem.

A wide variety of suggestions has, then, been made to account for avoidance of competition among detritivores, with their apparently rather generalized feeding habits. Experiments with single and mixed cultures of oribatid showed that vertical distribution was more restricted in mixtures, but changes in microhabitat distribution and gut contents were more difficult to interpret (Anderson, 1978a). Whatever the mechanisms of avoidance, detritivores have a part to play in decomposition, but just how important are they?

8.5   Relative roles of agents of decay

Except for the action of certain molluscs and insects which produce their own cellulases, major chemical changes in the composition of the necromass, especially its structural components such as lignin and cellulose, are effected by microbial enzymes. On the basis of evidence presented in Chapter 9 the microflora is believed to account for about 80% of the energy flow within decomposer systems, with a relatively minor contribution from detritivores. However, many of these animals are described as wasteful feeders, egesting 85% or more by weight of their food as faeces, and this might be expected to influence decomposition processes. Animals enhance microbial activity in a number of ways, for example by aiding spore dispersal or by exposing necromass tissues to the microflora [see Visser (1985) and Moore *et al.* (1988) for reviews of these interactions].

Spores and other propagules occur on and within the bodies of animals, and although some may be digested by particular species, others germinate freely. Some of these spores are transported from exhausted substrates to environments which are more favourable in terms of food availability or climate. Swift (1976) has suggested that this method of transport is particularly important to fungal species within the forest floor.

Organisms with limited mobility, such as bacteria, could be introduced to otherwise inaccessible substrates by the feeding of animals, especially within tissues (as shown during acorn decay). Comminution greatly increases the surface area of exposed tissue: it has been estimated that the conversion of a 60 mm long conifer needle into 10 μm³ fragments by phthiracaroid mites results in a 10 000-fold increase in surface area (Harding and Stuttard, 1974). These tissues could be exploited by microbes either within the animal (in the case of species which can tolerate conditions in the gut) or after egestion. Faeces form foci of high nutrient status, with moisture-holding and pH characteristics which are often more favourable to microbes than the ingested litter (Webb, 1977). Consequently, microbial numbers and activity (e.g. respiration) are normally higher in faecal material (Figure 8.9). There is now evidence of increased microbial activity, especially among bacteria, as a result of certain levels of grazing on faeces by arthropods. Browsing on litter might rejuvenate senescent microbial colonies, but could also destroy active hyphal tips.

These are some of the ways in which it has been proposed that detritivores help to unblock 'bottle-necks' in energy flow and nutrient cycling, thus contributing much more to soil metabolism than is indicated by their own energy requirements (Macfadyen, 1961). Nutrients are

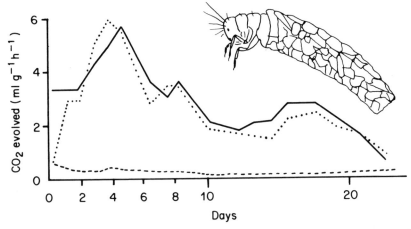

Figure 8.9 Microbial respiration from food and faeces of larvae of the terrestrial caddis, *Enoicyla pusilla*, at 25°C. The similarity in activity on faecal pellets (——) and on oak leaves (. . .) ground to the same fragment size as in the pellets contrasts with that on entire leaves (– – –). (From Drift and Witkamp, 1959.)

considered further in Chapter 9, and this section is concluded by reference to some quantitative studies of decomposition.

Decomposition rates are normally calculated from losses in weight of litter placed on or in the forest floor, whether loose, tethered or confined in mesh bags. Disappearance of fragments from bags or lines, or its transport deeper into the profile (as when dragged down by earthworms) results in exaggerated estimates. The relative importance of various agencies can be assessed by altering the influence of particular groups, although absolute exclusion is rarely achieved.

Naphthalene has been used to reduce animal numbers. At Oak Ridge (Figure 8.10b) the abundance of arthropods decreased by 82%, compared to control plots, Collembola being particularly affected. Bacterial populations in treated litter increased up to six-fold, probably because of the greater availability of corpses, but no changes were observed in amounts of fungal mycelium. Weight losses from litter bags in the two plots were not markedly different, in contrast to the results of Russian work using *Quercus robur* (Kurcheva, 1960), possibly reflecting differences in composition of the decomposer community, or a consequence of applying greater concentrations of naphthalene.

The significance of microarthropods in the decomposition of oak leaves was demonstrated at Rothamsted when leaf discs disappeared more rapidly in soils treated with DDT than in controls; the differential effect of the insecticide on certain predators enabled Collembola populations to increase markedly.

Litter bags with different mesh sizes provide a means of distinguishing between the effects of leaching, microorganisms, mesofauna and macrofauna; in making quantitative comparisons between results from different bags it is assumed that microbial decomposition and leaching are similar, whatever the mesh size. Results derived from this method are somewhat equivocal (Figure 8.10). Edwards and Heath (1963) recorded no loss in area of oak and beech leaf discs after 9 months in 3 μm mesh bags in pasture soil. Losses from 0.5 mm bags were associated with fragmentation by Collembola, Diptera larvae and enchytraeids, which were supplemented appreciably by lumbricid earthworms in the 7 mm bags. On the other hand, lumbricids caused little acceleration of decomposition of elm, ash, birch or lime.

In an Australian dry sclerophyll forest, with few earthworms or arthropods, leaching and microorganisms accounted for 70% or more of the weight loss from leaves of *Eucalyptus* spp. In wet forest, with abundant detritivores, the contribution of megascolecid earthworms and macroarthropods was similar to that of leaching plus microarthropods in the case of *E. delegatensis*, but not with *E. pauciflora*, which was apparently not attacked by earthworms (Wood, 1971).

Anderson (1973) found no significant differences between weight losses from bags of different mesh sizes, using *Fagus sylvatica* leaves in either of two woodland sites, or with *Castanea sativa* leaves in the site with a mor-like humus, suggesting that mycophagous species (such as various mites), woodlice, millipedes and *Lumbricus rubellus* made little contribution. Greater differences were recorded for leaves of *Castanea* in the site with mull-like characteristics, presumably due to *Lumbricus terrestris* (Figure 8.10c,d). The use of aerially suspended bags, in which microbial activity was assumed to be minimal, suggested that a very large proportion of the weight losses could result from leaching. There was no evidence of stimulation of decomposition by mycophagous or litter feeding microarthropods, although exploitation of their faeces was possibly hindered by the retarded descent of bags and contents into the F_2 sub-horizon. A review of 15 litter-bag studies by Seastedt (1984) suggested that on average microarthropods increase the litter decay rate by about a quarter.

Relative importance thus varies with the type of necromass and with the make-up of the decomposer community. Exclusion experiments yield the most extreme differences in mulls and similar sites where earthworms or other macrofauna are typically plentiful. Even in sites where animals are unimportant in the initial decomposition of litter, evidence of increased decomposition rates might be expected among detritivore faeces. Respiratory activity has been shown to increase as certain faeces age, correlated with a build-up of bacteria in pellets of *Enoicyla* (Figure 8.9)

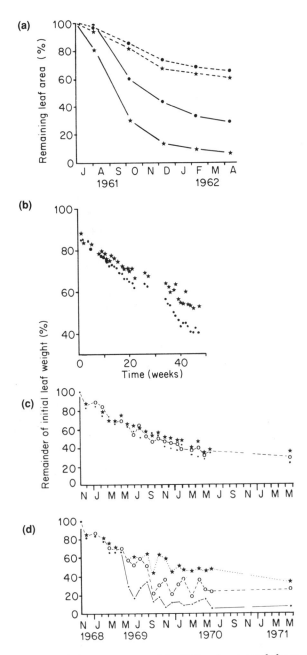

Figure 8.10 Rates of disappearance of leaf litter from mesh bags. (a) Leaf discs (2.5 cm diam.) of oak (*Quercus* sp., ★) and beech (*Fagus sylvatica*, ●) in 7 mm (——) and 0.5 mm (– –) mesh bags in newly cultivated pasture soil, with mull humus, Rothamsted. Leaves were picked from trees in July. (Redrawn from Edwards and Heath, 1963.) (b) Leaves of white oak (*Quercus alba*), collected from trees at leaf-fall, in 2 mm mesh bags in *Q. alba* litter treated with 100 g of naphthalene per m² (★) and untreated (●), Oak Ridge, Tennessee. (From Witkamp and Crossley, 1966.) (c), (d) Leaf discs (2.5 cm diam.) of freshly fallen sweet-chestnut (*Castanea sativa*) in 7 mm (●), 1 mm (○) and 48 µm (★) mesh bags in litter of (c) a *Castanea* site with mor-like moder, and (d) a *Fagus sylvatica* site with mull-like moder, both in Blean Woods, Kent. (Modified from Anderson, 1973.)

and of the pill millipede *Glomeris*. Nevertheless, decomposition rates of pellets of *Glomeris*, confined in bags at the litter surface, were similar to those of uningested leaves (Nicholson *et al.*, 1966); their fate might have been different deeper in the profile. Observations of faeces *in situ* (e.g. in sections) suggest that some, such as those of many termites and microbe-feeding Collembola, are much more resistant to microbial attack than those of many other litter ingesters (Harding and Stuttard, 1974).

The direct effect of most animals on litter decomposition in woodlands is therefore much smaller than that of the microflora and abiotic factors such as leaching. The influence of comminution and transport of materials into more climatically buffered regions of the forest floor is difficult to quantify, but is probably of considerable significance.

The most important interactions between soil organisms and plants occur in the rhizosphere, where nutrient fluxes are fuelled by the release of a substantial proportion of NPP in the turnover of fine roots, including mycorrhizas (section 9.6). Fauna may feed directly on roots or sloughed products, or they may regulate populations of microbes or of other animals. Examples of such relationships, mainly from grassland, are explored in Edwards (1988).

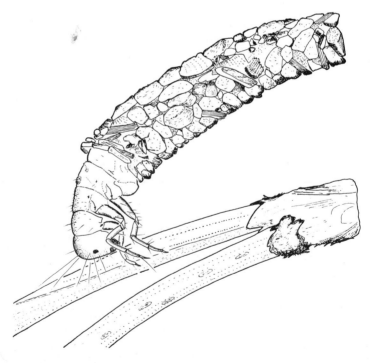

Larva of *Enoicyla pussilla* on pine needle at Chaddesley Woods, Worcs.

9
Energy and nutrients

9.1 Taking stock of woodlands

Foresters have traditionally assessed their forest stands, annual increments and yields in terms of trunk volume, whereas ecologists try to measure the weights and energy contents of various components of ecosystems, and also to quantify the major pathways of energy flow and nutrient cycling.

The heyday of ecosystem studies was during the decade of the International Biological Programme (IBP, 1964–74), whose Woodlands Biome programme was designed to quantify the productivity of a range of forest types under various environmental conditions and management regimes. The ultimate aim was to facilitate effective management of woodland resources on a long-term sustainable basis. Years of effort in 116 international research sites, predominantly in temperate and boreal regions, yielded the Woodlands Data Set, the components of which are summarized and analysed in Reichle (1981). Unfortunately, attempts to make comparisons and generalizations are bedevilled by the sheer diversity of site variables, including species, age, soil, climate, and whether plantation or natural stand, as well as by variation in the actual parameters recorded and in the techniques used. Mean values of 13 parameters for each of 11 forest types, derived from 68 sites, are presented in Table 1.3 of Reichle (1981), representing a major, if incomplete, source of information on the world's forest resource. In attempting to review current knowledge, we have used this information, together with data from some of the individual sites, and from studies other than IBP, but the problems inherent in comparing such data should always be borne in mind.

9.2 Biomass and productivity of autotrophs

Details of methods involved in woodland productivity studies are given by Newbould (1967) and Pardé (1980). Estimates of amounts of living

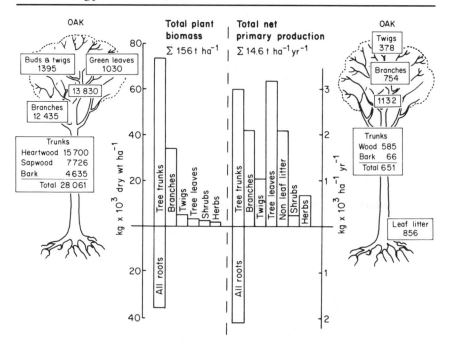

Figure 9.1 Estimated values for plant biomass (kg ha⁻¹; to left of centre) and for net primary production (kg ha⁻¹ yr⁻¹; to right) in a mixed oakwood at Virelles-Blaimont, Belgium. Data for oak are given in the tree outlines. Number of trees per hectare: *Carpinus betulus* 1135, *Quercus robur* 195, *Fagus sylvatica* 87, *Acer campestre* 69. Shrubs include *Crataegus oxyacanthoides*, *Cornus sanguinea*, *Corylus avellana* and *Carpinus betulus*. For description of herb layer and soil, see Figure 1.2. (From data of Froment *et al.*, 1971. © UNESCO 1971. Reproduced by permission of UNESCO.)

organic matter (**biomass**) present in trees, shrubs and herbs are complicated by the sheer bulk and heterogeneity of material. Conventionally, as in IBP studies, woody structures such as branches are included in biomass estimates only if they bear living buds; however, standing deadwood and, in some instances, litter, are sometimes included in totals of biomass. Rather than use destructive sampling, which involves removing and weighing all plant material from a site, alternative methods are normally employed, especially if estimates are to be repeated. For example, aerial woody biomass of trees can be calculated by measuring trunk girth at 1.3 m above ground level, and referring to a regression of, for example, dry weight against girth, derived from destructive sampling of part of the site. Sampling of roots is extremely laborious, and often a nominal value of *c.* 20–30% of aerial biomass is used.

The distribution of biomass in a Belgian broadleaved woodland IBP site (whose dominant trees were *c.* 75 years old) is shown in Figure 9.1.

Table 9.1 Estimates of aerial biomass and production in four forest types included in the IBP Woodlands Data Set; means and standard errors with number of stands in parentheses (data from Table 1.3 in Reichle, 1981)

	Stand Height (m)	Aerial standing crop (t ha^{-1})		NPP (t ha^{-1}y^{-1})
		Woody	Total	
Boreal conifers(9)	17.2 ± 1.7	124.4 ± 18.8	139.2 ± 18.7	5.2 ± 0.6
Temperate broadleaved deciduous(19)	20.8 ± 1.4	162.5 ± 25.6	173.5 ± 22.4	9.2 ± 0.7
Temperate coniferous evergreen(5)	21.0 ± 2.0	225.0 ± 54.9	214.4 ± 40.0	11.6 ± 2.4
Tropical broadleaved evergreen(4)	37.8 ± 0.3	371.3 ± 17.5	432.7 ± 31.1	15.5 ± 0.4

Mean values from 37 IBP sites are summarized in Table 9.1, indicating latitudinal trends. Among individual extremes, estimates from the temperate rain forests of the Pacific Northwest include 1600 t ha^{-1} for Douglas fir (*Pseudotsuga menziesii*) and 2300 t ha^{-1} for coastal redwood (*Sequoia sempervirens*), considerably higher than the 562 t ha^{-1} recorded for evergreen tropical forest at Banco, Ivory Coast.

The failure of the Woodland Data Set to name those sites where root biomass was actually measured, rather than estimated, precludes discussion of most of these IBP root data. Earlier, Rodin and Bazilevich (1967) indicated maxima of 70–100 t ha^{-1} for root biomass in broadleaved and subtropical forests, representing 15–33% of the total biomass. They also suggested that coniferous forests tended towards an equilibrium root: shoot ratio of 0.29:1 at an earlier stage of stand development (when total biomass exceeds 75 t ha^{-1}) than in the case of deciduous forests, which stabilize at about 0.25:1 as total biomass approaches 300 t ha^{-1}. This pattern of biomass distribution appears to be consistent over a wide range of forest types (Harris, 1981).

There is an even greater paucity of data relating to fine roots of trees (variously defined as having a diameter less than somewhere between 0.2 and 1.0 cm). Vogt *et al.* (1986) present data from 76 sites, of which 40% fail to distinguish between live and dead material. This data set is biased towards cold temperate and boreal conifers, and also tropical broadleaf evergreen sites, so that generalizations should be treated with caution. Evergreen conifers in boreal sites had only a quarter of the live fine root mass of their cold temperate counterparts (1.4 and 5.0 t ha^{-1}, respectively). The limited ability of larch to store food reserves may account for the very low value of 1.4 t ha^{-1} for total fine root mass of Asiatic larch in a cold temperate site; the availability of starch seems to be critical in determining the amount of roots that a tree can support.

In the warm temperate zone, totals for fine roots in evergreen stands averaged c. 18 t ha^{-1}, being similar for broadleaved or needle-leaved forests; corresponding data for deciduous stands were approximately 5 t ha^{-1} less. Evergreens can allocate a greater proportion of photosynthate to fine roots than deciduous trees, probably because of the lower annual costs of foliage production in evergreens, rather than due to their prolonged growing season.

Tropical forests have the highest mean value for total fine roots (40.7 t ha^{-1}), with individual maxima from the least fertile sites. The value of 54.7 t ha^{-1} for a Venezuelan caatinga site contrasts with 2.8 for a montane forest in New Guinea; it represents 40% of the total root biomass, and five times the leaf biomass (Vitousek and Sanford, 1986). This may largely reflect the expected tendency of root:shoot ratios, and of the amount of fine, active roots, to increase in sites where nutrients are less available (section 9.6).

Biomass is a static measure of the amount of matter present in living organisms at a particular time. Energy flows through this matter and production is added to it, so that biomass changes with time. Values for plants vary throughout the year, depending, for example, on the timing of incremental growth of stems and roots, and on the presence or absence of leaves or reproductive structures. As a stand matures, total biomass steadily increases to a maximum, with the weight of the trunks, which come to contain more and more dead heartwood, making a proportionately greater contribution in older trees. Consequently root:shoot ratios decline over the years towards the aforementioned equilibrium values. As crowns expand, the weight of leaves becomes maximal when the canopy closes, later declining as death and thinning reduce the number of trees.

Biomass on its own gives little indication of **productivity**, the rate of formation of organic matter. Estimates of woody biomass of trees, made a year apart, provide values for annual increment, but these rarely include roots, while ignoring the production of leaves, flowers, fruits, fine roots and secretions, most of which will usually have been lost to herbivores or to decomposition during the year. Assessment of **net primary production** (NPP, which represents the amount of assimilated organic matter remaining after the requirements of plant respiration have been met) should include estimates of these various components, which are added to the annual woody increment, together with values for the shrub, herb and ground layers. Consumption by herbivores, turnover of fine roots and losses in secretions are normally excluded from such calculations, which again often involve logarithmic regressions, e.g. NPP on trunk girth.

Estimated values for NPP in woodlands are shown in Table 9.1 and Figure 9.1. The mean values in Table 9.1 show a general increase to-

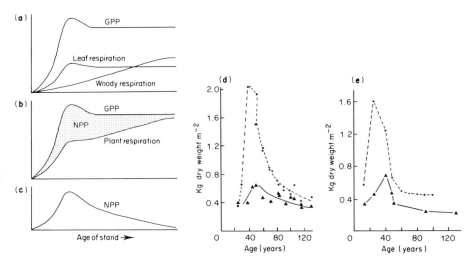

Figure 9.2 Relationships between productivity and age of stand. (a), (b), (c) Hypothetical curves showing changes in gross primary production (GPP) and net primary production (NPP), the difference representing respiration. (In Duvigneaud, 1971.) (d), (e) Net primary production (●) and annual litter fall (▲) in USSR. (d) Southern taiga spruce forests. (e) High oak forests, Voronezh. (From Rodin and Bazilevich, 1967.)

wards the equator, reflecting the influence of climate on production. The overlap between the ranges of the major forest biome types is partially explained by the effect of the age of stand. As shown in Figure 9.2, NPP increases during the early stages of stand development, but subsequently declines, as the green tissues gradually fail to synthesize enough material for respiration and growth, especially of woody tissues. The proportion of photosynthate respired progressively increases until, in theory, none is available for growth, when NPP falls to zero. Corresponding changes take place in the **production: biomass ratio** (P:B) and its converse, the **biomass accumulation ratio** (B:P). Mean values for NPP over a large area therefore depend in part on the relative distributions of climax and early successional stages.

Gross primary production (GPP), representing total assimilation of organic matter (i.e. NPP plus plant respiration), is extremely difficult to estimate, and there are relatively few published data (e.g. Kira, 1975; Reichle, 1981). Values range between c. 16 and 160 t ha⁻¹ y⁻¹, the highest being for Douglas fir in Oregon, followed by certain tropical rainforests in south-east Asia (UNESCO, 1978). Maxima for boreal conifers and for cool-temperate deciduous hardwoods seem to be c. 50 t ha⁻¹ y⁻¹. GPP has been shown to be related to leaf area duration (coefficient of proportionality = 0.85), assessed as leaf area index (LAI) times the

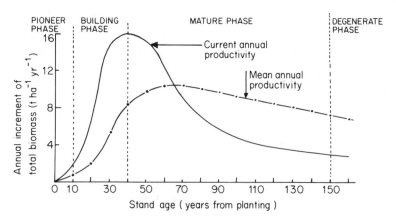

Figure 9.3 Productivity changes with time in an even-aged stand of Scots pine. Commercial plantations would be felled well before the degenerate stage. (Redrawn from Cousens, 1974.)

length of the growing season in months, for a wide range of broadleaved and conifer forests (Kira, 1975; Waring and Schlesinger, 1985). The proportion of GPP used in plant respiration varies between c. 30 and 75%, increasing with temperature and with biomass accumulation ratio (Whittaker and Marks, 1975). A 25-year-old Danish beech forest, losing c. 40% of photosynthate in respiration, had a comparable NPP value to that of mature lowland tropical rainforest, despite the GPP of the latter being more than double, due to respiratory losses of 75% (Kira, 1975).

Assessment of production in commercial forestry ignores many of the factors just considered, foresters being mainly concerned with the quality and rate of production of wood. This is for them an aspect of management and is discussed in section 10.4. Figure 9.3, which shows the annual increment of total biomass of *Pinus sylvestris* throughout the stand cycle, should be compared with Figure 10.5 which is concerned with volume increment of stemwood during the commercially viable part of that cycle.

9.3 Secondary producers: the herbivore subsystem

Values of biomass and production of woodland heterotrophs are even more difficult to estimate than for vegetation and have not been included in the synthesis of IBP woodland data (Reichle, 1981). However approximate, they are often used in attempting to quantify energy flow among consumers (for details of methodology, see Petrusewicz and Macfadyen, 1970; Phillipson, 1971; Whittaker, 1975). From figures published for the major forest types of the world, the total animal biomass

appears to be about a quarter of 1% of that of the world's forest vegetation. Only a small proportion of this biomass is represented by
vertebrates, which contribute less than 10% of the total in European
forests. The bulk is made up by invertebrates, most of which occur in
litter and soil.

Quantitative data on feeding by woodland herbivores refer mainly to
canopy-dwelling insects, especially phyllophagous (leaf-chewing) caterpillars and weevils. Some of these may occasionally cause widespread
and severe defoliation, as on oak and other hardwoods in many parts of
Britain in 1979 and 1980, but in general trees are subjected to chronic,
low-level herbivory. Evidence of such activity occurs as holes in leaves,
whose area is most readily measured at leaf-fall; in temperate broadleaved
trees such holes commonly represent about 5% of the canopy leaf area.
The corresponding consumption by insects (which may be less than the
final hole area, because of leaf expansion and necrosis subsequent to
feeding) is of the order of 1% of the net primary production of the trees
(Nielsen, 1978), and it is therefore usually assumed that the influence
on trees is negligible. Even if trees are more severely defoliated, certain
broadleaved species, notably oaks, often produce a second flush of foliage round about midsummer, although this in turn may be attacked,
e.g. oak by mildew and caterpillars of the Buff-tip, Phalera bucephala. This
lammas growth (sensu lato; section 6.7) is sometimes called regrowth
foliage, implying compensation for losses to herbivores, although the
relationship is far from clear (Crawley, 1983).

In July 1980 at Chaddesley NNR, Worcestershire, the canopy appeared
to be fully developed in individual oaks which had been completely
defoliated in May. Nevertheless, a study of Quercus robur at Wytham
showed that the summer-wood increment of individual trees was less
when caterpillars (mainly of winter moth and Tortrix) were abundant,
despite lammas growth (Figure 9.4a). Proportionate changes in growth
of five oak trees and in their associated caterpillar populations are related in Figure 9.4d. If a linear regression is assumed (but see Crawley,
1983), radial growth of summer wood would have been 73% greater
over the 8-year period if there had been no caterpillars. In energy terms,
such an increment loss is four times greater than the amount consumed,
indicating that lammas shoots do not fully compensate for the damage.
This is partially because of the timing of events: lammas growth in oaks
may not be complete until July, so that the LAI is less than it would
have been in undamaged canopy during the long days around midsummer. Consequently there is less photosynthate available for the
summer increment, which is dependent on the productivity of foliage
from May onwards (Figure 9.4b,c). In addition, some of the energy and
nutrients which would have been channelled into increment is expended
on lammas growth.

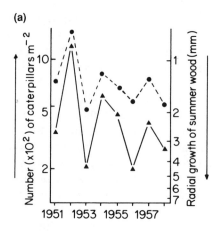

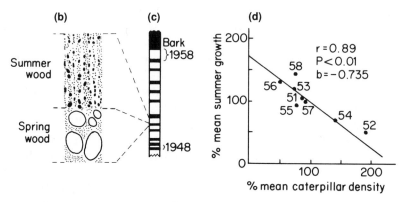

Figure 9.4 The effect of partial defoliation by caterpillars on timber production by *Quercus robur* in Wytham Wood, Oxford. (a) Abundance of caterpillars (▲) and, on inverted scale, radial growth (●) of summer wood by one of five studied trees, showing significant negative correlation (P < 0.01). (b) The appearance of summer and spring wood, the latter with very large vessels. (c) An enlargement of an increment core taken from a single tree, showing variation in size of annual rings. (d). Regression of proportionate change in summer growth and in caterpillar density, based on means for five trees. Each tree's growth (or caterpillar density) in each year was expressed as a percentage of the mean summer growth of that tree over eight years; the mean value, for the five trees, of each of the two percentages is plotted for each year. (From Varley and Gradwell, 1962.) A later, 20-year data set from these Wytham oaks suggests a much smaller and less significant effect on growth of summer wood (See Hodkinson and Hughes 1982).

Any surviving undamaged foliage, including lammas shoots, can enable reserves to be laid down which contribute to spring increment and to the flushing of buds; such trees can potentially survive successive defoliations, although with little increment, as evidenced by a series of narrow annual rings, for example in many parts of southern England in the period 1917–24. In fact these attacks leave trees more vulnerable; successive defoliations of oaks in Hayley Wood, Cambridgeshire, between 1916 and 1925, were followed by death of one in four trees from mildew, honey fungus or competition from other trees (Rackham, 1975). Defoliation, along with drought and atmospheric pollution, is one of the possible causes of crown dieback and stag-headedness, increasingly prevalent among Britain's ageing oaks.

Defoliation of beech by caterpillars of the pale tussock moth (*Dasychira pubibunda*) was found to have no measurable effect on incremental growth or production in a Danish site, probably because trees were attacked late in the summer, (Nilsson, 1978).

Deciduous conifers, such as *Larix* spp., also appear to be adapted to tolerate defoliation, which in effect brings about premature leaf-fall. Stands of larch in Holland have survived after defoliation for nine consecutive years by web-spinning larch sawfly caterpillars (*Cephalcia lariciphila*), but this species killed many British trees in 1978, whereas normally damage is masked by fresh growth in late summer. *Adelges laricis* appears to be largely responsible for dieback of European larch, especially in alpine provenances.

Timing is again important in **evergreen conifers**. Loss of needles may be less harmful if it occurs after the new buds have been formed, so that trees stripped of needles by pine looper caterpillars in late summer may be able to flush in the following year (but see section 6.6 for the increased susceptibility of such trees to attack by *Tomicus*). Pine beauty caterpillars, feeding in midsummer on all ages of needles, eventually kill the cambium so that buds fail to open. Among pine sawflies, larvae of *Diprion pini* also feed on various ages of needles, whereas *Neodiprion sertifer* leaves the current year's needles untouched; in Britain height growth may be checked by the first species, especially in young plantations, but in continental Europe both species are serious defoliators of pines. In North America the European spruce sawfly (*Gilpinia hercyniae*) feeds on older foliage, which has a lower photosynthetic capacity (Figure 2.15). It kills the crown of the tree from the bottom; no buds are killed directly and no adventitious buds arise. The young, photosynthetically efficient, foliage remains active for longer, and after a sawfly infestation the very small amount of foliage remaining is mainly new growth in the tip of the crown. Although severe defoliation of balsam fir by spruce budworm (*Choristoneura fumiferana*) can cause a decline in radial growth of as much as 90% in the following year, and death of some treetops within a couple of years, death of trees requires 4–5 consecutive years of almost 100% defoliation of new needles, which are the preferred food of larger larvae.

While loss of stem increment is usually proportional to the amount of defoliation, the actual outcome varies between the species, with different patterns of reallocation of resources following the disturbance of **source-sink** balances (Speight and Wainhouse, 1989; see also section 11.2).

Sap-sucking insects would be expected to have a greater effect on plant growth, per unit of energy ingested, than leaf-chewing species, since they remove not only photosynthate but also considerable quantities of water and nutrients. Components of injected saliva could affect growth rates, while some species may facilitate entry by pathogens (e.g. *Nectria* by *Cryptococcus fagisuga*). Among pest species, adelgids can defoliate conifers, as can the green spruce aphid (*Elatobium abietinum*) which was particularly abundant in Britain during the spring of 1989, following a mild winter. Premature leaf fall of the full complement of needles may check leader growth for at least two seasons.

Sycamores infested with *Drepanosiphum platanoidis* have smaller leaves than usual. Dixon (1971b) found that the average size of mature leaves in two 10 m trees was negatively correlated with the abundance of aphids in spring. The average population of aphids during a 7-year period was estimated to effect a 48% reduction in leaf area of one of the two trees, suggesting the possibility of competition between the requirements of growing leaves and aphids for translocated nutrients. However, estimates of the amounts of energy and nitrogen removed by the aphids (based on analysis of honeydew production) showed that these fell far short of the amounts represented by the reduction in leaf area: energy consumption, for example, was only one-third of the equivalent of the 'lost' leaf tissue. Dixon suggested that substances in aphid saliva, such as amino acids, indole acetic acid and phenolics, could reduce leaf growth, and a similar influence on xylem proliferation could provide a partial explanation for the observed reduction in the size of annual rings in infested trees, assuming that some is the result of reduced availability of photosynthate (cf. caterpillars feeding on oak leaves). In fact, the smaller leaves have a greater chlorophyll content, per unit area, than uninfested leaves, and infested leaves of saplings had a greater rate of net dry matter production. However, this is not reflected in the size of the annual rings, perhaps because this is a measure of the area of vessels required to supply a particular leaf area with water and salts. Whatever the reason, it has been estimated that the volume of stem wood could be increased by up to 280% per annum if there were no aphids.

Leaves of lime infested with *Eucallipterus tiliae* are of similar size to those of aphid-free plants, and there is no difference between the annual rings. Infested leaves contain less chlorophyll and fall earlier, suggesting damage by the aphids, and although aerial growth is un-affected, root growth may be completely inhibited; compensatory energy production is possible in the following year, when the leaves contain more chlorophyll, but only if aphids are absent (Llewellyn, 1972). Extra-polation of results from saplings indicates that the energy requirements of aphids on a mature lime (which may carry over a million aphids) are equivalent to 19% of net primary production (Llewellyn, 1975). Expressed in terms of ground area beneath the tree canopy, the energy consumed by aphids on lime is *c.* 20 times the estimated consumption by phyllo-phagous caterpillars on oak at Wytham.

The channelling of production into reproduction may be hindered by herbivores, either because destruction of foliage (e.g. by caterpillars on oak, Crawley, 1983) reduces seed production, or because the seeds are consumed. In a Danish beechwood an average of 36% of the annual production of beech-mast endosperm was consumed by seed worms (section 3.4), corresponding to less than 0.001% of net primary production (Nielsen, 1977). On the other hand, annual rings of *Fagus sylvatica* may be about half their normal width during a heavy mast year, showing how photosynthate is redistributed within the tree itself.

Seed production represents potential regeneration but in fact most of the energy in seeds or seedlings is often diverted into the production of heterotrophs. In the same Danish beechwood, heavy defoliation of ash saplings by the lilac leaf miner (*Caloptilia syringella*), combined with repeated removal of terminal buds by roe deer, resulted in negligible growth of saplings which, although of maximum height 0.5 m, were up to 36 years old (Nielsen, 1978). Quantitative estimates of consumption by vertebrate herbivores and their effects on seeds, saplings and mature trees are rare, but moose (*Alces alces*) were estimated to reduce produc-tion by young plants by 50% in a mixed forest near Moscow, mainly by decreasing growth rates rather than by direct losses due to consumption.

Whatever their effects on trees and other autotrophs, consumption by woodland herbivores probably rarely exceeds 10% of the net primary production. Even less is actually **assimilated**, and the proportion which becomes herbivore (or still less, carnivore) tissue is very small indeed. Values for assimilation/consumption efficiency (A/C) of woodland phyllophagous invertebrates are normally reckoned to be in the region of 20–30%, yielding 70–80% as faeces, destined for the decomposition subsystem. The corresponding value for vertebrate herbivores is about 50%, but in homoiotherms (warm-blooded animals) only 2% of this may pass into production, compared with 40% for invertebrates. In carnivores, A/C is typically in the region of 80%, but again homoiotherms 'lose' most of this energy in the form of respiratory heat.

Selected examples of **energy utilization** by woodland herbivores and carnivores are shown in Table 9.2. The values for *Operophtera* caterpillars

Table 9.2 The fate of ingested food in various woodland animals

Source	Species	Food	A/C*	P/A*
(a)	*Operophtera brumata* caterpillars	Hazel leaves	40	59
(b)	*Heterocampa guttivita* caterpillars	Hardwood leaves	14	40
(c)	*Drepanosiphum platanoidis* (aphid)	Sycamore leaves	9.5	44
(d)	*Mitopus morio* (harvestman)	Various arthropods	47–74	–
(e)	Salamanders	Insects	81	60
(f)	Shrews	Insects	90	2
(g)	Birds	Insects, fruits, seeds	30	2
(h)	Chipmunks	Foliage, seeds, insects	82	2

* From basic energy flow equations: C = A + Fu; A = P + R; where C = energy consumed (ingested), Fu = excreta, A = energy assimilated (i.e. digested and absorbed), P = energy in assimilated food used for tissue production, R = heat lost in respiration.

Data from: (a) Smith (1972); (b, e–h) Hubbard Brook Forest, New Hampshire; Gosz *et al.* (1978); (c) Dixon (1971a); (d) Phillipson (1960).

indicate that these short-lived leaf feeders, which represent the sole growth phase of the individual, convert their food into body tissue with minimal loss as faeces or as metabolic activity. The much lower assimilation efficiency of aphids reflects the necessity to ingest an excess volume of phloem sap, primarily to obtain enough nitrogen. Lime aphids may not be particularly productive, but they bring about a massive flow of energy, turning over the energy equivalent of their average standing crop nearly 500 times during the year; most of this sap is siphoned on to leaves or litter below as honeydew.

Among the major vertebrates at Hubbard Brook, production/assimilation, P/A, values for shrews, birds and chipmunks indicate how little of the assimilated energy ends up as production in homoiotherms, in contrast to the situation in salamanders, whose metabolic rate is reduced during the winter, and which are more efficient converters of food into biomass. The maintenance of a high and relatively constant body temperature throughout the year in an animal as small as a shrew (adult weight 15 g) necessitates catching about 140 items of invertebrate prey per day, or one every 5 minutes during 12 hours' activity; this is equivalent to an intake of about one kilocalorie (4.2 kJ) of food energy per gram body weight per day.

The proverbial slowness of sloths reflects a low metabolic rate, which is further reduced at night when their body temperature is lowered (Jenik, 1979). Adult three-toed sloths (*Bradypus tridactylus*) spend most of their time among tree foliage, moving an average of 38 m a day. An adult consumes *c.* 5 g of fresh foliage per kilogram of body weight per day, an intake far below that needed by other mammals, such as monkeys, in tropical forests. Low levels of food consumption are undoubtedly among the factors which enable quite large sloth populations to exist in fairly limited territories.

Within the whole woodland ecosystem the amount of energy represented by production of herbivores and carnivores is far less than the 10% of net primary production which enters the grazing chain, most of which is 'lost', either from the chain, as faeces, or from the whole ecosystem, as respiratory heat. However, faeces, mucus, cast skins and dead bodies add to the decomposition subsystem a contribution which, although small compared to plant necromass, may be qualitatively and temporally important to particular decomposers.

9.4 The decomposition subsystem

Quantitative estimates of woodland necromass and its turnover are bedevilled by sampling problems, especially for roots and woody material, and also by the varying usage of terms such as litter, humus and soil

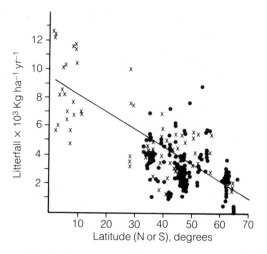

Figure 9.5 Litterfall for broadleaved (x) and needle-leaved (●) forests in relation to latitude. The regression line refers to broadleaves only. (From Vogt *et al.*, 1986.)

organic matter. Most data refer to input in the form of aerial plant litter, thus ignoring roots, standing deadwood, exudates, leachates, animal remains and frass. If all contributions from plants are included, then annual necromass input is closely related to net primary production; in climax forest, the two values theoretically differ only by the amount of plant production dissipated by the herbivore subsystem. Estimates of **annual litter fall** range from *c.* 1 t ha^{-1} in arctic-alpine forests to as much as 30 t ha^{-1} in certain tropical forests. A typical value for a temperate deciduous woodland is *c.* 4 t ha^{-1} y^{-1} (for reviews, see Bray and Gorham, 1964; Reichle, 1981; Anderson and Swift, 1983; Vogt *et al.*, 1986).

Although this tendency towards less litter at higher latitudes (Figure 9.5) follows the gradient of primary productivity, the role of roots could complicate the issue in terms of total necromass. In fact, the few estimates that have been made of the turnover of fine roots suggest that, at least in temperate woodlands, the input of organic matter from roots exceeds that from aerial litter by as much as three-fold (Harris, 1981; Waring and Schlesinger, 1985).

Litter input increases during the initial ageing of a stand (Figure 9.2), but once the canopy has closed there may be little difference in litter production by stands with very different densities or ages of trees. Variation from year to year within a particular site (Figure 9.6) results from the influence of factors such as the weather, pests and diseases on primary production; heavy investment in fruiting (e.g. mast years) or

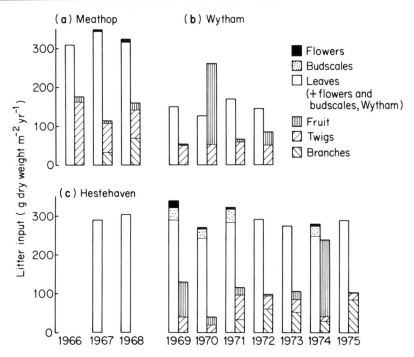

Figure 9.6 Annual variation in litter input in three temperate broad-leaved woodlands. (a) Meathop, Cumbria, England: oak, ash, birch, hazel and syca-more. (Data from Sykes and Bunce, 1970.) (b) Brogden's Belt, Wytham, Oxford, England: *Fagus sylvatica*. (Data from Phillipson *et al.*, 1975.) (c) Hestehaven, Denmark: *Fagus sylvatica*. (Data from Nielsen, 1977.)

storm damage to trunks and branches make a substantial contribution to variation.

In cool temperate regions the most conspicuous seasonal variation in leaf-fall occurs in deciduous trees, with major input in the autumn ('the fall'). Storms or herbivores can cause the addition of green leaf material to the litter, and green leaves may also accompany the abscission of inflorescences (e.g. *Castanea sativa*). Temperate evergreen species typi-cally shed leaves throughout the year, either irregularly (e.g. *Picea abies*), or with a peak in spring (e.g. *Quercus ilex* in the Mediterranean region, Figure 9.7a) or autumn (e.g. *Pinus sylvestris*), while *Cryptomeria japonica* is an example of a bimodal species in this respect. Certain tropical species also show bimodal behaviour, the leaves of *Terminalia catappa* being com-pletely and quickly replaced by fresh growth twice a year. Even in those tropical rainforests where rainfall is almost uniformly distributed throughout the year there are numerous species which are deciduous, although some may be bare of leaves for only a few days. Leaf-fall is

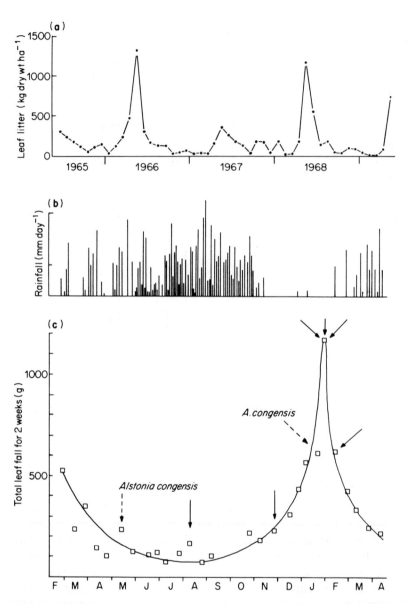

Figure 9.7 Seasonal aspects of litter production. (a) Leaf fall in the evergreen *Quercus ilex* at Roquet, near Montpellier (from Lossaint, P. and Rapp, M.; in Duvigneaud, 1971. © UNESCO 1971. Reproduced by permission of UNESCO.) (b), (c) Rainfall and leaf fall in a tropical forest at Ibadan, Nigeria. Arrows indicate maximum leaf-shedding for seven tree species, including bimodal *Alstonia congensis* (From Madge, 1965.) The cannon-ball tree (*Couroupita guianensis*) in an even better example of a bimodal litter shedding distribution. Twice a year it loses all its leaves in 3–4 days, only to replace them completely within a week.

typically continuous in tropical forests, often with a sequential pattern of different species (Figure 9.7c).

Input of non-leaf litter, especially bud scales, flowers and fruits, is also highly seasonal. The timing of litter fall is of significance, not only to the protection and development of seeds, but also to the activities of the decomposer community and the timing of nutrient release. The contribution made by the different parts of the plants to total litter input varies widely with species and age, the percentage of non-leaf litter tending to increase as stands mature. Coppicing rejuvenates stands, so that leaf-litter production may be unusually heavy under coppice. Although fruits sometimes constitute nearly half the litter fall, woody remains usually make up the bulk of non-leaf litter, which itself forms on average c. 30% of the litter. In terms of necromass, woody tissues probably represent a greater proportion than this; Swift (1977a) has suggested that twigs and branches may contribute c. 40% of aerial production, with further input, so difficult to quantify, from standing dead trunks and dead roots. The percentage of aerial litter stemming from the field and ground layers may be as high as 50%, but 10% is probably a more typical value.

Whereas annual production of necromass is assumed to increase from the poles to the equator, the standing crop of the forest floor to which it is added generally shows the reverse, implying a low organic matter content for tropical forest soils in general. The latter, in fact, show a wide range of values (Anderson and Swift, 1983), but the conventional wisdom is that plant litter decomposes more rapidly in the humid tropics than elsewhere.

Generally, quantitative estimates of decomposition are made on aerial litter, leaves or wood, rather than on total necromass, although net primary production can be used as a measure of total input in climax woodlands with negligible herbivory. Decomposition often refers only to visible depletion from the forest floor, i.e. until components lose their identity, thus ignoring the later, and usually slower, stages of decay of fine fragments and humus. Estimates of turnover coefficients and turnover times (k and $1/k$), based on ratios of input to necromass standing crop, depend on the latter being in a steady state; in reality, even in apparently mature woodland there can be considerable variation, both temporal and spatial, in the amounts of necromass present, so necessitating long-term studies of individual sites. Mean litter turnover times from the IBP Woodlands Data Set (Cole and Rapp, 1981) range from 353 years for boreal conifers to 17 and 4 years for temperate conifers and broadleaves, respectively; in certain tropical rainforests values may be less than 6 months (Anderson and Swift, 1983).

Some of the overlap between tropical and temperate values is a result of variation in resource quality, especially chemical composition, between the various types of necromass, but decomposability will also depend on site variables, such as moisture regimes, soil fertility and the composition of the decomposer community. By considering comparable sites along a tropical-temperate gradient, Jordan (1989b) concluded that the generalization that process rates, including decomposition, are higher in the tropics, seemed justified.

The use of mesh bags to assess the quantitative role of the various agents of decomposition has already been discussed (section 8.5). Alternatively, the contributions made by different groups of organisms can be estimated in terms of one or more of the components of the basic energy-flow equations, referred to in Table 9.2. Laboratory and field estimates of the various parameters, and their extrapolation to woodland populations, including IBP sites, are reviewed and critically analysed by Petersen and Luxton (1982). Soil animals generally consume the equivalent of 1–10% of their dry body weight of food per day, and for most species investigated the proportion of the litter input to their site which they are estimated to consume annually shows a similar range. Certain termites, Diptera larvae and oribatid mites have been reported to ingest 30–50% of the input to particular sites, and mull sites are characterized by a discontinuous litter layer for part of the year, implying removal of much of the input, usually by lumbricid earthworms. In some cases, studies have suggested that there may only be enough litter to satisfy the requirements of one or two groups of animals, e.g. lumbricids (Satchell, 1967), but not all of them, let alone the fungi and bacteria. This apparent paradox is partially explained by the widespread incidence of *coprophagy*. This involves the reworking of litter which is ingested but only partially digested, the resultant faeces being re-ingested, with the possibility of repetition of the process through a succession of organisms (including coprophilous members of the microflora). This reprocessing can occur only if digestion is less than perfect, as is implied by the generalization that detritivores are wasteful feeders, passing out 80–95% of their ingested food as faeces. In fact, the range of assimilation efficiencies (A/C) reported in the literature is very wide, from 1 to *c.* 70%. It used to be thought (Harding and Stuttard, 1974) that microphytophagous feeders had considerably higher values than macrophytophages, but such generalizations are no longer tenable, although oligochaete worms seem to be generally less efficient than other groups. Variability of data for the same species may partially reflect the quality of the food offered: values of 69% for *Steganacarus magnus*, compared to earlier estimates of 10–15%, could be due to differing degrees of fungal conditioning of their leaf litter food. Whatever their efficiency, many detritivores concentrate on a particular food source for considerable periods of time (months in the case of woody substrates such as beech cupules), producing veritable dung heaps, ripe for further exploitation.

Secondary production of soil animals, representing tissues of growing individuals or reproductive products, is usually assessed as the difference between the energy assimilated and that lost in respiration. Ecological growth efficiencies [production/consumption (P/C)] are usually less than 20%, while the production/respiration, P/R, ratio is generally low (<0.5) except at climatic extremes and among certain populations of micro-arthropods, tipulid larvae and millipedes, where food reserves are apparently built up against times when food is scarce. The majority of values of $P/\overline{B}$ (relative productivity) lie between 1 and 2, the chief exception being nematodes, with ratios of 4–7. Large oribatid mites, with greater fecundity, can be separated from others in terms of this production turnover, the threshold value of $P/\overline{B}$ being 2.5.

Because of the relative ease of determining respiration rates in the laboratory, and since it usually represents the most important pathway for energy loss, respiratory metabolism is the most widely used measure of a population's contribution to energy flow. Regression equations based on such determinations for various species over a range of temperatures and individual body weights can be used in conjunction with quantitative estimates of the composition of woodland populations sampled throughout the year. Numerous data relating to particular groups now exist, but only in relatively few cases do we have such information for whole detritivore communities (or at least for the majority of groups present). Petersen and Luxton (1982) present details for several woodland sites, all from temperate regions (Table 9.3). Oligochaete worms generally make the greatest contribution to community respiration in these sites, but with varying proportions from the component families, Lumbricidae and Enchytraeidae. The one exception is the *Liriodendron* site at Oak Ridge, where mites contribute almost half the total; otherwise, either mites or nematodes are usually in second place, after the segmented worms. These data lend some support to the generalization that the scarcity of larger decomposers, such as lumbricids, woodlice and millipedes, in mor (as compared to mull) sites, is compensated by relatively greater activity among the smaller detritivores.

Comparable data from non-temperate sites are virtually non-existent. It has been suggested that values for the biomass of soil animals are maximal in temperate and subtropical hardwood zones, declining towards the taiga and in tropical forests. Contrary to earlier hypotheses concerning latitudinal variation in biomass, Anderson and Swift (1983) emphasize that rainforests generally appear to have a lower soil fauna biomass than temperate deciduous forests; litter-bag studies of decomposition suggest compensatory activity among the soil fauna and microflora in terms of the overall pattern of decomposition in different sites. However, until figures for community metabolism are available from such sites, ideas about relative contributions must remain speculative. Meanwhile, on

Table 9.3   Respiratory metabolism (kJ m^{-2} y^{-1}) of major groups of soil animals in IBP woodland sites (from Petersen and Luxton, 1982)

	Deciduous mull			Deciduous mor	Coniferous mor
	Meathop UK	Hestehaven Denmark	Oak Ridge USA	Netherlands	S. Finland
Lumbricidae	654	150	563	–	66
Enchytraeidae	651	62	107	251	42
Nematoda	237	42	596	209	16
Mollusca	2	?	15	–	–
Collembola	12	19	134	15	11
Acari	15	32	1303	54	24
Diptera larvae	99	30	125	13	11
Diplopoda	2	?	22	8	–

the basis of a handful of estimates from temperate forests (including Meathop, Figure 9.9), we are left with the conventional wisdom that 80–90% of total decomposer respiration is accounted for by the microflora, implying minimal direct input from the soil fauna to community metabolism.

9.5   Energy flow through woodland ecosystems

As a result of extensive research, especially during the IBP, certain generalizations can be made about the importance of particular pools and fluxes of matter and energy in a range of forest types, especially from temperate and boreal regions. However, with the exception of a few long-term, intensively studied sites, such as Hubbard Brook Forest, New Hampshire, we are far from fulfilling the expectations of Tansley (1935) and Elton (1966) that in future ecologists would quantify the dynamic relations between all the components of a particular ecosystem. It is therefore conceivable that certain assumptions incorporated into simulation models, e.g. of tropical rainforests, may prove inadequate at a time when ecosystem studies are increasingly moving from a descriptive to a predictive phase (Waring, 1989). Predictions of the effects of perturbation are particularly relevant in an era of forest decline, deforestation and probable climatic changes (enhanced global warming).

The following examples from temperate woodlands are incomplete as ecosystem studies, but give some idea of the relative importance of the different subsystems in terms of energy distribution.

Energy flow values for Hubbard Brook Forest, New Hampshire, are

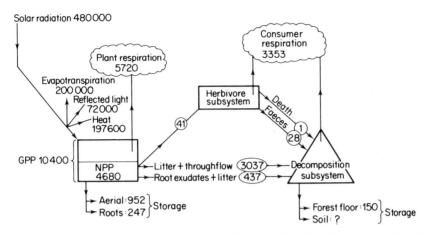

Figure 9.8 Energy flow diagram for an undisturbed hardwood area in Hubbard Brook Forest, New Hampshire; values in kcal m^{-2} yr^{-1}. (Data from Gosz *et al.*, 1978.)

summarized in Figure 9.8. Biomass was still being accumulated in this subclimax forest, so that only 72% of NPP was added to the decomposition subsystem, 84% of this in the form of aerial litter; there was very little information regarding the decomposers.

Details of energy flow are available for two English woodlands, relating particularly to the herbivore subsystem and the decomposers, respectively. Data for Wytham Wood (Figure 9.9) were derived largely from studies of population dynamics of various herbivores and carnivores (sections 7.2 and 7.4), from which estimates were made for consumption and production. Consumption by caterpillars was equivalent to *c.* 1.4% of the NPP of trees, but the quantitative importance of other herbivores was not assessed. Secondary production declined markedly as energy passed through the trophic web: 40 kJ m^{-2} y^{-1} for *Tortrix* and *Operophtera* caterpillars, 1.2 for mice and voles, 0.17 for titmice and 0.01 for tawny owls which, with weasels, were the top carnivores. This value for owls represented 0.00004% of the NPP of trees and shrubs, compared with 50% which entered the decomposition subsystem.

Studies at Meathop Wood, the IBP site in Cumbria, concentrated on primary production and decomposition (Figure 9.9). Intake by herbivores was equivalent to 0.01% of the NPP of trees, indicating the lesser importance of this subsystem than at Wytham. Annual increment represented 46% of NPP and input to the soil 54%. Assessments of decomposer respiration indicated the importance of segmented worms among the animals, but estimates of microbial activity should be viewed with care since they are particularly liable to error.

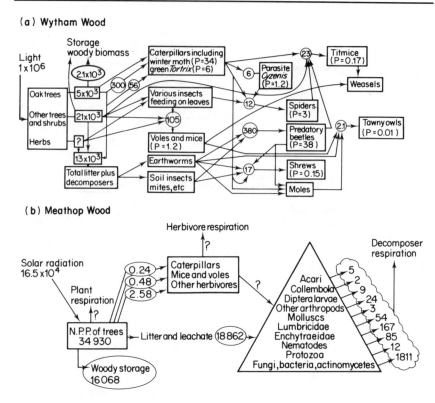

(a) Wytham Wood

(b) Meathop Wood

Figure 9.9 Simplified energy flow diagrams for two English broadleaved sites (kJ m^{-2} yr^{-1}; values in boxes are for net production, those in circles are for energy input, and those in 'clouds' for respiration). (a) Wytham Wood, Oxford; there are no data for decomposers. (b) Meathop Wood, Cumbria; data are not available for carnivores. (From The Open University, *Energy flow through ecosystems* (S323 Unit 5: Whole ecosystems). © 1974, The Open University Press. With additional data for Meathop decomposers from Satchell, in Duvigneaud, 1971; see Table 9.3 for amended values).

9.6   Nutrient cycling in temperate, boreal and tropical woodlands

As a forest develops, nutrients are fed into the vegetation/soil system from the atmosphere and by the weathering of rocks, and are in turn lost to the atmosphere, to streams and to groundwater. Nutrients are gradually accumulated by this system and tend to be distributed among various compartments in ways considered to be characteristic of particular biomes. For example, of the total amount of essential nutrients such as nitrogen, phosphorus and potassium, the proportion in the above-ground vegetation increases on passing from boreal to tropical latitudes, with a concomitant decrease in the relative importance of the forest floor. This leads to the generally accepted, and much repeated, view that in tropical forests most of the nutrients are stored in the plant biomass (by implication mainly aboveground), in contrast to other regions where the major pool is in the soil. Comparisons are complicated by definitions of compartments and of total amounts: some authors ignore the soil beneath the forest floor, while soil phosphorus, for example, may be quoted as available or total. A survey of data from two dozen tropical and temperate forests by Jordan (1985) revealed that the proportion of calcium and potassium stored in vegetation and litter ('the biomass') was very high in tropical rainforests, but mainly because of low levels in the soil. Corresponding data for nitrogen showed that in all cases there were larger stocks in the soil than in the biomass, supporting the contention by Waring and Schlesinger (1985) that in most forests the largest nutrient storage pool is in soil organic matter.

These major, long-lived nutrient pools are of particular relevance to perturbations, such as clear felling and climatic changes. Here we are largely concerned with the more rapid turnover, at the seasonal level, of nutrients moving between soil, vegetation and forest floor in particular woodlands, the so-called **intrasystem transfers**.

Although a certain amount of nutrient exchange occurs between foliage and the atmosphere, the major pathways involve the soil–root interface. Ions reach the root surface either along concentration gradients or in the mass flow of water associated with transpiration. Mass flow often fails to meet plant demands for nitrogen, phosphorus and potassium, whose supply is therefore dominated by diffusion (Waring and Schlesinger, 1985). Foraging for scarce nutrients, notably phosphates, depends on the development of permeating roots; the actual uptake of many ions is an active process, requiring energy.

Foraging efficiency can be increased, especially in soils of low fertility, by increasing the root: shoot ratio, and by supplementing the interface area by the hyphal component of mycorrhizas, the symbiotic associations of fungi and roots which probably exist in the majority of woodland plants. Most trees in northern temperate regions, together with southern hemisphere genera such as *Nothofagus* and *Eucalyptus*, plus members of various families in the tropics, including all dipterocarps, possess sheathing (ectotrophic) mycorrhizas. These develop as a weft of hyphae over the surface of short fine roots, in which the production of root hairs is suppressed (Jackson and Mason, 1984). Hyphae radiate out from the sheath into the soil either as individuals or, in some species, as aggregated cords or rhizomorphs. Inside the root, a complex system of branching hyphae (the Hartig net) constitutes an enormous area for exchange of materials with the cortical cells. An individual tree may support several species of fungi, many of which give a characteristic form to the infected roots; those of beech, birch and many conifers are often much branched, but others are club-shaped. Several of the fungi involved have large fruiting bodies, such as those of the conspicuous fly agaric (*Amanita muscaria*, often associated with birch or pine), ceps (*Boletus* spp.) and earthballs, or of the more elusive truffles (*Tuber* spp.).

Vesicular–arbuscular (V–A; sometimes called endotrophic) mycorrhizal infections, which involve species of the zygomycete family Endogonaceae, do not normally affect root morphology. A coarse, branching mycelium in the soil is connected at penetration points to hyphae in the root cortex. These produce storage vesicles and intracellular arbuscules, whose fine branches are presumably the main sites of nutrient exchange. V–A fungi are associated with the majority of crop plants and tropical trees (especially those in mature forests, rather than early successional species), but also with sweet gum, the tulip tree, maples, alders and poplars.

Hyphae not only exploit a greater volume of soil than the roots alone, but also store nutrients which subsequently can be passed to the tree. This is especially significant with nutrients such as phosphorus, where availability in the soil solution varies seasonally. There is some evidence to suggest that certain mycorrhizas may be directly involved in decomposition (Alexander, 1989), but the majority seem to depend on their higher-plant partner for carbon compounds. The metabolic costs of this transfer may be high (15% of the net primary production of certain pines is diverted into mycorrhizal biomass) but this is presumably more than compensated by the more efficient utilization of soil nutrients.

Many species of tree seedlings have been shown to grow more quickly when infected with mycorrhizal fungi, especially in nutrient-poor soils. There is also evidence of reduced susceptibility to pathogens, such as *Phytophthora cinnamomi*, and of greater tolerance of adverse abiotic conditions, such as low pH, toxic chemicals and water stress.

The failure of many exotic pines in plantations, and of attempts to revegetate areas of industrial dereliction, such as mining spoil, highlighted the importance of the mycorrhizal complex. Natural infection can be supplemented by importing soil or seedlings from healthy stands, but increasing attention is being paid to the possibility of inoculating cultures of selected species of ectotrophic fungi into nursery beds or the compost of containerized seedlings. This may be particularly beneficial in the establishment, for example, of Sitka spruce on deep acid peats.

Nitrogen is another element which often limits forest production, and which can also be taken up by mycorrhizas. Ammonium ions, which are less mobile than nitrates, and also potentially toxic at relatively low concentrations, are the preferred source for many forest species, especially in sites where microbial nitrification is inhibited.

Because they represent by far the greatest proportion of the biomass of a mature tree, woody tissues of trunks, branches and roots amass the bulk of the nutrients accumulated over a lifetime. The actual concentrations of nutrient elements are generally very much lower in wood than in leaves, fine roots and reproductive structures, and it is to these particularly active regions that most of the nutrient uptake is translocated. Internal nutrient cycling is closely linked to the growth, death and replacement of these organs, for which most information relates to leaves.

Changes in nutrient concentration during leaf expansion are complicated by the build-up of carbohydrates, including cellulose, but nitrogen, for example, is required for the synthesis of chlorophyll and of photosynthetic enzymes; root enzymes include those involved in the active uptake of nutrients. After expansion, the concentrations of certain elements, such as nitrogen, phosphorus and potassium, often peak during the summer, but then decline towards autumn, whereas calcium typically continues to increase throughout the growing season. Some of this decline is due to leaching from the leaf surface, especially of highly mobile ions such as potassium, which may be hastened during leaf senescence and also by the activities of defoliators. The remainder represents nutrients which are withdrawn from the senescing leaves and translocated into perennial tissues, such as living wood. This process of nutrient conservation provides a store, within the tree, for subsequent flushes of foliage.

Rainwater passing through the canopy (throughfall) and down the branches and trunks (stem flow) may become enriched in nutrients leached or exuded from within the foliage, as well as from pollutants scavenged by the canopy. Although the vast majority of the aerial input of leached nutrients is in throughfall, stem flow constitutes a more concentrated source around the base of the tree. Below ground inputs from root leachates and exudates are usually ignored when quantifying nutrient cycling, and the same applies to the contribution of fine roots to the necromass nutrient pool.

The few studies which have taken into account fine roots, including mycorrhizas, indicate that their turnover can comprise the largest input of nutrients to the soil. Otherwise, most data relate to aerial litter, especially leaves, where seasonal patterns linked to a peak abscission period may be complicated by summer storms or by herbivory, occurring before nutrient conservation has begun. Significant contributions to nutrient inputs may also come from the ground flora and from non-leaf litter, such as bud scales, flowers and fruits. Woody necromass, whether as a standing dead hulk or as fallen branches or trunks, has a low nutrient content per unit volume, but its sheer bulk constitutes an appreciable nutrient pool within a woodland, from which nutrients are normally released only slowly.

Studies of decomposition of branches in the forest floor enable us to

Table 9.4 Nutrients in decomposing hardwood branches from Meathop Wood, Cumbria (from Swift, 1977b)

| | Concentration (μg cm^{-3}) | | | |
	N	P	K	Ca
Living branches	1057	70	767	1935
Wood plus decay fungi	1304	54	141	3871
Wood invaded by fauna	710	28	21	1777

visualize some of the interactions between biotic and abiotic factors in this subsystem (Table 9.4). The steady decline in potassium content presumably reflects leaching, which was probably accelerated by the tunnelling activities of tipulid larvae. The reductions in all elements which followed invasion of the wood could be partially due to subsequent migration of animals into the surrounding litter. Conversely, the increases in nitrogen and calcium which accompanied fungal attack could imply translocation via hyphae from richer sources, fungi being well known for their abilities to accumulate various elements (Wallwork, 1983). At the time when animals began to invade branches decayed by a single species of fungus, nearly all the nitrogen and phosphorus were within the hyphae, which then acted as a nutrient sink. Absolute increases in nitrogen content, commonly observed in decomposing litter, may sometimes indicate fixation of atmospheric nitrogen, for example by bacteria inside logs.

Changes in carbon: nutrient ratios during decomposition, recorded from a wide range of litters, including needles and broad leaves, indicate that while carbon is lost through microbial respiration, nitrogen, phosphorus and sulphur are incorporated into microbial biomass. Potassium, calcium, magnesium and manganese are not normally in limiting supply, and so are released into the soil solution, although calcium may sometimes be immobilized as oxalate in hyphae.

The proportion of the underground nutrient pool immobilized in microbial biomass is generally small (Anderson and Domsch, 1980). Since the biomass of detritivores is orders of magnitude smaller, their role would appear to be insignificant. Wallwork (1983) reviews the situation in oribatid mites and suggests that release of nutrients from microbial tissues may be a key industry among fungivorous oribatids and Collembola. The stimulation of losses of mobile ions such as potassium and calcium from litter bags with animal access (Seastedt, 1984), normally attributed to greater leaching, could also involve stripping of associated microflora. Evidence of faunal–microbial interactions has been explored by Anderson and his co-workers, with particular reference to

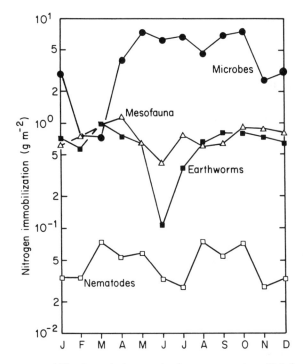

Figure 9.10 Immobilization of nitrogen in decomposers in a *Liriodendron tulipifera* forest floor at Oak Ridge, Tennessee. The period of maximum root growth (January–March), when the roots absorb most nitrogen, coincides with minimal microbial immobilization. (From Ausmus, Edwards and Witkamp, in Anderson and Macfadyen, 1976.)

nitrogen (Anderson *et al.*, 1985). Microcosm experiments revealed that the presence of millipedes increased losses of N from oak/beech litter by as much as 15 times compared to controls. This was probably largely due to unblocking of microbial sinks, although other experiments also implicated stimulation of microbial activity, as well as the animals' own excretion. Whatever the processes, the end result was that most of the liberated nitrogen was taken up by tree roots, rather than transferring to the humus pool, where turnover rates are very slow. Finally, the seasonal distribution of nutrients is also of significance. Studies at Oak Ridge showed that maximum root growth in a *Liriodendron* forest occurred between January and March, this period of maximal uptake of nutrients coinciding with minimal microbial immobilization. Conversely, losses to groundwater during periods of low uptake were minimized by high immobilization (Figure 9.10). Further buffering of the system was provided by soil animals, supporting the idea of their role in regulating the balance of nutrient availability.

Together, detritivores and microflora can provide innumerable pools and pathways for the conservation and timely release of nutrients; such complexity may impart a characteristic stability to nutrient cycling in temperate woodlands (Witkamp and Ausmus, 1976). In forests generally, a significant proportion of the energy captured in photosynthesis is allocated to heterotrophs, especially it now seems via fine roots, thus ensuring that nutrients are effectively recycled, and so helping to maintain the high productivity of forest ecosystems.

Temperate broadleaved woodlands

The IBP site at Meathop Wood, Cumbria, can be taken as an example to illustrate some of the features of nutrient cycling in this type of woodland. Provisional data are summarized in Figure 9.11, showing the major pools and fluxes, while more recent values are shown in Table 9.5. Nutrient conservation, involving reabsorption before leaf-fall, is particularly marked for nitrogen, phosphorus and potassium, constituting 19–32% of their respective annual requirements. Note that 87% of the requirement for nitrogen goes into production of foliage and incremental growth of larger roots. There are no data for fine roots, whose turnover probably contributes considerably more nutrients to the soil than are derived from aerial litter and leachates; studies in a *Liriodendron* forest in Tennessee suggest ratios for below: above-ground inputs of 2 and 7 for nitrogen and potassium, respectively. Above-ground returns are dominated by leaching for potassium and magnesium, but by litter for calcium, nitrogen and phosphorus, with *c.* 30% of the litter inputs of nitrogen and phosphorus being represented by material other than leaves, twigs and branches. The fall of budscales and catkins in spring, followed by peduncles, with or without acorns, accounts for the fact that in the nearby oakwood at Bogle Crag more phosphorus, and almost as much nitrogen and potassium, are added to the litter in the spring and summer as during the autumn. Significant contributions are also made by herbaceous litter, notably nitrogen and potassium from bracken. Elsewhere it has been shown that spring-green plants can improve the seasonal availability of important ions. When the leaves of *Allium ursinum* decay (Ernst, 1979), their large nitrate pool is immediately made available to summer-green plants without involving the action of nitrifying bacteria.

Nutrient conservation may be prevented if oaks are defoliated by caterpillars such as *Tortrix viridana* before nutrients are translocated out of the leaves. Complete loss of leaves at Meathop represents the removal of amounts of nitrogen and phosphorus, and to a lesser extent of potassium, equivalent to the requirements of 4 years' incremental growth.

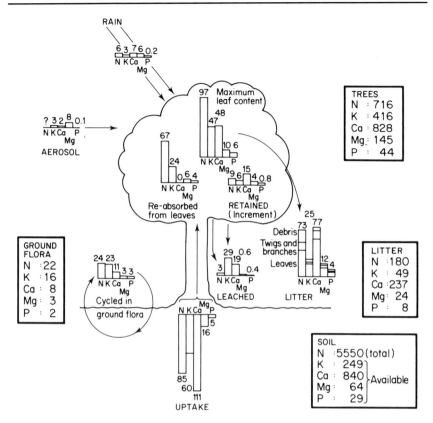

Figure 9.11   Mean nutrient contents (kg ha^{-1}; in boxes) and nutrient fluxes (kg ha^{-1} yr^{-1}) in Meathop Wood, Cumbria; the canopy is dominated by oak, ash and birch, with an understorey of hazel and other shrubs, and a rich herb layer. A brown earth with mull humus, developed on glacial drift, has a pH of 4.1–7.5, depending on the depth of soil over the underlying Carboniferous Limestone. The mean annual rainfall is 1240 mm. Net uptake by trees is the sum of amounts in increment and those cycled in litter and leachates (e.g. for N, 85 = 9 + 73 + 3). (Modified from Brown, 1974; see also Table 9.5).

Lammas growth necessitates further supplies of nutrients, which may drain considerably the trees' reserves, especially if defoliations occur in successive years. Studies in the Central Chernozem Reserve, USSR, revealed that litter decomposition rates increased by two-fold in *Tortrix* outbreak areas (Zlotin and Khodashova, 1980); this was attributed to more solar radiation reaching the forest floor, and to the inputs of more readily decomposable frass and green leaf fragments, with considerably higher contents of nitrogen and phosphorus than in autumn leaf fall. In this Russian woodland the flush of released nutrients was exploited by

Table 9.5 Distribution and circulation of nutrients in Meathop Wood, Cumbria; IBP data from Cole and Rapp (1981), after Waring and Schlesinger (1985)

	N	P	K	Ca
In plant biomass (kg ha^{-1})	509	34	369	781
Tree foliage	86	5	39	42
Trunks and branches	192	12	174	437
Tree roots	223	12	113	235
Field layer	8	5	43	67
Annual requirement, R (kg ha^{-1} y^{-1})	114	6.2	85.5	97
Tree foliage	86	4.6	39	42
Woody increment in				
Trunks and branches	6	0.3	5	13
Roots	13	0.6	6.5	14
Leaching				
Throughfall	8.8	0.7	30	23
Stem flow	0.4	0.03	5	5
Annual uptake, U (kg ha^{-1} y^{-1})	92	4.2	65.5	138
Woody increment	19.4	0.9	11.5	27
Leaching	9.2	0.7	35	28
Aerial litter fall	63.5	2.6	19	83
Reabsorption (R–U)	22	2	20	–
Reabsorption/R(%)	19	32	23	–
Foliage and root growth/R(%)	87	83	53	58
Litter fall/total return(%)	87	79	35	75

grasses rather than by the oaks, which appear not to have been compensated for their losses.

At all IBP sites the annual uptake of calcium exceeds its requirement. At Meathop, calcium is taken up, shed in leaf litter without any reabsorption, and lost from the soil in greater amounts than any other nutrient. By contrast, phosphorus and nitrogen are tightly cycled, with negligible losses to drainage waters of phosphorus and only 12.6 kg ha^{-1} y^{-1} for nitrogen. The latter is only partially compensated by atmospheric inputs, but the deficit is more than made up by nitrogen fixation in the soil. Phosphorus deficiency is much more likely to limit growth, especially at Bogle Crag. The value for nitrogen reabsorption, as a percentage of total annual requirement, is similar to that of a mixed hardwoods site at Hubbard Brook, but lower than the average of 23% for the 14 temperate deciduous IBP sites listed by Cole and Rapp (1981); values ranged from 3 to 65%, compared with 8–44% (average 22%) for phosphorus. It should be noted here that reabsorption values in the literature, based on these IBP data, can vary with definitions of uptake and requirement, e.g. whether the latter includes leachates. The exceptionally low value of 3% for nitrogen was from an *Alnus rubra* site, presumably reflecting the ready availability of nitrogen fixed by symbiotic *Frankia*, which enables this species to continue photosynthesizing until the leaves freeze. Stands of red alder can fix as much as 80 kg ha^{-1} y^{-1}, compared to values of 0–3 for nitrogen fixation by free-living bacteria in most temperate forest soils.

Temperate conifers

Studies of nutrient cycles and budgets in coniferous plantations have been reviewed by Miller (1979) who, with his co-workers, has made a particular study of Corsican pine (*Pinus nigra* ssp. *laricio*) planted in 1928 on the sparsely vegetated Culbin sand dunes in north-east Scotland. As well as yielding data on the cycling of various nutrients, these studies have stressed the importance of the changes which occur at various stages of a rotation, and the implications of these for the use of fertilizers in forestry. This idea of changing relationships during stand development is, of course, not unique to conifers.

During the early stages, before canopy closure, nutrients are required not only for the rapidly increasing biomass, especially of foliage and roots, but also to establish the intrasystem cycle. Accumulation of capital in the young trees and in the litter-based humus is largely dependent on uptake of nutrients from the mineral soil, whose availability may be reduced through competition from species such as ling and bracken. With canopy closure, the rate of foliar accumulation declines, and so do demands on the soil, as cycles between trees and humus, and by nutrient reabsorption, become fully charged. Fluxes and accumulation rates of various nutrients needed to maintain maximal growth rates of 20 m^3 ha^{-1} y^{-1} in 40-year old Corsican pine are shown in Table 9.6, together with the major sinks and sources. These data indicate that half or more of the nutrients required for new growth are derived from senescing tissues, especially needles, from which they are withdrawn for temporary storage into twigs, roots and remaining needles, until required in the following growing season. In the case of potassium and magnesium, atmospheric inputs almost balance immobilization in new tissues and humus, so that the trees absorb very little of these elements from the mineral soil, which is itself low in these nutrients. On the other hand, immobilization of nitrogen and phosphorus in slowly decomposing litter implies that these cycles need to be continually recharged from the soil, uptake of phosphorus being enhanced by mycorrhizas. Nitrogen is

Table 9.6 Nutrient cycling within Corsican pine stand (40 years, height 11 m) in Northern Scotland (from Miller, 1981b)

| | Rate of process (kg ha^{-1} y^{-1}) | | | | |
	N	P	K	Ca	Mg
Overview					
Uptake	69	6.0	28	34	10.9
Accumulation in trees	18	2.1	11	12	3.6
Litter fall	51	3.9	11	21	4.3
Crown leaching	trace	trace	6	1	3.0
Accumulation in humus	12	1.5	1	5	1.6
Sinks					
In new needles	92	9.4	45		
To replace lost structures	20	1.7	4		
Net structural increase	26	3.0	13		
Total requirement	138	14.1	62		
Sources					
Uptake from soil	69	6.0	28		
Reabsorption from needles	61	7.2	32		
Reabsorption from other tissues	8	0.9	2		
Total	138	14.1	62		

particularly likely to become deficient during the later stages of a rotation, leading to mobilization of reserves from older tissues, whose sacrifice is marked by a progressive decline in incremental growth, a feature of the pines at Culbin. Although it is generally assumed that nitrogen fixation is negligible in British coniferous plantations, an appreciable contribution to nutrient budgets may be made by the fully developed canopy scavenging substances, including pollutants, from the atmosphere. Stands under maritime influence may derive much of their chloride and magnesium from this source, while various sulphur and nitrogen based components of acid precipitation are particularly efficiently intercepted by conifer canopies. For example, planting of Sitka spruce on 6×10^4 ha of moorland at Kielder in northern Britain has increased annual inputs of nitrogen from the atmosphere by 90% to a total of 23.4 kg ha^{-1}; uptake of gaseous NO_2, HNO_3 and NH_3 is encouraged by the aerodynamic roughness of the forest canopy, with deposition as cloud water being particularly significant above 400 m (Fowler et al., 1989). If the underlying soils are insufficiently buffered, this effect can lead to the acidification of freshwater systems.

Miller concluded that the high productivity of P. nigra on low-fertility soils at Culbin, comparable to that of agricultural crops at that latitude, depended largely on efficient internal conservation of nutrients, supplemented by atmospheric inputs and backed by accumulated nutrient capital.

Gosz (1981) reviews some of the implications of variations in nitrogen availability for processes in the forest floor and soil of coniferous ecosystems. Conifer species in nitrogen-deficient sites have a low uptake requirement and high internal redistribution, supplemented by longer persistence of their leaves. Shed leaves are not only low in nitrogen but also relatively rich in polyphenols, lignin and organic acids, a combination which is inimicable to decomposers. Microbial activity may be further suppressed by antibiotics emanating from abundant mycorrhizas which enable the conifers to exploit the low levels of nutrients resulting from slow rates of mineralization. This scenario is typical of mor humus sites, where slow cycling of nutrients is largely independent of the mineral soil. At the other extreme, sites which are high in available nitrogen are characterized by conifer species which require more nitrogen, much of which is probably absorbed from the mineral soil by non-mycorrhizal roots. Nitrogen levels in plant tissues are higher than in mor sites, with less internal redistribution, and litter is more readily decomposed and mineralized, so sustaining high levels of nitrogen availability in mull sites.

Such markedly different strategies may be obscured by reliance on mean values. IBP results from 13 temperate coniferous sites show that, on average, amounts of nitrogen taken up are identical to the trees' requirements, with no evidence of reabsorption before needle fall (Cole and Rapp, 1981). In fact, half of the sites resemble the Culbin Sands situation, with requirements greater than uptake.

Average IBP values for annual cycling of the major nutrients, whether calculated as uptake, requirement or return, indicate that rates in coniferous stands are roughly half those of deciduous sites. Cole and Rapp attributed this to the fact that most conifers retain individual needles for several years, rather than to nutrient reabsorption. Internal conservation may also be enhanced in conifers by lower losses to leaching (Parker, 1983).

Whatever the mechanism, **nutrient-use efficiency**, defined as net aerial production per unit uptake of nutrient, is on average higher in conifers than in deciduous stands for all major nutrients except phosphorus (Cole and Rapp, 1981). The cost of concentrating nutrients such as nitrogen into a smaller leaf area, associated with higher specific leaf weights for conifers, is greater than for deciduous species (Gosz, 1981), but is spread out over the longer life-span of persistent foliage, so contributing to a greater yield of photosynthate per unit investment of nutrient. The ability of conifers to produce a given amount of biomass from less nutrients may account for their success in nutrient-poor sites (see also section 2.6).

Boreal forests

The low productivity of boreal forests is a consequence of the relatively short period in the year when temperatures favour biological activity. This applies not only to photosynthesis but particularly to nutrient fluxes below ground, where the situation may be further exacerbated by high moisture levels in forest bog sites. Black spruce forests, dominated by *Picea mariana* and mosses such as *Hylocomium splendens*, are the most widespread vegetation type in the Alaskan taiga, and are considered to be the most nutrient-limited and least productive taiga forest type (Flanagan and Van Cleve, 1977). The correspondingly low inputs of litter are of very low quality, so that the forest floor has a higher concentration of lignin and lower concentrations of all macronutrients than any other forest type (Bonan and Shugart, 1989). Thick mats of mosses or lichens which characterize the forest floor of many boreal sites insulate the lower layers, while permafrost restricts the downward extent of biological activity, as well as preventing leaching losses. Poor litter quality, low temperatures and often acid conditions, all mitigate against decomposition and mineralization. A large proportion of the nutrient capital is thus bound up in slowly decaying organic remains; calculated turnover times for individual nutrients in certain sites run into thousands of years. Research in black spruce sites suggests that microbial activity, especially by fungi in the litter, leads to immobilization of nitrogen and phosphorus within hyphae. Subsequent increases in respiration rates of roots, which are concentrated deeper in the forest floor, are presumed to mark mycorrhizal uptake of nutrients, released during declines in microbial biomass. The role of microfauna, enchytraeids and microarthropods in such processes is unknown. Weathering is minimal at such low temperatures and atmospheric inputs are generally unlikely to be supplemented by anthropogenic pollution. Mosses absorb nutrients such as NH_4^+ from precipitation, while cyanobacteria (whether free-living or associated with mosses or lichens) are the most important nitrogen fixers; in both cases the nitrogen is only released after the characteristically slow processes of decomposition. Trees which are adapted to these very low levels of nutrient availability achieve high values for nutrient-use efficiency (Table 9.7). In the case of *P. mariana* this is correlated with the extreme longevity of needles (25–30 years) and with the fact that even the oldest needles make a positive contribution to the carbon balance.

Severe nutrient immobilization occurs in stands dominated by *Picea sitchensis* at Glacier Bay, Alaska (section 5.4). Pure stands of this species

Table 9.7 Nutrient-use efficiency: aerial productivity (kg ha^{-1} y^{-1}) per unit of nitrogen uptake (from Cole and Rapp, 1981)

	Sites	Mean productivity
Temperate deciduous	14	138
Temperate coniferous	13	184
Boreal coniferous	3	236

are very infrequent in South-east Alaska where *Tsuga*, which is abundant in mature forest, may play an important role in mixed stands by extracting nutrients from organic substrates.

Tropical forests

When considering the various patterns of nutrient cycling in woodlands it may be tempting to try to characterize each major forest type by a single, representative pattern. Such an approach ignores the diversity imposed by variables such as climate, tree species, age of stand and soil fertility. This is no less true of tropical forests, yet authors continue to present a stereotype, based on sites of exceptionally low nutrient status, as the typical tropical situation. Recent reviews by Jordan (1985) and Proctor (1989) exemplify some of the diversity, and reveal that the amount of research on the least fertile soils is out of proportion when they make up only a tenth of the total area of the moist lowland tropics. The important review by Vitousek and Sanford (1986) redresses the balance in the case of moist tropical and montane rainforests, for which the authors explore the influence of soil fertility on nutrient cycling. In summarizing their findings, we will emphasize those processes which are commonly considered to be tropical characteristics.

Site data were arranged into three major categories for the lowlands and one for montane soils. The moderately fertile category includes a wide range of soil types, which generally have the greatest agricultural potential. They often occur in regions of relatively low annual rainfall (1500–2000 mm) or are of volcanic origin, and together make up 15% of the total moist lowland tropical area. The associated forests are highly productive and have relatively high concentrations of the major nutrients in their leaves (e.g. mean values for the various species in each site of 2–2.5% nitrogen) and in litter fall. As in the lowland tropics in general, throughfall is relatively unimportant except for potassium, where it represents the major input to the forest floor; the sparse data available suggest that more potassium and calcium may be lost in throughfall in more fertile sites. Rapid cycling of carbon-bound nitrogen would be expected in the light of high litter turnover rates, which correlate with nitrogen concentrations in lowland litter over a broad geographical range (Anderson and Swift, 1983). Rates of nitrogen mineralization and nitrate production are higher than in temperate forests, supporting the idea of rapid circulation of large amounts of nutrients in these moderately fertile sites.

Oxisols and **ultisols** are the most widespread soil types of the lowland tropics, especially in Africa and the Neotropics, comprising almost two-thirds of the total. Both are strongly weathered iron sesquioxide-rich soils of moderate to strong acidity with low cation exchange capacity, formed in warm, moist to seasonally dry climates. Agriculturally they range from mildly to severely infertile. Tree leaves are moderately high in nitrogen (site means 1.3–1.9%) but low in phosphorus and calcium, values of 0.1–0.2% for the latter in two Venezuelan oxisol sites being the lowest ever encountered by Vitousek and Sanford. The low availability of phosphorus in these soils apparently results from intensive long-term weathering and from adsorption by sesquioxide clays. As an example of tight cycling of scarce nutrients, two neotropical sites had values of $c.$ 70% for retranslocation of phosphorus before leaf-fall. Detailed studies in a Venezuelan oxisol site suggest that annual inputs of phosphorus and nitrogen from the turnover of fine roots may be, respectively, 5 and 13 times as great as from fine litter fall, if we assume that there is no retranslocation of nutrients from these roots.

Montane rainforests develop on a variety of soils, but with a greater proportion that are geomorphologically young than in lowland sites. Moderate to high fertility is associated with volcanically derived soils or with input from rocks to shallow soils subject to frequent landslips. Nutrient levels in foliage are generally lower than would be expected in lowland sites of comparable fertility (e.g. 0.6–1.8% nitrogen), as are inputs of nitrogen to the forest floor. In a lower montane site in New Guinea one-quarter of this input occurred in throughfall, much apparently due to fixation in the phyllosphere (Edwards, 1982). Release of nitrogen from large quantities of organic matter in the soil was hindered not only by reduced microbial activity at temperatures on average 5–10°C lower than in lowland rainforests, but also by complexing of organic matter with soil colloids. Shortage of available nitrogen may therefore be characteristic of montane sites, in distinction to the lowland tropics, where phosphorus is more often limiting. Nutrients are also scarce in the ecotone between lower and upper montane forests, which often coincides with the base of the cloud cap. Anaerobic conditions in the sodden forest floor inhibit decomposition, leading to peat formation, which is further encouraged by the growth of *Sphagnum*. The process of nutrient immobilization may be reinforced by the growth of conifers with recalcitrant litter (Whitmore, 1990).

Soils with virtually no agricultural potential, mainly because of their exceptionally low nutrient status, include the **spodosol/psamment** group of sandy soils developed on old river terraces and in highly weathered upland sites. Heath forest is an example of associated vegetation, represented by Amazonian caatinga in the Rio Negro basin and by kerangas ('land that will not grow rice') in Borneo; it occurs on very acid (pH < 4.0) poorly buffered podzols, with a bleached silica sand layer which may be several metres thick (Whitmore, 1990).

Such sites epitomize the dilemma posed by potentially massive losses of nutrients through processes such as leaching, driven by high temperatures and rainfall, from soils with little or no input from weathering. In fact, recorded losses to drainage waters are minimal, implying the ultimate development, foreshadowed in oxisol and ultisol sites, of mechanisms to retain the scarce supply of nutrients within the plant–humus system. Various characteristics of foliage (none being unique to the tropics), which may aid nutrient conservation include slow turnover rates for evergreen leaves, drip tips, and low specific leaf areas of tough, thick leaves, which may counteract leaching. The deterrence of herbivores and pathogens by toughness or by secondary compounds may assume extra significance among communities on, for example, nutrient-poor white sand soils, where plants can ill afford to lose precious nutrients to consumers (Janzen, 1974). High levels of these compounds may also serve to conserve nutrients by retarding decomposition and inhibiting nitrification; litter turnover (k_L = 0.76, implying more than a year to disappear) in the caatinga site of the San Carlos project in Venezuelan Amazonia is slower than would be expected with average temperatures of 26°C and annual rainfall of 3500 mm. Tannins and other phenolics ('humic acids') percolate rapidly through the coarse sands, unimpeded by absorptive clays, and impart their stamp on the clear, almost lifeless blackwater rivers such as the Rio Negro (Jordan, 1989a).

Foliar nitrogen levels of vegetation on these sandy soils are generally lower than those on oxisols and ultisols, but low values for phosphorus overlap with those of more fertile sites. Despite high values for dry mass: nutrient ratios in fine litterfall, there is little evidence of particularly marked retranslocation. The relatively high value calculated for nitrogen-use efficiency in the San Carlos spodosol may indicate that nitrogen, rather than phosphorus, limits productivity in such sites (Jordan, 1989a).

The key components of the nutrient retention system are the least conspicuous, being situated underground. High values for root biomass and, especially, for root: shoot biomass ratios, which characterize vegetation on nutrient-poor soils world-wide, are the hallmark of the least productive tropical sites. In extreme cases root: shoot ratios may be as high as 7:1, and there is some evidence of a much greater biomass of functionally active fine roots in more infertile sites, where their biomass exceeds that of leaves. In all of the sites detailed by Vitousek and Sanford (1986) there is more nitrogen in the roots than above ground.

This large mass of roots is concentrated at the soil surface, sometimes as a discrete root mat, well placed to intercept throughfall and to prevent nutrients released from necromass from reaching the mineral soil and so becoming virtually unavailable. The efficacy of this absorptive barrier was demonstrated by applying ^{32}P to a root mat at San Carlos, and finding that only 0.1% reached the mineral soil (Stark and Jordan, 1978). Bypassing of this soil sink is facilitated by the ability of mycorrhizas to provide a throughway (**direct cycling**) from decomposition sites into roots, as well as by more circuitous pathways within the microflora/detritivore web.

Whitmore (1989) refers to the gradual dismantling of the twin pillars of conventional wisdom regarding nutrients in tropical rainforests: that most nutrients are in the above-ground biomass, and that decomposition rates are always higher than in temperate forests. In summarizing the wide range of tropical situations, we must not lose sight of one fundamental truth: the continuing fertility of these rainforests depends on maintaining the integrity of their living components, especially those below ground. This applies particularly to forests, such as those in much of Central Amazonia, whose nutrient cycles are largely independent of deeply weathered soil.

9.7   The influence of man

Sustainable utilization of forests requires careful consideration of the constraints of nutrient budgets, not only in terms of inputs and losses but of the key components of nutrient cycling in a particular site. Any imbalance or major disturbance may necessitate remedial action.

The disruptive effects of clear felling were dramatically demonstrated at Hubbard Brook, New Hampshire, where all the vegetation in one watershed was felled and slashed but not removed (Likens et al., 1970). Herbicides were applied for 2 years to prevent regrowth or colonization. Stream outflow from the cleared site during the following 3 years was about one-third greater than from a comparable control watershed largely because of a corresponding reduction in evapo-transpiration (which normally equalled 41% of precipitation). Losses of major cations were markedly increased, while nitrate-nitrogen was also lost rather than gained by the system. These changes reflected greater concentrations in the soil solution, nutrients no longer being taken up by the vegetation. Nitrifying bacteria, which are normally inhibited in forest soils, increased

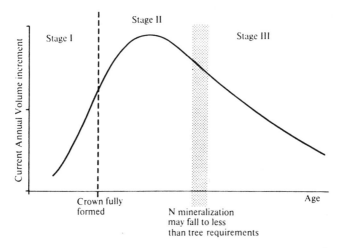

Figure 9.12 The three nutritional stages in the life of a coniferous forest stand (Miller, 1981a).

appreciably, the resultant nitrate ions being readily leached, whereas previously ammonium ions had been absorbed by roots or adsorbed to soil colloids. Increasing acidity, resulting from decomposition of slash, reduced the base exchange capacity of the soil colloids, so facilitating the leaching of potassium, calcium and magnesium.

The vital link in the cycle can be restored by natural regeneration or by replanting, while natural inputs may be supplemented by the use of fertilizers.

Fertilizer applications

On the basis of investigations into nutrient cycling and fertilizer response in pine, Miller (1981a) proposed three concepts to encourage rational use of fertilizers in forestry. Firstly, benefits normally accrue to trees rather than to site, so that permanent improvements in the nutrient status of a site are only likely if a disproportionately large amount of nutrient is applied, e.g. as sewage sludge, relative to nutrient capital already in the soil. Secondly, response in the form of enhanced growth is best described and costed in terms of time saved to achieve the desired crop, i.e. a reduction in rotation length. Thirdly, in the light of the changes in demand for nutrients with time three distinct nutritional stages can be recognized (Figure 9.12), although stage III, characterized by late rotation nitrogen deficiency, may not be reached in nitrogen-rich sites. Fertilizers are most likely to be of value during stage I, when the subsequent development of the stand is most susceptible to manipulation, and towards the end of the rotation, with a minor role in between, to accelerate recovery of the canopy from the effects of thinning or insect defoliation.

The nutrients most commonly supplied in fertilizers by foresters throughout the world are nitrogen and phosphorus, the latter being deficient in soils over much of Britain. *Thuja plicata* and *Abies grandis* require soils rich in calcium and magnesium, but calcium is rarely deficient in plantations of other conifers. Some broadleaves, such as *Eucalyptus grandis*, have a high calcium requirement. Local deficiencies of micronutrients such as copper and manganese are normally readily rectified.

McIntosh (1984) showed that analysis of foliar nutrient concentrations in the top whorl of *Pinus sylvestris* provided a sound guide as to which established stands should receive nitrogen fertilizer. Recently developed bioassay methods to detect phosphorus deficiency work well with feeder roots of pines and of *Picea sitchensis*. These laboratory tests are based on the rate of metabolic uptake of elements labelled with appropriate isotopes, e.g. radioactive ^{32}P: roots from deficient plants show a 'hunger response' with a greater rate of uptake. Binns *et al.* (1980) provide an illustrated guide to deficiency symptoms in the conifers of British forests. They emphasize that making good an evident deficiency of one element may induce another deficiency symptom resulting from an unsatisfactory new balance between the nutrient elements.

Quite mature trees may benefit; in one case 90-year-old beech responded very positively to NPK and lime. Higher responses are usually obtained on sites of lower quality (Figure 9.13), but even on the better sites the improvement obtained normally justifies the cost of fertilizer application.

In North America nitrogen is far more important in limiting tree growth than any other mineral nutrient element (Daniel *et al.*, 1979). This is as true for the northern hardwood forests as it is for Douglas fir in the Pacific Northwest. Western hemlock (*Tsuga heterophylla*) is, however, an example of a species whose growth is not normally improved by additional nitrogen. Nitrogen-tolerant species, i.e. those with modest soil nitrogen requirements (red and white oak, red maple and aspen), compete well on sites with low nitrogen status, while nitrogen demanders, such as white ash, the basswood and the tulip tree (*Liriodendron tulipifera*), are at an advantage on soils with abundant nitrogen. Engelmann spruce responds vigorously to ammonium nitrogen but not to nitrate. At high latitudes and altitudes the amount of nitrogen available to trees is frequently low because so much is 'locked up' in the litter and organic matter on the forest floor.

Fertilizer applications influence many aspects of the forest ecosystem. Understorey plants and 'weed trees' often grow much more vigorously

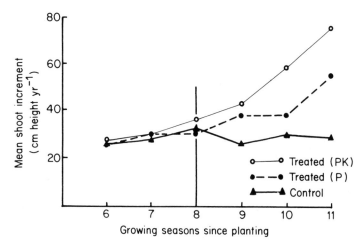

Figure 9.13 Response of Sitka spruce (*Picea sitchensis*) to remedial addition of fertilizer in the autumn of 1973, eight years after planting. The single aerial applications to the trees, whose canopies had not closed, were at the rates of 375 kg unground phosphate rock per hectare (P), or this amount of phosphate together with 200 kg potassium chloride per hectare (PK). Control values are for trees grown for eleven years without this remedial fertilization. All trees were on unflushed deep peat in Galloway, Southern Scotland. The cost of the fertilizer application, in terms of 'years saved' before cropping, was more than recovered within three years of treatment. The same was true of Sitka spruce on a variety of other soils at various elevations in this area. (Redrawn from McIntosh, 1978; by courtesy of the Royal Scottish Forestry Society.)

as a result, so that mechanical weeding or the application of herbicides becomes necessary. Insect populations are affected in complex ways, as are the pattern of disease in forest trees and the development of mycorrhizas. With young stands in particular the increase in total leaf biomass, and consequently in transpiration, may be sufficient to reduce the run-off into streams. The major consequence of the entry of fertilizer into aquatic systems associated with forest is **eutrophication**, in the sense of enrichment of waters and increased primary production. The respiratory requirements of associated microorganisms and other heterotrophs may lead to oxygen depletion, especially in the deeper regions of thermally stratified lakes (Richards, 1985).

The tragedy of the tropics

The necessity to maintain natural cycles is particularly pertinent to those tropical rainforests which are largely independent of their nutrient-poor mineral soils, and where economics generally preclude the use of fertilizers. Selective logging may cause relatively little depletion of nutrient capital, but subsequent recovery of the ecosystem is likely to be hindered if the soil subsystem is disrupted (Whitmore, 1990). Clear felling to provide agricultural land not only destroys the all-important mycorrhizal networks but may also result in soil erosion and the loss of surface organic matter. Traditional shifting agriculture exploits the greater soil fertility resulting from slash-and-burn to grow food crops for 1 or 2 years. The subsequent sharp decline in yields is due to depletion or non-availability of soil nutrients such as phosphorus (Jordan, 1989a), possibly through combination with Al^{3+}, released as the soil becomes more acid; crop plants are also particularly sensitive to aluminium toxicity (Whitmore, 1990). Weeds, pests and diseases also take their toll. Secondary succession during the bush fallow period involved native woody species, more tolerant of acidity and aluminium, and better adapted to the low soil fertility. Deeper roots and mycorrhizas enable them to act as 'nutrient pumps', priming the buildup of nutrients in their biomass and so into the litter and soil organic matter. After 8–10 years nutrient levels are sufficient to warrant a further round of slash-and-burn. This ideal situation can usually only support 10–20 people km^{-2}, (Whitmore, 1990). Population pressure and commercial exploitation of forests have led to prolonging of the period of cultivation or to shorter fallow periods, with diminishing returns, compacted soils and degeneration to scrub savanna. Shifting cultivation is thus a major cause of forest destruction, which could be avoided by organizing the intercropping of food crops and forest trees or tree crops. This practice of **agroforestry** (section 10.7) perpetuates nutrient pumping and the inputs of nutrients and organic matter from an abundant litter fall. Such sustainable systems are in marked contrast to the short-term exploitation of vast areas of Amazonian forest.

10
Working forests

10.1 Resource potential and forest products

The global potential for sustainable exploitation of the renewable resources of forests and woodlands has been drastically diminished, not only by deforestation of vast areas, but also by misuse and mismanagement of much of the remainder. Tangible productivity – whether as timber, fuelwood, food or medicines – is jeopardized, as are the less readily quantifiable benefits, such as conservation of genetic resources, amelioration of soil erosion, and influences on hydrology and climate. Increasing attention is therefore being paid to the manifold roles played by forests, both regionally and globally. Multipurpose use, practised for thousands of years by indigenous forest-dwellers, has been re-invented by forestry experts and adopted, at least in principle, throughout the world.

Nevertheless, the prime call on forests is still to produce timber and fuelwood. By 2025 world demand for wood is expected to rise by as much as 150% from its 1980 level of 3 billion tonnes per annum, while the UK's requirements may have doubled. Even when the major objective is timber production, there are many alternative strategies. Is the aim to gain short-term financial advantage, whether to the individual or to a nation's balance of payments, or should consideration be given to the longer term implications of managing hardwoods?

Forestry Commission planting grants differentiate in favour of hardwoods over conifers, as do extended income supplements under the Farm Woodland Scheme, but many small private owners regard their trees less as a continuous source of income than as a capital reserve, to be realized when the price is right in a widely fluctuating market. At the other extreme the International Paper Company, which has developed fast-growing 'supertrees' from the southern pines (e.g. *Pinus taeda*), affords an example of a large corporation with planned forest cropping programmes extending many decades into the future. The increasing demand for wood and the financial necessity to obtain a return on investment as soon as possible are both strong inducements to reduce

the period between planting and felling; in future the number of coniferous trees in commercial forests allowed to reach beyond the age of
60 years is likely to be low, apart from those in cold areas.

Practical constraints may reduce timber production to a level less than
that theoretically possible for a given site, since many forests are on
relatively poor unfertilized soils, or experience severe climates. This
sometimes has advantages; slowly grown timber from northern forests
often works better when planed or chiselled than that grown in warmer
countries. Trees at higher latitudes are subjected to greater exposure,
lower temperatures, and receive less light in winter, all of which lessen
their productivity. Considerable capital investments in modern machinery, purchase of stock and construction of forest roads are required,
while highly trained staff, fuel and fertilizers are expensive. Ethical
problems may also be involved. Phosphates for food production are
already in very short supply; is their use to increase timber production
justified? Each forest is unique in the environmental, financial and even
political conditions which must be considered by the forester managing
it. The silvicultural system used (section 1.7), whether it be coppicing,
clear felling, shelterwood or selection, must be appropriate to the general management objectives. The most important decisions of all are the
choice of trees to replant, and whether to maintain a complex forest
with many species or a simple monoculture. Here a detailed knowledge
of indigenous pests and diseases, and of potential exotic invaders, is
most important, particularly with regard to planting pattern.

Even if the type of site and the tree species to be planted are fixed,
the forester can greatly influence the form of the trees and the quality
of the wood they produce. Such interactions between the forester's
options and the quality of the crop (Figure 10.1) are of crucial importance.

Clearly planning for future forests should be related to estimated demand for wood and forest products. The Forestry Commission's review
'The wood production outlook in Britain' (1977) proposed that some of
the increasing demand for wood be met by creating up to 1.8 million
hectares of new plantations. This option would almost double the wooded
area of Britain. It was assumed at the time that most of this planting
would be carried out in low-grade areas of upland Scotland, with minimal
losses to Britain's agricultural production, but with far-reaching implications for visual and recreational amenities, water catchments and
conservation. The Nature Conservancy Council later became increasingly
concerned over conflicts between afforestation and conservation interests in areas of national or even international importance for wildlife,
such as the peatlands (Flow Country) of Caithness and Sutherland
(Ratcliffe and Oswald, 1987). The claims and counter-claims of foresters,
economists and conservationists over this region provide a salutary reminder that translating the concept of multipurpose use into practice

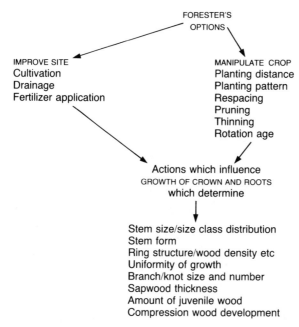

Figure 10.1   The forester's options and their effects on tree growth and timber development. (From Brazier, 1979. Crown copyright. Reproduced by permission of the Building Research Establishment, Princes Risborough Laboratory.)

may, in particular instances, be fraught with considerable difficulty when apparently irreconcilable interests are involved.

More recently the planting programme has been modified by proposals under MAFF's 1987 ALURE scheme (Alternative Land Use and Rural Enterprise), which aim to encourage planting 'down the hill', i.e. away from the uplands (Harding, 1987). The Farm Woodland Scheme provides grant-aid for planting up to 12 000 ha per annum of farmland, with greater incentives for broadleaves and for better land. The set-aside scheme, which started in Britain in 1988, aims primarily to reduce surpluses of arable products but one of the options is tree-planting on set-aside land. Forestry is in fact the major contender as an alternative land-use for the 2–3 million hectares which may be taken out of agricultural production in Britain by the turn of the century. Plans are already being drawn up to create large-scale mixed forests in the lowlands of Scotland and in the English Midlands.

As the management of woodlands and forests becomes more intensive, and as ever more of the tree – roots, stump and bark, together with the mineral nutrients which it contains – is removed in forest products, so foresters require an increasingly accurate knowledge of the ecology of woodland processes. The same is true of those endeavouring to conserve woodlands of aesthetic and scientific value: the complex structure and dynamic nature of woodland ecosystems continue to pose management problems and provide an enduring challenge to ecologists.

10.2 Multipurpose forests

In 1960 the US Congress enacted the Multiple-Use-Sustained-Yield Act. This redefined the purposes of the national forests as the 'most judicious use of the land' for its renewable resources, referring specifically to outdoor recreation, soil, timber, watershed, wildlife and fishing. The implementation of this policy became the responsibility of the US Forest Service, whose history and functions are described by Frome (1984). The act also contained a clause declaring its provisions consistent with the establishment and maintenance of wilderness areas.

These are broadly the objectives of multipurpose forests everywhere, though the main emphasis will vary from one area to another, with provision for scientific study areas being a main function in one, while another may be extensively employed for recreational use by walkers, riders, campers and fishermen. In North America timber companies have frequently clashed with conservation interests in attempts, often successful, to fell large stocks of huge overmature trees, such as Douglas fir, whose biomass per hectare is high but whose net productivity is low. Determined attempts are still being made to preserve some of the last remaining areas of mature old growth forests in the Pacific Northwest and elsewhere. Clear felling of such areas is aesthetically disastrous, entails vastly accelerated soil erosion, and frequently interferes with fisheries because of increased sedimentation in streams and lakes. In the early part of this century the Forest Service advocated selection systems, but after the Second World War came to support clear cutting using heavy machinery which could fell up to 1800 trees a day. After the 1960s the flaws inherent in this policy became manifest and the advantages of selection systems in sensitive areas are again recognized.

In Britain the loss of ancient semi-natural woodland, often as a result of felling broadleaves and replanting with conifer, has been a graver problem (Peterken, 1981). In continental Europe countries such as France and Germany possess much greater areas of traditional multipurpose forests, often managed by coppicing or selectively with natural regeneration. Elsewhere, in Holland for example, wooded areas are often so small that truly multipurpose use is not appropriate.

Some British multipurpose forests

With the exception of the magnificent Caledonian pinewoods, such as the Black Wood of Rannoch (Bunce, 1977), most British native forests are deciduous and at best semi-natural, having been continuously subject to the influence of agricultural man for up to 6000 years. The New Forest, Hampshire, is a striking remnant of the very considerable forests which still remained when the Romans first landed in 55 BC. It was designated as a royal hunting preserve about 1079 AD and now has an area of 37 700 ha, 18 000 ha of which are enclosed woods or farmland. The remainder is unenclosed, 4500 ha being woodland and the remainder acid grassland, heathland, wet heath and bog.

The forest was intended to shelter a considerable deer population in country open enough for the chase and it was kept in order, for use by visiting monarchs, by a considerable local population and governed by forest laws which were not totally abrogated until 1971. Local commoners had various rights of access to unenclosed land for fuel and animal fodder. The principal rights were **rights of common**: of **pasture** (grazing for cattle, horses and exceptionally sheep), **estovers** (wood fuel), **turbary** (turf or peat), **marl** (limey clay to improve land), **fern** (bracken as bedding for stock), and **pannage** (for pigs in the autumn when they would eat the green acorns harmful to ruminants). In theory, though seldom in practice, the right of common of pasture was withdrawn during the winter heyning and for the fence month, when the deer dropped their fawns. In time, access was assumed by all citizens, though there is in fact no public right of access to Crown land in the New Forest, and in the Review Group Report for 1988 the Open Spaces Society and the Ramblers Association were unsuccessful in getting such a right recommended. However, it is impossible to exclude humans from unenclosed land and most of the gates to enclosed Crown woods carry a notice 'The Public may use this access for the peaceful enjoyment of air and exercise'.

Any attempt to enclose land reduced the value of the rights and was resisted by the commoners, but enclosure was essential to ensure regeneration of trees when the Forest was needed as a source of wood and timber. The Act for Inclosing of Woods, 1483, authorized enclosures for 7 years 'pur la salvation de lour germe' – [for the preserving of their young spring (the coppice shoots)]. The Act for the Increase and Preservation of Timber, 1698, also made the first statutory recognition of common rights. Interest in the Forest as an amenity for walking and recreation of visitors and commuters brought by the London and South Western Railway Company was encouraged by John Wise in his attractive book 'The New Forest, Its History and Scenery' (1883).

The responsibility for management of the Forest was vested by the

Crown in the Board of Woods and passed to the Forestry Commission in 1924. Acts of Parliament in 1877, 1949 and 1964 recognized the multipurpose nature of the Forest for timber production, commoning, transport routes (major roads and pipelines), nature conservation, and recreation. The prime recreation is day visiting by car, with 4–5 million day visits annually, mostly for less than 3 hours. As a result of the implementation of proposals in a Conservation Report in 1971, these visitors are well provided for and their impact on the Forest has been sensitively and successfully minimized: roadside parking is prevented by ditching and providing screened off-road parking, often in old marl pits, with litter bins and sometimes picnic tables. Other recreation includes walking, painting, photography, wildlife study, hunting, shooting, fishing, orienteering, model aircraft flying and considerable use by youth organizations. The recreation now causing most concern is riding in groups: shod horses damage vegetation by widening paths under damp conditions. The greatest threat to the Forest is the decline of commoning: commoning of cattle or ponies is run at a loss unless cost of labour, rent and capital is ignored.

The New Forest shows the importance of a large area of open ground for the full enjoyment of a multipurpose forest: it may be contrasted with the Forest of Dean where the open area is more restricted. On a smaller scale, Cannock Chase in Staffordshire has open heathland, which is almost absent from the Wyre Forest.

The Wyre Forest (2430 ha), on the Shropshire/Worcestershire border, has been managed for its timber for over 900 years: as early as the twelfth century the extraction of oak was strictly regulated and required royal approval. Silvicultural selection caused it to become an oak forest managed as coppice-with-standards, species such as elm, wild service tree (*Sorbus torminalis*) and gean (*Prunus avium*) being discouraged. About the middle of the sixteenth century virtually all timber quality oaks are said to have been lost as a result of indiscriminate cutting for timber and a demand for charcoal that was to become so intense that by 1650 almost the entire forest was reduced to coppice. After the industrial transition from charcoal to coal the forest began to recover, many oak coppice stools being singled and allowed to grow into high forest (Hobson and Packham, 1991).

By the end of the First World War much of the ownership of the forest had become fragmented and local forest industries had declined. In 1926 the Forestry Commission acquired much of the forest, beginning a period of rapid redevelopment in which the forest was planted with a diversity of conifers. In the early years much of the forest was put down to larch, most of which has now been felled after reaching forester's maturity (section 10.4) but this tree continues to regenerate naturally. *Pseudotsuga menziesii* and *Tsuga heterophylla* have been very

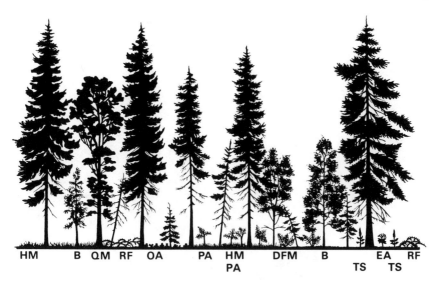

| HM | | B | QM | RF | OA | | PA | HM | | DFM | | B | | | EA | RF |
| | | | | | | | | PA | | | | | | TS | TS | |

Figure 10.2 Mixed conifer-broadleaved area of the Wyre Forest. Douglas fir planted by the Forestry Commission is regenerating naturally from seed, but has suffered losses due to windblow in some places. Sessile oak (QM, *Quercus petraea*) has regrown from a singled coppice stool, and birch (B, *Betula* spp.) established from windblown seed in regeneration gaps is here outcompeting young Douglas fir. HM, creeping soft-grass and other low growing herbs; RF, bramble; OA, wood sorrel; PA, bracken; DFM, male fern; TS, wood sage; EA, wood spurge.

successful; pines have grown well in the drier, shallower soils of the plateau. In many areas there is now an interesting mix of conifers and broadleaves (Figure 10.2). The relatively small size of the individual plantings and cuttings ('penny packets' as they are known locally), together with the retention of large areas of ancient singled oak make this area one of great conservation value and some 76 ha of the Forest have been designated as a National Nature Reserve. The oaks of Wyre are of generally poor genetic quality, often diseased, and frequently fail to regenerate naturally largely because of browsing by fallow deer. The flora is very varied and sunny glades have been established to encourage the adder (*Vipera berus*) population.

Forest road networks, glades and the careful management of verges all encourage visitors from the Midlands conurbation. A visitor centre explains the ecology of the forest and there is a range of forest walks, a deer hide, and fishing along the River Severn. In recent years the Forestry Commission (FC) has taken much more interest in the recreational value of its forests; they commissioned a study of six FC forests by Willis and Benson (1989) who employed a travel-cost method of evaluation.

Figure 10.3   The multistemmed Seckley Beech, Wyre Forest. The tree is shown early in 1988, a year before it was blown down. It possessed at least 27 trunks and appeared to have arisen from a number of saplings which fused together and gave rise to a considerably greater number of stems after coppicing. Sectioning of one of the major stems showed it to be over 140 years old but the whole beech probably lived for at least 400 years. (Photograph by John R. Packham.)

The nature conservation philosophy of the FC has evolved over a long period (Steele, 1972); the Commission now has statutory responsibilities for conservation under the Wildlife and Countryside Act 1981 and its amending Act 1985. Boyd (1987) describes the application of this philosophy to management, dealing with such matters as **transitions** (at forest edges, along the rides, glades and outer margins); **roads** as a source of mineral enrichment, **road margins** which form seed-beds for volunteer tree species (especially birch, sallow and alder), and the importance of **streams and open water**. Equally important is management of the **forest matrix** to encourage and control various animal communities, the use of conifer/broadleaf admixtures, and the need for both short- and long-term research.

Protection of the nature conservation interest

Experience in both USA and Britain shows that the battle between conservation and commercial interests never ends. In its early years the

UK Forestry Commission was responsible for many visually intrusive plantings of massed conifers, often of a single exotic species. Gradually its handling of broad landscape features, detailed forest architecture, wildlife issues and provision of recreational facilities improved, reaching an excellent standard by the mid-1980s. No sooner had this occurred than the Government prepared to privatize the FC, so national non-governmental organizations like the Woodland Trust will become even more important.

In the USA great things were expected of the 1964 Wilderness Act, which established the National Wilderness Preservation system and enabled roadless areas to be designated so as to assist the conservation of rare animals and plants as well as retaining truly primitive areas. The Wilderness Protection Act, which became law in 1982, was designed to protect wilderness and wilderness study areas from exploitation for energy, minerals and commercial development until the year 2000. Ironically, 1982 was also the year that President Reagan established the Property Review Board to dispose of 'unneeded' Federal land and made the American Forest Service responsible for identifying such lands.

Commercial interests and governments concerned more with material benefits are always likely to act against the conservation interest. This is why organizations like RSPB and the RSNC in the UK, and the Sierra Club in the USA are so important. Clearly commercial interests, particularly in terms of timber extraction, are of importance in multipurpose forests, but there must also be a strong voice for the environmental point of view. In 1982 the Forest Service planned hydroelectric schemes, roads, mines, geothermal leases, oil wells and timber sales even in wilderness areas. This was opposed by many local groups such as the state Natural Resources Department in California.

Control must be exercised over many aspects of public use including off-road vehicles (**ORVs**): motor cycles and four wheel drive machines in summer, and snowmobiles in winter. Special interest groups, especially the gun lobby, are often highly organized and difficult to control. There is also the danger that forests and other natural areas will be degraded by excessive use: in many American forest areas this problem has been tackled by a system of quotas and entry permits.

In Sweden the right of common access (Everyman's right, Allemansrättan) is embodied in the Nature Conservancy Act (1964), and allows the picking of wild flowers, mushrooms and berries where these are not protected. This right allows anyone to move freely in the countryside, and is accompanied by a very responsible attitude to nature. Countries with higher population densities could not allow the picking of wild flowers, but should move towards easier access and a truly responsible and understanding approach to working forests.

10.3   Genetics, choice of species and tree breeding

Provenance testing

By collecting seed from many different areas and raising seedlings from them under the same conditions, it is possible to determine which of the wide variations seen amongst natural populations of trees are environmentally induced and which are caused by genetic differences. The 1929 unreplicated experiment concerning four provenances of *Picea sitchensis* laid down in Radnor Forest, Wales, using seed from the Queen Charlotte Islands, California, and intermediate areas, provided a striking demonstration of clinal variation in growth rates. It also showed the importance of using a seed provenance within a few degrees of latitude of the British site. Such **provenance studies** are worthwhile because some tree species with a wide natural distribution grow in a range of climates and on different soil types in their country of origin. Trials are particularly necessary with a species such as *Pinus contorta* which has a wide natural range, but a discontinuous distribution. Suitable choices of seed origins for the main forest species in Britain have been worked out in great detail (Lines, 1987); the data available for *P. contorta* enable the forester to select a compromise between growth rate, stem form and exposure tolerance suitable for particular purposes. If Alaskan or North Coastal stock is grown as a nurse with Sitka spruce, self-thinning is likely to produce a final crop of pure spruce. Skeena river stock, on the other hand, gives good form and high volume production in pure stands of lodgepole pine. *Pseudotsuga menziesii* occurs extensively in North America, displaying marked ecotypic variation. In Britain the most satisfactory trees grow from coastal material from Washington State. When planted in Britain the Rocky Mountain provenance of this tree is slow growing and liable to fungal attack (section 4.3).

The major regional tree exchange relationships are described by Wright (1976) who points out that the Pacific Northwest (Oregon, Washington and British Columbia) has many valuable timber trees (*Picea sitchensis, Pinus contorta, Pseudotsuga, Thuja* and *Tsuga*) which prove useful in Europe and elsewhere (Figure 1.5). Many British forests now consist mainly of exotic conifers. In North America *Quercus robur* from England grows twice as fast as American white oak (*Q. alba*), has a good growth form, and produces very large crops of acorns palatable to wildlife. This tree may well be planted in quantity in areas such as Michigan and Indiana.

The use of exotic trees in an afforestation programme requires great care in selecting the best parent trees from the most appropriate provenance. Earlier in this century many collections of North American conifer seeds were despatched to Europe by local collectors, but more recently various national institutions and international organizations have collected the seed and tested the progeny themselves. The Forestry Commission's first 50 years' experience with such plantings, since its establishment in 1919, is of particular interest, being concerned largely with investigations of North American conifers (Wood, 1974). There have also been extensive trials of European pines, Japanese larch, of various hybrids, and of other trees including Chilean *Nothofagus. N. obliqua* and *N. procera* are broadleaved trees capable of growing as fast as conifers on sites outside the normal range of broadleaved species in Britain. The latter seeds prolifically and may well naturalize itself. The experience gained with fast-growing broadleaved trees, including alders and poplars, will be most useful if cultivation of hardwoods for pulp becomes an economic proposition. It has also been shown that a small number of species of *Eucalyptus* can be plantation grown in those areas of Britain where winter minimum temperatures are not too low.

Edwards and Howell (1962) produced a simple and effective **rule for provenance trials of exotic trees,** which should never be judged on a single seed source. A reasonably safe preliminary judgement can be made with trials of seedlings from:

1. the best climatic match in the natural range with the proposed planting region;
2. the part of the range where the species grows best; and
3. a third source representing another logical approach to the climate of the new site, perhaps an area with a similar temperature regime but a rather different rainfall.

When trees are planted in regions to which they are native it is often possible to increase growth rate by using seeds from lower latitudes: fast-growing German *Pinus sylvestris* can be planted successfully in parts of southern Sweden. With microsites that experience exceptional frosting cold injury can be lessened by using stock from areas to the north.

Ecological amplitude of Scots and Monterey pines

Pinus sylvestris is the most common and important timber tree in Eurasia and has been more intensively studied from the provenance standpoint than any other tree in the world (Wright, 1976), the earliest tests being made in Finland during the eighteenth century. The species shows much variation having many ecotypes, which in turn give rise to different ecads under various environmental conditions. Its native range extends from Spain and Scotland to eastern Siberia. It has been extensively planted for centuries in France, Germany and Czeckoslovakia, and is now widely grown as an exotic in Canada and northeastern USA. The

Forestry Commission found early on that the progeny of the old Breckland hedgerow *P. sylvestris* grew well not only in Thetford Forest, the largest English forest, which was established in 1922 when agricultural land in East Anglia could be bought cheaply, but also under markedly different conditions in Scotland. Since that time attempts have been made to discover widely tolerant forms of other species.

Langlet grew plants in a nursery to the age of 2–3 years from seed collected in 582 localities in Sweden, concentrating on characteristics concerned with winter hardiness (section 3.5) such as the fresh weight/dry weight ratio of their foliage, sugar content of the leaves and changes in moisture content with season. Many published reports concern trees that are 10–20 years old. Nanson (1968) found that results obtained in Belgian provenance tests at 10–15 years generally agreed with those obtained for the same trees at the age of 50+ years. Tests with ecotypes of a species with such a large natural range must be conducted at a number of stations; central European trees are consistently winter-killed when tested in central Sweden.

Natural selection in widely different climates has resulted in great genetic diversity in *Pinus sylvestris*. The fastest growing varieties have a growth rate four times that of the slowest, some suffer 10–15% mortality if planted in areas infested by pine root weevil, while with others 60–70% die. Trees of northern origin have leaves with the lowest fresh weight/dry weight ratios; an appreciable portion of their leaf dry weight is sugar which helps to make them frost-hardy. The more succulent foliage of trees from Spain provides a reservoir against desiccation in an arid climate. The foliage of trees of northern provenance turns yellow in winter and the Spanish variety *iberica*, with a good green colour in winter, is the exotic now frequently grown for use as a Christmas tree in the USA. Provenance tests in Europe and North America have had great value in providing trees with improved form and growth rates for a wide variety of sites.

Monterey pine (*Pinus radiata*), which on a world scale is planted more extensively as an exotic than any other tree, has a limited natural range consisting of six small areas in California and Baja California. All have cool, dry summers, warm moist winters and an absence of extreme cold. The lowest recorded temperature at the Monterey Peninsula site is −7°C and the average annual precipitation 380 mm. The minimum winter temperature is clearly critical; −10°C has proved fatal in Britain and the tree meets with little success on the eastern edges of continents. Most of the areas in which it succeeds (Chile, New Zealand, South Africa, Western and Southern Australia) are on the west of a continental mass and provide a maritime climate under which the tree produces quite outstanding yields. A plantation only 37 years old contained trees 49 m high in New Zealand. The original seed lots of *P. radiata* appear to have been gathered from a range of parents in more than one site and mixed. The resulting vigour is partly responsible for the tree's rapid growth as an exotic. The long-fibred wood is suitable for pulping and construction work, but the trees retain dead branches and produce knotty lumber unless pruned. *P. radiata* and the much more crooked *P. attenuata* (knobcone pine) occasionally hybridize on the Monterey Peninsula from whence some of the original seed may have come (Wright, 1976). The crooked form of some of the southern hemisphere populations of Monterey pine may result partially from introgression between the two species. In Queensland and Argentina, where summer rainfall is plentiful, disease is a problem with *P. radiata*.

Clonal forestry in spruce

When new trees are grown from cuttings the percentage of failures increases as the stock plants age. In Britain, first cycle cuttings of *Picea sitchensis* are normally taken from 2-year-old stock plants grown in polythene greenhouses; plants grown outdoors yield fewer cuttings. With suitable nutrient regimes juvenility in fifth cycle (10-year-old) Sitka spruce is maintained well enough for a reasonable yield of apparently satisfactory young plants. However, there is clear evidence from elsewhere that cuttings from older stock plants show topophysis, their main stems growing at an angle oblique to the ground like branches instead of being erect, while height growth is sometimes retarded. **Topophysis** is the long-term persistence of age or position effects in trees grown from cuttings or grafts. Scions taken from the upper branches of spruce and many other conifers retain their flower-producing ability after grafting; breeders frequently utilize this positive application of topophysis.

Roulund (1981) reviewed the problems of clonal forestry in spruce, many of which concern the physiology of rooted cuttings, and their influence on breeding strategy. All propagation programmes must be closely connected with the breeding programmes which produce improved forms from which further selection is possible. Clonal propagation can be employed by use of multiclonal varieties or bulk propagation. **Multiclonal varieties** are based on selection and testing of individual clones and must contain sufficient clones to diminish risk. They contribute a great deal of information to the breeding programme and can be so selected as to be very well suited to specific regions. **Bulk propagation** involves the replication of selected provenances or progenies without individual selection or testing of separate clones, producing populations of large genetic variability with production rates that are good but not the highest achievable.

Micropropagation, the use of shoot apex cultures for the rapid clonal production of plants, has tremendous potential for producing very large numbers of clonal trees (Evers *et al.*, 1988). In recent years micropropagation has been applied to spruce; the resulting trees are not significantly different from those produced from seedlings.

Breeding programmes

It is more difficult to breed trees (Figure 10.4) than smaller plants; the life cycle takes longer and seed is not so easy to collect. Even so, modern methods have created shortened natural generation times; heavy cone crops have been borne on 6-year-old *Picea sitchensis* grafts induced to flower early by growing them for 12 months inside a plastic greenhouse. Flowering can be consistently induced by injecting small potted Sitka spruce grafts with gibberellin $A_{4/7}$ under conditions of heat and drought (Philipson, 1983). The basic principles are well described in a letter from the International Paper Company describing their 'supertrees' which are specially selected and bred southern pines: 'In creating this tree, we selected a breeding stock of individual trees with a variety of desirable characteristics such as rapid volume growth, straight stem, disease resistance. Then we carefully selected their offspring and their offspring's offspring, to encourage their desirable characteristics.' Other features often considered include branch habit and angle, chemical characteristics of the wood, wood density and in the case of loblolly pine (*Pinus taeda*), fastest growing of the southern pines, resistance to fusiform rust. In Britain initial selection of 'plus' trees of *Picea sitchensis* on the basis of form and vigour is 1 in 750 000 of the seeds originally sown. Only 15% of the individuals selected in this way reach the national average for wood density.

Polyploids are often used in plant breeding programmes. Natural polyploids are rare amongst gymnosperms, the most notable being *Sequoia sempervirens* ($6n = 66$) in contrast to *Sequoiadendron giganteum* ($2n = 22$), which is closely related. Artificial tetraploids in pines, spruces and larches are mainly of interest as dwarf plants for horticultural use. Polyploids are common amongst angiosperm trees and while many are not commercially useful others, such as the octoploid forms of *Betula papyrifera* and *Alnus japonica* are large, fast-growing representatives of their genera. A naturally occurring clone of *Populus tremula* with exceptionally large leaves and unusually rapid growth was the first triploid forest tree to be discovered. A number of fast-growing triploid clones of *Populus*, *Alnus* and *Ulmus* have been bred since.

Hybrid vigour and other advantages often result when tree species or genera are crossed. Leyland cypress (× *Cupressocyparis leylandii*), now much planted as an extremely hardy, fast-growing ornamental tree, is

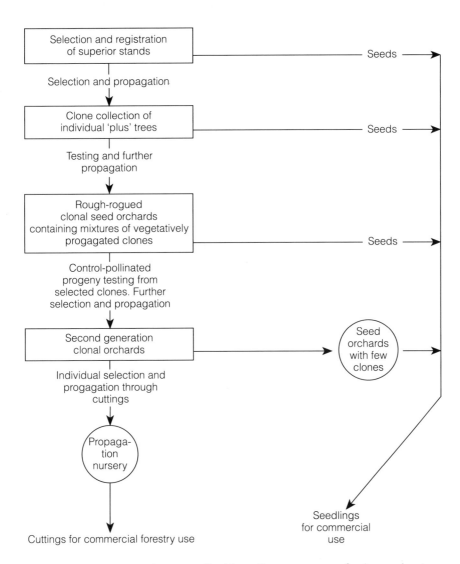

Figure 10.4 Diagram of a generalized breeding programme for improving tree stocks. The programme commences with mass selection methods to obtain better seed or sources of cuttings for practical forestry use, and provides young trees of steadily improving quality as it proceeds. When free pollination is allowed in the seed orchards many offspring have one parent, usually female, in common. These are half-**sibs**; full-sibs are trees which share both parents. In more intensive programmes controlled pollination is accomplished by enclosing young female cones, or female portions of the flowers, in polythene bags and introducing pollen artificially. Vegetative propagation from cuttings (*Thuja plicata, Picea sitchensis, x Cupressocyparis leylandii*) often occurs more readily when the mist technique is used. Grafting is used extensively to propagate clones of more difficult trees (*Pinus, Pseudotsuga menziesii*). Most practical planting of hardwoods and conifers is with seedlings. Cuttings for commercial use are very popular for a few conifers and a number of hardwoods such as willows and poplars. Their use is essential with triploid forms of poplars, alders, elms, apples and pears. In addition to the procedures shown in the diagram, a full breeding programme will allow for some hybridization and recurrent selection work.

an intergeneric cross which has arisen more than once. 'Haggerston Grey' arose in 1888 and had Nootka cypress (*Chamaecyparis nootkatensis*) as the female parent and Monterey cypress (*Cupressus macrocarpa*) as the male. *Larix* × *eurolepis* is a hybrid which has inherited the good form and vigorous growth of the European larch (*L. decidua*) together with resistance to canker fungus (*Lachnellula willkommii*) of Japanese larch (*L. kaempferi*). Amongst spruces *P. sitchensis* × *P. glauca* is a promising hybrid.

A further example of an interspecific cross giving hybrids having the desirable properties of both parents is between pitch pine (*Pinus rigida*), which has a rough form but grows extremely well when naturalized in Korea, and loblolly pine (*P. taeda*) which survives in Korea only along the south coast. The initial (F_1) cross had the cold resistance of pitch pine and the clean bole of loblolly pine. The backcross pitch pine × (pitch pine × loblolly pine) ($A \times F_1$) was even more cold resistant and retained excellent vigour and form; it was from this basis that the Koreans began the largest conifer hybrid programme in the world. Pollen production in conifers may vary markedly from year to year: stored pollen can be used to produce crosses in years when production is low.

When clonal orchards of most forest trees, such as larch, Scots pine and Douglas fir, are being established, the method of vegetative propagation employed is **grafting**, though **cuttings** are used with a few species (Hong, 1975). **Cleft**, **side** and **bark grafting** are most commonly used with trees; all entail attaching the **scion**, a small severed shoot, to the **rootstock** so that the cambial layers of the two are in contact. Successful grafting is relatively easy with potted conifers under glass, but in Britain quite a high proportion of takes can also be obtained under field conditions if the grafting is done in April.

In successful grafts the union is confined to one or two overlapping tiers of cells on either side of the graft. If there is a slight incompatibility between the scion and the stock, the cells of each do not interlace to form a mechanically strong joint and sudden severe strains may cause breaks at the point of grafting. With greater incompatibility the stem immediately above the graft often grows to a much greater diameter than that beneath it. This is not uncommon is Douglas fir where 2-year-old grafts from replicate graftings can be examined for evidence of wound-xylem areas in the union. Such an early check is not possible in *Pinus radiata* where the incompatibility arises in the phloem. This poses a serious problem as the trees get older; up to 50% of 14-year-old grafted trees may be severely affected. **Root grafting** can be used to solve incompatibility problems with such species. In this method scions are grafted onto excavated lateral roots of the same tree; the plants produced have genetically identical roots and shoots. Triclonal grafting is often employed with commercial rubber trees (*Hevea brasiliensis*) to unite strong rootstocks, stems with many latex-yielding tubes in the phloem, and disease-resistant crowns.

Tree-breeders usually endeavour to make the best use of **genetic variation** at four levels: within genera, within species, within ecotypes, and even within a stand, which is the smallest unit which can provide a base for a breeding programme. Initial selection to ensure that the programme starts with genetically superior stock is essential, otherwise years of work may be largely wasted attempting to improve basically mediocre material. Selection may be direct, e.g. for the tallest trees or those least damaged by a particular insect, or indirect, which might involve selecting trees with the highest photosynthetic rate in the hope that this is causally related to growth rate. In practice measurements of features such as height are usually made on young trees: juvenile selection is vital if breeding programmes are not to be impossibly long. Figure 10.4 gives a very simple view of a modern tree improvement programme which has to provide seeds or cuttings for effective short-term production, with a marked risk of inbreeding, but should also develop a broad-based genetic population for future research by continuous breeding. Weir and Zobel (1975) describe the use of such advanced-generation seed orchards with *Pinus taeda*. In the long term the British *P. sitchensis* improvement programme obtains highly superior seed from clones derived from a system of recurrent matings combined with family selection (Faulkner, 1987).

It is of great importance to the health of future forests that we create **tree banks** including genotypes of forms not currently in commercial use. If the genetic bases of forestry stocks became too narrow they might lack the variation necessary to meet changed conditions. The Food and Agriculture Organization (FAO) has strongly recommended that conservation of forest gene resources within their natural range should be carried out *in situ* where this is practicable, taking care to prevent contamination by pollen from adjacent stands. This process has been followed very thoroughly in Finland, where the conservation and administrative procedures enable foresters to obtain fresh seed from individual trees of proven worth as parents. The genetic resources of forest ecosystems are extraordinarily complex (Stern and Roche, 1974): in managing them we are not conserving a static state but endeavouring to contain a dynamic system of which our understanding is still imperfect (Namkoong, 1986).

10.4 Growth, thinning and economic factors

Plans for the development of an individual forest are now usually based on detailed measurements of the growth of particular tree species in similar sites, and computer modelling is often employed to examine alternative strategies. Inflation makes precise financial forecasts exceedingly difficult; although it is certain that very large quantities of wood will be needed in the future, the degree to which market forces will increase its value is unknown. All that can be done is to make forestry operations as efficient as possible, though subsidies and taxation anomalies often minimize the direct effects of economic competition. A summary of good forestry practice, with methods of establishing, maintaining and harvesting forest crops, is given by Hibberd (1986). In commercial plantings, the forester endeavours to obtain either maximum profitability or the maximum volume of the highest quality wood (the two aims do not always entail the same management), while restraining operational costs as far as possible. In such a long-term operation he must consider the whole life of the tree, unlike the practice of forest managers in some developing countries where semi-natural forests are often felled in what are virtually exploitation situations. Over the short term it is often possible to make good forecasts of the costs of such operations as logging and log transport. Comparative costings of various systems of felling, skidding logs out with crawlers or wheeled vehicles, road construction, and different methods of organizing labour, help to avoid losses arising from failure to adopt more efficient modern methods or machinery.

The duration of the period between tree planting and the felling of the main crop has profound financial implications. In some sites particular species grow relatively rapidly; it takes *Pinus radiata* only 35 years to develop into a large tree in several regions of the southern hemisphere, whereas *P. nigra* is usually far smaller when harvested after a 55-year rotation in England. As £100 of planting costs would, if invested at 8% compound, accumulate to £1478.5 after 35 years, but to £6891.4 in a further 20 years, there is a great financial incentive to obtain revenue from thinnings and to achieve as short a rotation time as possible.

Assessment of production in commercial forestry

Many of the concepts involved in assessing tree growth are set out in forest management tables such as those used by the Forestry Commission (Hamilton and Christie, 1971). Trees increase in fresh weight, dry weight, height and volume, of which the last two are the easiest to measure. Measurable volume, conventionally taken as being that of stemwood exceeding 7 cm diameter overbark, is the most meaningful measure in economic terms and the pattern of growth in an even-aged stand is usually described in terms of annual volume increment. Figure 10.5 shows that the **current annual increment** (**CAI**) increases for some

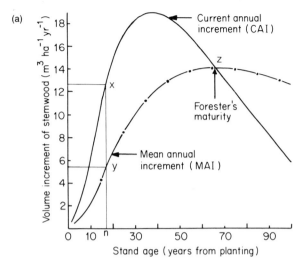

Figure 10. 5   Patterns of volume increment in an even-aged stand of Scots pine (Pinus sylvestris) belonging to yield class 14. For symbols see text. Drawn from the data of Hamilton and Christie, 1971. (Figures 10.5 and 10.10 appear by courtesy of the Forestry Commission.)

years after planting and subsequently declines. CAI at n years is x m³ ha⁻¹ while y is the **mean annual increment (MAI)**, the average rate of increase from planting to a given point in time. At **forester's maturity,** where the two curves cross (z), the stand has the maximum average rate of volume increment which can be achieved by a particular species on the site concerned. Theoretically the maximum average rate of volume production could be maintained in perpetuity by repeatedly felling the stand at this age and replanting with the same species. In practice soil deterioration, buildup of pests and pathogens and other factors may cause yields to fall gradually over a number of rotations.

The **yield class system** is based on maximum MAI, i.e. the maximum average rate of volume production reached by a given species on a particular site, irrespective of when this occurs. Maximum MAI under British conditions may be as low as 4 m³ ha⁻¹ for many hardwoods, larch and pines, but may exceed 30 m³ ha⁻¹ for grand fir. Each **yield class** represents the number of cubic metres (to the nearest even number) of timber produced per hectare at maximum MAI. Thus the maximum MAI for a stand of trees in yield class 14 is greater than 13 m³ ha⁻¹ but less than 15 m³ ha⁻¹.

In any species, MAI reaches its maximum value at successively greater ages in the slower growing stands, as the yield class curves for oak (Figure 10.6) illustrate. Maximum mean annual increments of different species are often of the same magnitude but reached at quite different

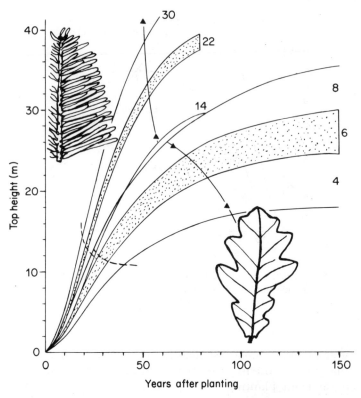

Figure 10.6 General yield class curves for grand fir (*Abies grandis*) and native oak (*Quercus robur* and *Q. petraea*) under British conditions. The stippled areas represent the median yield classes for each species; the other figures indicate the lowest and highest general yield classes normally encountered. – – – Time of first thinning, ▲–▲ age of maximum MAI. (Curves redrawn from Hamilton and Christie, 1971.)

times in the life of the tree, e.g. 35–40 years for poplar, 60 for Douglas fir and 80 for Norway spruce (trees in yield class 12 under British conditions).

The direct measurement of tree volume is time consuming. Fortunately there is a fairly close relationship between top height and the total cumulative volume production of a stand. **Top height** of a stand is the mean height of the 100 trees of the largest diameter at breast height (1.3 m) per hectare. Top height/age curves (i.e. **general yield class curves**), such as those given in Figure 10.6, have been worked out for all the major species planted in Britain. By entering top height in the appropriate year after planting, the practical forester can ascertain the general yield class of the stand. The volume of standing timber can then be determined graphically by entering yield class and basal area in the appropriate standard volume chart (Hamilton, 1985). **Local yield classes** may be used for sites where, for example, a particular species at a given spacing tends to form trunks rather thinner or thicker in relation to top height than is normal.

Figure 10.6 contrasts the general yield class curves of oak with those of grand fir, which under British conditions grows faster than any other tree for which Hamilton & Christie (1971) give data. Oak is a magnificent broadleaved deciduous tree, whose growth encourages the development of a diverse insect fauna and an attractive herb layer, but it grows far more slowly than most conifers and in Britain few commercial timber growers now plant it. Even with conifers, costing at compound interest over 50 years or so influences whether and when operations such as brashing and spraying are carried out: in many cases owners have to cut their losses by 'premature felling'. Timber, especially in northern countries, tends to be a long-term, low-return crop; indeed a cynic could say that the main advantage of softwoods in Britain is that they are less unprofitable than hardwoods.

Influence of tree spacing

As tree spacing has considerable influence on planting costs, weed problems, and the volume and quality of wood obtained from a forest, many experiments have been carried out to determine its effects upon crop characteristics and yield. Analyses of spacing trials conducted by the Forestry Commission with Pinus sylvestris, Picea abies, P. sitchensis, Larix decidua, L. kaempferi and Pseudotsuga menziesii are given by Hamilton and Christie (1974; Figures 10.7–10.9). Many of these experiments involved initial spacing of the seedlings at 3, 4.5, 6, or 8 ft (0.9, 1.4, 1.8 or 2.4 m), both between and along the rows. The number of seedlings planted per hectare were 11 960, 5315, 2990 and 1682, respectively.

With closer spacing, competition between young trees begins earlier and mortality prior to the thinning stage is much greater (Figure 10.7). There were slight variations between the species in the trials but roughly 90% of the seedlings in the 2.4 m spacing plots survived at the 10 m top height stage, whereas almost half of those in the 0.9 m plots were dead when the survivors reached this height. However, weeds were suppressed earlier, as the canopy closes much more rapidly when the trees are planted close together.

Trunks of trees planted at the wider spacings grow to a greater diameter than those planted close together (Figure 10.8), but taper rate in the lower part of the stem is much lower in the latter. Greater volumes of wood are produced if trees are planted close together, and Figure 10.9

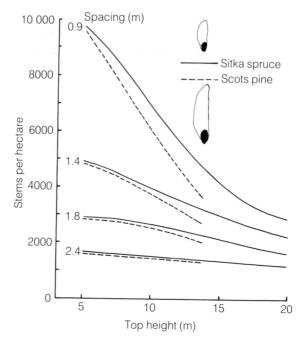

Figure 10.7   Surviving stem numbers in unthinned Sitka spruce (*Picea sitchensis*) and Scots pine (*Pinus sylvestris*) for four initial spacings, together with scale outlines of the winged seeds. Norway spruce (*Picea abies*) has survivorship trends very similar to those of Sitka spruce. (Curves redrawn from Hamilton and Christie, 1974.)

shows the extra amounts of wood obtained if trees are planted at spacings closer than 2.4 m. The effect is not so obvious if a minimum 7 cm dbh limit is applied to exclude stems that are technically not of merchantable quality, though astute foresters often market them for use in rustic work. The **basal area** of a tree is obtained by measuring the overbark cross-sectional area of the trunk 1.3 m above ground level. Basal areas per hectare, like total volume, are considerably greater when trees are closely spaced.

The measurement of standing and felled timber, known as **forest mensuration**, is very important when timber is being bought and sold (Hamilton, 1985). The sizes and shapes of trunks strongly affect the value of a given volume of timber; wide trunks with low taper produce the greatest area of usable planks per unit volume of uncut wood. Isolated trees in particular often have a form poorly suited to economic conversion to sawn timber. Large limbs often arise relatively low on the main trunk, and in species such as *Thuja plicata* may even bow down to the ground and become rooted.

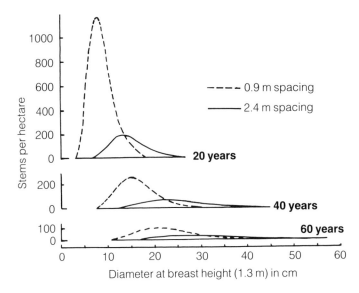

Figure 10.8 Distribution of stem numbers by dbh classes for 0.9 and 2.4 m spacings at 20 year intervals, in unthinned Sitka spruce, yield class 16. (Redrawn from Hamilton and Christie, 1974.)

The basis used for measurements of height of trees planted at different spacings needs to be carefully considered; the average height of all the trees present is not a valid measure since in closely spaced trees the shortest individuals are about to be suppressed and the forester's interest lies in those which will form the mature stand. In this respect top height is the most useful measure. In the experiments described above marginally greater top heights were associated with closer spacings.

In general, experiments carried out elsewhere with various species confirm the main results of the analyses made by Hamilton and Christie (1974), though Heding (1969) found survival of *Picea abies* as a percentage of initial stocking to be notably higher at comparable top heights. Reports of the effect of spacing upon height growth vary and sometimes concern the mean height of the total population, in which case the heights given for closely planted areas are low as a higher proportion of trees is restricted to the lower canopy. However, there is no general agreement regarding the effect of spacing on dominant or top height; Cromer and Pawsey (1957) report increases in dominant height of *Pinus radiata* with increased spacing.

Mensuration research and computer modelling can yield considerable dividends particularly in situations where management is becoming more intensive, as in the very large *Pinus radiata* forests established in New Zealand. These exotics are being thinned and pruned more drastically

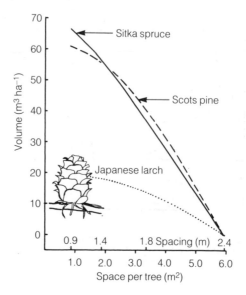

Figure 10.9 Total volume production of different spacings, relative to that of 2.4 m spacing, at age of maximal mean annual increment in Scots pine, Sitka spruce and Japanese larch. The curve for Norway spruce is very similar to those of Sitka spruce and Scots pine. The diagram shows the **additional volumes** of timber obtained by growing trees at spacings closer than 2.4 m. (Curves redrawn from Hamilton and Christie, 1974.) Note the reflexed cone scales of the Japanese larch.

than in the past. Fertilizer application is becoming more widespread and improved genotypes are coming into general use; there is also a need to establish more precisely the volume losses caused by pests and disease. At the age of 10 years, when dominant height is about 12 m, most stands will have received their last pre-commercial thinning and have been high pruned to 6 m. With such rapid growth, poor management can cause considerable volume losses so the New Zealand Forestry Service has for many years provided a computer-backed mensuration service, established forest planning models, and run regular training programmes for foresters.

Thinning

Thinning promotes the growth of the remaining trees, provides intermediate financial returns, and may increase the total yield of usable timber over the life of the stand. The major factors which constitute a thinning regime concern its intensity, type and timing. **Thinning in-**

tensity is the rate at which timber is removed per year, e.g. 10 m^3 ha^{-1} y^{-1}. It should not be confused with **thinning yield**, the actual volume removed in any one thinning. Given a **thinning cycle** (time between one thinning and the next) of 5 years, this would be 50 m^3 ha^{-1} in the case being discussed. At low thinning intensities stands of normal initial spacing become so overstocked that cumulative production of usable timber is reduced, suppressed trees dying before harvesting. Though the remaining trees respond by increasing diameter more rapidly than if left unthinned, they do not use the extra growing space fully if thinning intensity is excessive. In practice cumulative volume production of usable timber remains the same over a wide range of thinning intensities. The **marginal thinning intensity** is the maximum thinning which can be used without incurring loss in this total volume. Once the stand is sufficiently well developed for thinning to commence this intensity, until the stand reaches maximum mean MAI, is about 70% of maximum mean annual volume increment per year (Rollinson, 1985). For a stand with a yield class of 10, for example, the marginal thinning intensity is 7 m^3 ha^{-1} y^{-1}, which would give a thinning yield of 42 m^3 ha^{-1} with a 6-year thinning cycle.

Selective thinning involves the removal or retention of trees on their individual merits. If the type of thinning is systematic, e.g. by harvesting of rows, strips or chevrons, competing dominants will be removed along with subordinate trees. The thinning cycle adopted influences the profitability of the operation as long cycles involving heavy individual thinnings are the most profitable. They may, however, increase the risk of windblow and loss of volume. Higher yield classes within particular species normally reach maximum MAI relatively early, and as thinning usually ceases a few years before this, their **thinning period** ends earlier than in lower yield classes, though their thinning cycles are shorter for a given volume removed.

As the cost of timber and other forest products continues to rise, foresters in the developed countries are attempting to use more and more of the total production. This tendency is exemplified by the Swedish forest industry's Whole Tree Utilization Project which uses stumps, twigs and small trees formerly abandoned in the woods after felling operations; it is estimated that stumps from fellings could provide 12–13% of the raw material required by the Swedish wood pulp industry. Some conifer bark is made into very useful compost. When pulverized and allowed to undergo natural thermophilic bacterial fermentation, bark loses volatile oils harmful to growing plants and its pH is raised. Such complete use of trees is of obvious economic importance and, by reducing the substrate available to fungal pathogens like *Heterobasidion* and *Armillaria*, may lower the incidence of disease. On the other hand, the removal of nutrients from the forest ecosystem is considerably accelerated.

10.5   Woodland maintenance and cropping

Clear felling (section 1.7) is the method of forest cropping most widely used today. In North America vast areas of cut-over forest were formerly left in the hope that adequate regeneration would occur naturally, but the secondary forests often contained a low proportion of commercially valuable species and were sometimes slow to develop. Felling is now usually followed by planting, as in Europe; the higher costs are justified by shorter rotation times and improved productivity. Dead stumps often harbour pathogenic fungi and harmful insects, and in extreme cases may be extracted – in very intensive cropping systems even the main roots are harvested – or a delay allowed before replanting. The ground is often ploughed, drained and scarified to provide a good seed bed. Difficulties associated with a young woodland may include the development of a luxuriant understorey, climbers such as traveller's joy (*Clematis vitalba*) and honeysuckle (*Lonicera periclymenum*), and undesirable tree species, which require weeding, and the cost of thinning coppice shoots of any regenerating trees to one per stump. If the forest is clear-cut in patches the remaining trees are more susceptible to windblow, and the creation of large blocks of trees at a similar stage, including the cut-over areas and impenetrable young stands in the thicket stage, lowers landscape quality. The young seedlings must also be capable of growing well in the open, as do those of many pine species. In this silvicultural system all the mature trees are extracted together so there is no danger of damaging new plantings as there is with shelter wood and selection systems, though severe soil erosion can occur.

It is particularly important to prevent soil washing into major watercourses by leaving fringe timber to act as a buffer along their margins, while felling alternate strips greatly encourages natural regeneration. **Clear felling** often involves the use of very large numbers of **artificially grown seedlings or cuttings** (section 10.3), and this in turn frequently lowers the genetic diversity of the forest. In Europe **selection systems** remain useful in areas of outstanding natural beauty and for protection forests, while **shelter-wood** methods are generally preferred for *Fagus sylvatica*; they are used for *Picea abies* in parts of southern Germany, Switzerland and Czeckoslovakia.

Trees may modify the soil on which they grow very appreciably: adjacent Scottish plots under *Pinus sylvestris* and *Betula* often tend to form mor and mull humus, respectively. Loss of water by transpiration from tree foliage is often a factor in preventing gleying in rainy areas. When areas of *Pinus sylvestris* near Loch Maree, Scotland, were clear felled and abandoned the ground quickly became much damper and rushes invaded. Conversely, there has been a striking improvement in the permeability and aeration of the deep peat upon which *Pinus contorta* has been grown in Scotland; irreversible drying has caused deep shrinkage cracks.

The sequence of operations involved in removing a tree crop and establishing another on the same site is quite complex, even when the area is clear felled. It also involves making a number of management decisions whose outcome may be greatly influenced by factors, such as the weather and disease patterns, which cannot accurately be foreseen. In Britain, in years with adequate summer rain, a decision to plant seedlings relatively late on may be justified by a high 'take' and an extra season's growth. However, summer drought after late planting of bare root seedlings can result in a mortality of 60% or more, and much extra expense in subsequent 'beating up' operations to replace the dead seedlings in later years. The success of the new crop is influenced by the method of clearing, treatment of residual stumps, branches and foliage, preparation of the seed bed, the condition of the young trees when planted, soil conditions, fertilizer treatments, competition from other plants and the extent to which forest pests and diseases can be controlled either chemically or by sound silvicultural practice.

The development of wheeled machines whose jaws grip the base of the tree trunk, which is rapidly sawn or sheared, and then transported on a skidder, has made it possible to crop trees of moderate size selectively. It is now very rare for axes to be used for large-scale felling. With the huge trees of the Pacific Northwest large winches are used to operate running skylines and high lead systems with which very large trunks can be hauled out. In areas free from strong winds captive balloons can be used to lift logs which are again winched from the felling site to the loading deck.

Clearing felled areas often involves the use of fire (section 10.6), though the fungus *Rhizina undulata*, which in Britain can cause group dying of *Picea sitchensis*, tends to grow on the sites of quite small brushwood fires. When stumps and stems are pushed into lines and left as rucks, as is done in some British plantations, there is clearly a danger of encouraging rabbits and small mammals, as well as leaving a substrate for pathogens and insect pests. If pine stumps are left in the ground it is often worthwhile to paint them with a sealant or other preparation to reduce the incidence of fungal disease in the new crop (section 6.3). On unburnt areas brash and ground cover are often chopped with rotary cutters before planting.

The use of heavy machinery to remove felled timber and to grade restored land, such as old gravel workings, may cause soil compaction too severe to remedy adequately by conventional ploughing and the Forestry Commission now tine to a depth of 0.7 m with a multishank ripper. Rotary mouldboard ploughs have proved effective in improving the mixing of the organic and mineral layers of the soil. In damp sites the practice is to plough and plant seedlings on the tops of the ridges, whereas in drier areas such as Cannock Chase *Pinus nigra* grows best in the clean groove left by a double-throw plough. In very damp sites the whole area is drained by deep cross ditches which are often almost perpendicular to the ridges on which the seedlings are planted.

For many years foresters have improved the germination rates of seeds by stratifying them in moist sand and chilling them prior to sowing. With many conifers a naked pre-chilling treatment has been shown to produce a greater number of usable seedlings at the end of the season than does one involving sand. The lethal effects of cold fungus (*Geniculodendron pyriforme*) during the moist pre-chilling of *Pinus contorta*, *Pseudotsuga menziesii* and *Picea sitchensis* can be avoided by pre-soaking in a 0.5% Thiram suspension for 48 hours. Better germination rates can also be achieved by lightly raking the seed bed instead of leaving it compacted, and also by covering the seeds with coarse light-coloured sand instead of grit. Both practices improve seed–soil contact.

Trees such as *Pinus sylvestris* grow well when planted out directly as seedlings, often after being dipped in insecticidal solutions. Species with compact fibrous root systems are favoured; those with relatively few long thin roots, such as *Pinus nigra*, are not. However, in Britain *P. nigra* showed almost 100% take when grown in Japanese paperpots in a polythene greenhouse for 20 weeks and hardened off outside for another 4 weeks. A pointed tool enables the pots to be planted quickly even on sites covered with chopped brash. The method avoids damaging the roots and provides them with a known environment; drought losses are greatly reduced. Transparent or translucent tubes are now frequently employed in establishing broadleaved trees. Such **tree shelters** enhance growth, afford protection from browsing deer and allow herbicide spraying (Tuley, 1983).

Weed control assists the establishment and rapid growth of young trees; it is difficult to overestimate the importance of adequate initial clearance of clear-felled sites or of cleaning during the thicket or small-pole-stage phases (Davies, 1987, Figure 10.10). It is less expensive to establish good control from the start, with regular inspection to check on the weed development, than have an extensive weeding operation after rank growth has overtopped the young trees, which rarely need weeding once they have reached a height of 1.2 m. Physical methods such as ploughing, mulching and hoeing have their place, but chemical control by herbicides (employing tree shields where necessary) is more effective in keeping down the weeds which might otherwise grow vigorously until the tree crop closes canopy.

The use of herbicides and insecticides is rigorously controlled by law

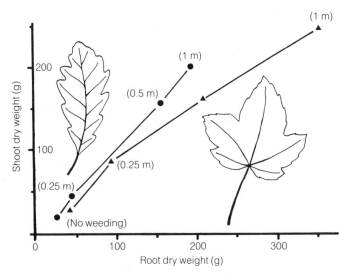

Figure 10.10 The effect of eliminating weeds from areas of different diameter on the growth of British oak (*Quercus* sp.) and sycamore (*Acer pseudoplatanus*). Areas around the young transplants were treated with paraquat and glyphosate for three years. Note also the higher shoot: root ratio of oak. Accidental herbicide damage to the oak in 0.25 m treatment partially balanced relief from weed interference. (Redrawn from Davies, 1987.)

(Control of Pesticides Regulations, 1986) so the appropriate authorities (e.g. Evans, 1984; Sale *et al.*, 1986) should be consulted as to which chemicals are legally available, the weeds and pests they control, and the season at which they should be applied. Brushwood stumps can be killed by painting cut surfaces with 2,4-D, 2,4,5-T or glyphosate. 2,4, 5-T also controls broadleaved woody weeds when sprayed from the air at ultra low volume; this method was first banned by countries such as Sweden. If a heavy growth of bracken (*Pteridium aquilinum*) already exists it should be sprayed with asulam before planting tree seedlings, while in other cases herbicides may best be applied at the post-planting stage. In general herbicides are either absorbed from the soil (dichlobenil and propyzamide) and are thus residual in the short term, or act by contact with green tissue (asulam, glyphosate and paraquat). Careful tests are needed to ensure that the herbicides used do not damage the tree crop at the concentrations employed. The residual herbicide simazine prevents re-invasion of weed-free soil by killing germinating seeds. Stems and root collars of conifer seedlings are frequently attacked by pine weevils (*Hylobius*) and beetles (*Hylastes*), which breed in old stumps. Young plants are often dipped in insecticides to help reduce damage by these animals, but tolerance tests must be made; root collar swellings with poor development of root systems have been reported in *Picea sitchensis* after treatment with both gamma HCH and gammacol.

Foliage analysis is often used to determine whether fertilizer treatments are necessary; in Britain phosphate levels are usually adequate but some pockets of severe phosphate deficiency exist (section 9.6). *Picea sitchensis* may be held in 'check' by *Calluna vulgaris*, and the trees often require two applications of nitrogenous fertilizer to enable them to close canopy. The alternative is to spray the *Calluna* with glyphosate.

Traditional methods of woodmanship (Figure 1.6) maintained stands of trees without necessarily employing seedings. Coppiced areas retained their genetic diversity over long periods of time and soil erosion was minimized because tree roots remained undisturbed. **Coppicing** was practised in Roman times and in the nineteenth century well-kept coppices existed in many parts of Europe and even in Tibet. The crops of poles and brushwood which they yielded were distinguished, as wood, from the timber obtained from large trees as planks, beams, and posts. By growing coppice-with-standards it is possible to obtain both products, and standard trees with their wide crowns, and frequently crooked trunks were formerly the source of the knees, catheads, and other shapes of timber required for wooden ships. Patterns of cutting, and the length of the coppice cycle, need careful consideration in modern woodlands managed along traditional lines for reasons of conservation.

10.6   Fire: factor, hazard and management tool

In many areas the aim has been to prevent all fires; such policies of fire protection can swiftly alter vegetation, while in the northern Wisconsin pine barrens Vogl (1970) has shown that the frequency of fires helps to determine the species composition of the conifers. Fire also helps recycle nutrients retained in the ground layer and in long-lasting plant litter. Grasslands are often burnt to prevent the development of scrub, to recycle nutrients and to encourage fresh growth. The Plains Indians used fire to keep much of the North American prairie free from trees; forest invaded vast areas of this land after the destruction of these peoples and their sustaining buffalo herds. *Juniperus virginiana* and a number of fire-tender hardwoods became much more common following the advent of fire protection. Many even-aged forests date back to the last fire; the relatively few older trees probably represent the original seed sources.

In central North America white pine (*Pinus strobus*), red pine (*P. resinosa*) and jack pine (*P. banksiana*) frequently became dominant after fire, and it is considered that these species do not re-establish extensively after logging or in natural forest unless burning occurs. Fire spreads rapidly

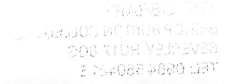

through the crowns and the well-aerated, non-compacted litter of red pine is a great hazard in pure cultures of this tree, though less so where it is mixed with white pine and shade-tolerant hardwoods. Red pine bark is thick, corky and resistant to fire, so trees which would be killed by crown fires often survive ground fires. The cones of red and white pine are easily destroyed by crown fire or scorch, whereas those of most jack pines are **serotinous** (closed by resin) and remain on the live trees for many years. Heat destroys the resin enabling the cone scales to open and release the seed quickly after fire. Some cones of lodgepole pine (*P. contorta*) are serotinous while others are not. Black spruce (*Picea mariana*) has persistent semi-serotinous cones that are opened by heat, but the germination of this species occurs mainly in the second or third years after burning, whereas jack pine germinates most strongly in the first year. In Australia fire is particularly important in the regeneration of *Eucalyptus regnans* (section 3.4), and in south-east Labrador it creates and perpetuates mosaic patterns involving paper birch (section 5.3).

Fire is a major factor in perpetuating savanna in areas where the climate would permit the growth of tropical forest. The life-form spectrum for the Olokemeji Forest, Nigeria (Figure 10.11a) is typical of tropical forests in that over 90% of the plant species are phanerophytes (Hopkins, 1965), other life forms being rare and grasses virtually absent. The area is relatively dry which is why geophytes are more abundant than is usual in tropical forests. In the grassy savanna woodland only 5 km away (Figure 10.11b) less than one-third of the species are phanerophytes, one-quarter therophytes, and over one-fifth hemicryptophytes. The percentage of geophytes is four times greater than in the forest. The majority of trees and shrubs in tropical forests have very thin bark so that the cambium is easily damaged. That of trees and shrubs of the savanna, however, is usually more than 1 cm thick, a factor which leads to their resistance to and selection by fire. In addition the enormous root systems of many woody savanna species sucker readily when the shoots have been burnt to ground level. The Olokemeji Forest reserve was used in an experiment on the fire protection of derived savanna. In 1929 the site was burnt, coppiced and cleared of trees. Three plots were established and their vegetation was analysed in 1957. The plot burnt annually in March, at the end of the dry season, had become a tree savanna whose fire-scarred trees were widely spaced. In the plot burnt annually at the start of the dry season there was a fairly mature savanna woodland with some fire-tender forest trees protected inside fire-resistant clumps of closed woodland. Grasses were almost absent from the fire protected plot in which there was a well-developed canopy of trees and shrubs – forest vegetation was re-establishing swiftly.

The litter of such trees as *Eucalyptus* and *Pinus ponderosa* helps to ensure high flammability in their forests, in which fairly intense fires are

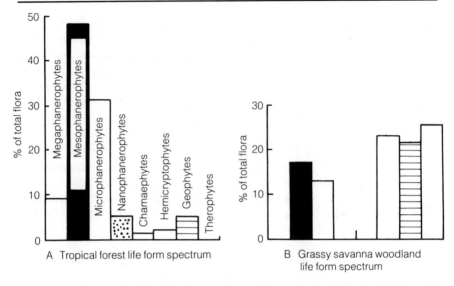

Figure 10.11 Contrasting life form spectra for two stands in the Olokemeji Forest Reserve, south-western Nigeria. The various life forms follow the same sequence in both diagrams. (Redrawn from Hopkins, 1965.)

inevitable given an ignition source and suitable weather. Mutch (1970) suggests that fire-dependent plant communities burn more readily than communities which are not fire-dependent because the species which they contain have been subjected to processes of natural selection which have led to their materials burning more easily, as well as giving them a greater capacity for surviving fire.

In recent years the extent of tropical rainforests has been greatly diminished by felling and by fires started by man; huge areas of bamboo forests and grass have resulted from shifting agriculture. In Asia aggressive fire-resistant species such as teak (*Tectona*) and sal (*Shorea robusta*) have colonized large areas; indeed fire is essential in the natural regeneration of the moister types of Burmese teak forest. In forests in which teak occurs, the monocarpic bamboo undergrowth dies after flowering, and if this litter is burnt teak regenerates well and may come to form up to 50% of the tree stock. Teak and sal both produce a whippy shoot which dies back later in the season. The rootstock grows in size and vigour, and finally a more robust leading shoot develops which is capable of surviving fire and other hazards.

Fire affects the invertebrate populations and alters the physical and chemical characteristics of the forest floor. The resulting seed bed is often very suitable for the growth of conifer seedlings, and in California *Sequoia sempervirens* produces 5–10 times as many seedlings per unit area where slash burning is medium to heavy as it does in unburnt or lightly burnt places. The tree has thick fire-resistant bark so some seed parents are likely to survive, and the species can also regenerate from dormant buds just above the massive lateral roots. The removal of heavy canopy by fire also helps the growth of young seedlings. Fire destroys or reduces the thick dry litter and humus which normally prevent the roots of many conifer seedlings reaching the moisture in the underlying mineral soil. Because their vertical roots grow slowly many conifer seedlings, such as those of most pines, do badly on soils with a thick humus layer, whereas the roots of some hardwood seedlings can penetrate 12–15 cm before the first leaves begin to function.

Deliberate burning is used as a management tool in many parts of the world, particularly with dwarf shrub communities in which *Calluna* is dominant. Fire has been used deliberately for many decades by North American foresters; the costs and techniques are well established. Broadcast burns lessen future fire hazards and help reduce vegetation cover which physically impedes tree regeneration. They also influence the field layer. In an American experiment a burnt area was later covered by the shrub *Ceanothus intergerrimus*; an adjacent space in which woody rubbish was simply windrowed was revegetated by *Arctostaphylos viscida*. Great care has to be employed to prevent fires getting out of control and to avoid altering the environment so as to favour undesirable species. In Australia, aerial incendiaries are used to start low intensity fires which burn off forest fuel accumulations; in Tasmania ground ignition techniques are used in the *Eucalyptus* forests. In Scandinavia the slash and upper portions of the humus layer are often burnt to release immobile nitrogen from the cold moist soils. Dry sites with thin humus layers are not burnt.

'Let-burn' policies have now been adopted by the American National Park Service with regard to low intensity fires started by lightning in high altitude *Sequoiadendron giganteum* and *Abies magnifica* (red fir) forests. These help maintain the appearance and species composition of the forests as they were prior to the exclusion of fire. Where such natural fires do not periodically sweep through unharvested forest, enormous accumulations of dead trees and other fuels build up on the forest floor. If, or rather when, these eventually ignite the result is catastrophic, particularly as relatively young conifers act as fuel for 'ladder fires' which carry flames up into the canopies of the largest trees. Let-burn policies came under intense criticism in the summer of 1988 when, after the driest summer on record, fire swept through one-third of the 2.2 million acres of Yellowstone National Park. The policy adopted in 1972 had been to put out fires caused by man, but leave those started by lightning. In 1988 forty-four of the 52 lightning fires went out by themselves without burning one acre. After 15 July attempts were made to stop all fires but burning continued to the end of October after heavy snows occurred, though the 9000 firefighters saved virtually all the important buildings in the park.

Subsequent recommendations were that major fires should be controlled in future, and the possibility of using prescribed fires to stop the buildup of forest fuels be investigated in developed areas. This will reduce inconvenience to visitors and residents; it also recognizes that the total area of North American wilderness is now too small to allow fire to range unchecked (Romme and Despain, 1989). As the first US National Park, Yellowstone is very much in the public eye. In the summer of 1988 the 720 000 acres burnt here attracted far more attention than the more than 2 million acres burnt elsewhere in continental USA and the 2.2 million acres burnt in Alaska, but the total losses were enormous.

The ground vegetation of some British woodlands is highly flammable, particularly in spring. Combined with a windy climate, this makes for the rapid spread of **ground fires**. Following severe drought **crown fires** occurred in many parts of Britain in late summer 1976, but such fires are unusual as the relative humidity of the atmosphere is usually rather high. In countries with fairly stable weather conditions the moisture content of the potential fuel (plant litter, brash, etc.) is often used to calculate a **hazard rating**.

Destructive fires are always liable to develop or get out of control in droughts, particularly in high winds. Forests should be designed to reduce fire hazards and fire plans devised so that the most effective sequence of operations is followed if an outbreak occurs (Teasdale, 1986). In Britain belts of larch or broadleaved trees tend to arrest fires, while the complete brashing of all trees to a height of 2 m in a strip 10–20 m wide may prevent fires getting into the tree canopy from neighbouring ground vegetation by the forest edge. Conveniently placed supplies of water, beaters and other fire-fighting equipment are essential. Much can be done to reduce the hazards by using suitable spacings for access points and firebreaks, employing appropriate mixtures of tree species, and keeping a careful watch, sometimes from observation towers and aircraft which in North America are also used to drop many tons of water, foam and other fire retardants. Roads act as excellent firebreaks and enable fires to be fought while they are still small. Herbicides such as paraquat provide a fairly cheap method of eliminating weeds in firebreaks which are unsuitable for cultivation or mowing.

10.7   Agroforestry

ICRAF, the International Council for Research in Agroforestry, Nairobi, Kenya, defines **agroforestry** as a collective word for all land-use practices and systems in which woody perennials are deliberately grown on the same land management unit as annual crops and/or animals. Increased use of agroforestry is now seen as having great advantages in improving production while protecting the soil (Gholz, 1987).

Agriculture frequently involves the destruction of woodland, so it is easy to assume that it cannot co-exist with forestry. However, the earliest forms of agriculture were often practised in temporary clearings in woodland. Iversen (1941) identified the use of such clearances, **landnam**, from a study of the pollen and charcoal layers of peats and soils formed in Denmark in Neolithic times (section 5.6). The perpetuation of forest cover was required for soil conservation and nutrient recycling. Swidden, or 'slash-and-burn' methods of shifting agriculture remain common in wooded areas of Africa and Asia but are now less in tune with the environment because land shortage often reduces the period of forest fallow. Rappaport (1971) describes the care with which the Tsembaga in New Guinea protect regenerating tree seedlings, **duk mi** 'mother of gardens', amongst the crops of yams, cassavas and bananas within their swidden clearings.

It is more difficult to re-establish trees in areas which have been completely cleared for farming or become denuded wasteland but this promises great benefits, especially if it can be more widely adopted in semi-arid areas. The **taungya** system was developed from a shifting agriculture system in Burma to establish teak forest, under the direction of Sir Dietrich Brandis in the mid-nineteenth century. It is used to rehabilitate degraded land, and to raise teak and eucalypts in Kerala, India, for example, where the tree component is interplanted with cassava and rice. In experiments at Trichur, rice conserved mineral nutrients and soil better than cassava, though cassava with 10% grass strips was the least erosive of several treatments (Khurana and Kohli, 1987). Taungya differs from agroforestry in that the trees take over from the crops and people must be moved on to a new area intended for rehabilitation.

Large-scale state or commercial forestry has the disadvantage that local people are deprived of land and inevitably become poachers and encroachers, whereas they will tend and value the trees amongst their crops. Social forestry is geared to the needs of the local community, providing forest products for local industry rather than for an external organization. Agroforestry may be regarded as a form of social forestry, in which crops for human food and animal fodder are interplanted with trees. A distinction can be made between **sylvo-pastoralism**, where animals graze amongst the trees, and **agri-silviculture**, in which forage is harvested for feeding elsewhere: both systems were practised in the New Forest, England, until the Deer Removal Act of 1851. The benefits of the trees can include the production of fuel, fodder, food, fertilizer, fibre and fruits (the six F's of conservation forestry). They may also be a source of medicine or even insecticides (e.g. from *Azadirachta indica*, the neem tree). Trees may rehabilitate degraded land, help to conserve soil, and act as biological pumps to stop the water table of irrigated land reaching the surface and causing salination. They also provide shade and reduce extremes in the microclimate.

Agroforestry systems include trees grown in hedgerows round crops, mixed cropping of trees scattered amongst crops, and alleys of crops between rows or strips of trees. The trees themselves may be 2 m apart, or even closer. The rows are 4–8 m apart and arranged to run from east to west to minimize shading of the alleys left between them. Harrison (1987) regards agroforestry as vital to a sustainable increase in food production in Africa, especially in the form of alley cropping as a replacement for fallow. In work at the International Institute for Tropical Agriculture at Ibadan, Nigeria, maize has been grown between hedgerows of the leguminous tree *Leucaena leucocephala*, subabul, pruned back to 1 m during the growing season, but allowed to grow to as high as 4 m during the dry season. The tree produces leaves and twigs used as fodder or mulch, rich in fixed nitrogen, as well as stakes and fuelwood. The maize yielded 83% more over 3 years than a control plot. Subabul possesses many of the characteristics needed in an agroforestry tree. It is shade tolerant in its early stages, deep rooted, fast growing, with light branching but a compact crown, and deciduous, with a leafless period when maximum crop growth is required. It yields proteinaceous leaf fodder palatable to mammals but repellent to birds (Khosla *et al.*, 1986). In the Kandi region of the Punjab, it has been harvested on a rotation of 3–4 years for fuel wood of high calorific value. If the leaves are not needed for fodder, their litter improves tilth, water-holding capacity, and fixed nitrogen content, besides reducing the alkalinity of the soil. Members of the leguminous genus *Prosopis* can be of value in agroforestry in arid regions: *Prosopis cineraria*, khejri, did not reduce grain yield of pearl millet at a density of 50 trees per hectare and increased fodder yield by 31% at 100 trees per hectare under semi-arid conditions in Rajasthan.

Intercropping can often yield more total biomass per hectare because of differing moisture and nutrient needs, the extended growing season of perennials, and the use of nitrogen-fixing trees. Advantage is being taken of the complementarity of different life forms and there is a recreation of some of the stratification of natural woodlands. However, in some cases crop yields are reduced and most tree products are not as valuable as cereal grains or pulses. The greatest advantage may be that the system can be sustained for long periods without soil degradation. Both tree and crop species being grown together need careful selection to avoid allelopathic effects: an experiment at Samaru, Nigeria, showed that maize and sorghum were more resistant to eucalypt leachates than

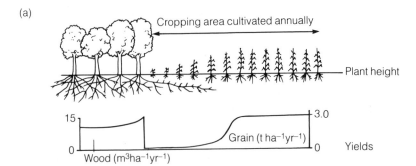

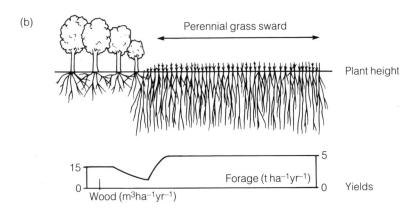

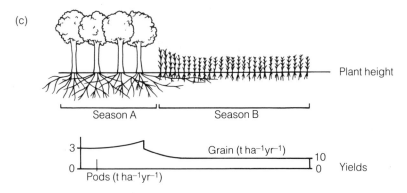

Figure 10.12 Sections through tree/crop interfaces showing effects of competition between root systems and canopies. (a) An extended tree/crop interface, in which the tree species dominates the crop. (b) The growth of the tree is inhibited by a deep-rooted perennial grass. (c) The growth of the tree species and the crop are separated in time and both benefit. (Redrawn from Huxley, 1983).

other crops. Trees should not compete with the crops for light and nutrients yet recycle nutrients effectively: the architecture of their canopy and root systems must be carefully considered. The interface between trees and crops is shown in Figure 10.12. Crop root systems are usually dense and strongly stratified, in contrast to tree root systems, which are more diffuse and extensive. Competition between them can be reduced by careful timing of tree pruning and crop planting, or by the choice of species, for example, crops such as millet and sorghum are better close to *Leucena* or *Sesbania* trees than grain legumes whose flower initiation and fruit set is reduced by shading. If the overall biological effect is positive, the interface can be maximized, as in mixed cropping: if it is negative, the interface must be minimized, as in zonal or alley cropping.

Since the tree is expected to provide fodder as well as fuel, it must be palatable, lopped at the right time and fed to the animals which make best use of it. Pods and seeds (e.g. *Acacia* and *Prosopis*) may be valuable. The value of foliage will depend on the content of crude protein, condensed tannin, and toxic substances. For example, a sample of *Leucaena leucocephala* was found to have a total tannin content of 2.8% of leaf dry matter, of which a proportion could be hydrolysed by the gut of ruminants, but 1.5% was condensed tannin and would bind with the proteins of the leaf, making them unavailable to the animal. Although this is not an unusually high level of condensed tannin (compared with *Robinia pseudacacia*, 2.3% or *Quercus incana*, 2.5%), *Leucaena* also contains mimosine, a toxic non-nutritive amino acid. The harmful effects of such foliage may be reduced by feeding it as part of a mixture of species, while in the longer term genetic improvement may be used to produce strains with less tannin and toxins.

Though agroforestry is most commonly practised in the tropics it can be adapted to many other parts of the world, including the UK, where various agroforestry regimes are now being tested experimentally in Devon, Wales, Northern Ireland and Scotland.

11
Contemporary problems and the future of forests

11.1 Vanishing forests: desertification and climatic change

Desertification is a continuous process of degradation in soil and vegetation caused by land misuse in dry areas. Degradation of soil includes erosion, compaction, waterlogging, salination and alkalination; degradation of vegetation includes reductions in biomass, productivity and diversity. The vegetation becomes inferior in density and ability to stabilize the land, e.g. perennial grasses are replaced by less productive annual species. Grainger (1990) lists the four main types of poor land use in drylands as over-cultivation, over-grazing, incorrect irrigation and deforestation. These are a direct result of human impact. Desertification is associated with drought, but drought alone would cause a temporary and reversible reduction in yields of crops and livestock. Because of familiarity with biblical accounts of drought (I Kings 18 and Ruth 1) and predictions of the seven years of plenty and the seven years of famine (Genesis 41), drought is often regarded as an unusual occurrence, rather than a restriction on the carrying capacity of the land. Over-use and over-grazing is then attempted in an effort to compensate for falling yields.

World dryland is centred on subtropical latitudes of about 30° (somewhat greater than those of the Tropics of Cancer and Capricorn) where regions of subsiding air cannot provide rain by the condensation of water vapour. They cover about one-third of the Earth's land surface, principally in Africa and Asia. Rainfall may be up to 500 mm y^{-1}, but agriculture is affected by its seasonal nature, unreliability and rate of evaporation as much as by its quantity. Rainfed crop culture requires at least 350 mm y^{-1} but traditionally pastoralism has been the dominant land use in areas with up to 600 mm y^{-1}. Drylands suitable for some form of agriculture would be expected to support a natural vegetation of open woodland with thorny shrubs and trees of moderate height scattered in grassland. The trees include many leguminous genera, especially **Acacia** in Africa and **Prosopis** in Asia. Their leaves, seeds and pods are a source of fodder, especially towards the end of the dry season, and their dead branches are a source of fuelwood. Increased herds and population lead to more intense culling of the trees or to their clearance for agriculture, sometimes for growing cashcrops. The pressure on the remaining woodland then becomes intense, especially for fuelwood and around centres of population. When trees are removed, land is exposed to baking sun and eroding wind and rain, leading to dust storms and ensuing desertification.

Desertification of lowland areas forces farmers to deforest and cultivate mountainous slopes which then deteriorate rapidly. Rainfall is no longer released gradually from vegetational cover and flash floods are followed by dried-up watercourses, in place of more evenly flowing streams. Deforestation can have effects far away where the sediment load of rivers silts up reservoirs and irrigation channels and diverts the course of rivers. The shortage of fuelwood leads to the burning of animal dung which is no longer available for use as a fertilizer.

If land is used within its carrying capacity desertification can be controlled, but over a long period, not just during the most favourable years, and by employing a system that makes production resilient to climatic variation. Traditional cropping methods achieved this, often by the inclusion of a long fallow period. The fallow land itself could be very productive, being grazed, with the nutrient benefit of animal dung, or even left long enough for trees to recolonize. Grainger (1990) describes the fallow of Kordofan in the Sudan where *Acacia senegal* invaded and after five years could be tapped for gum arabic for a further seven years, meanwhile protecting the land from erosion, building up a litter layer, fixing nitrogen through the bacteria in its root nodules and finally yielding fuelwood.

Desertification can be corrected by making land use sustainable and needs programmes whose social and political implications have been thoroughly explored. Large-scale tree plantations may fail where growth rates are reduced by poor soils and low rainfall. Social forestry may succeed where local people are involved and likely to protect trees along roads and channels outside the regular forest areas.

Climatic change may alter the rate at which desertification occurs, or the likelihood of it occurring. It may also have other effects on vegetation and tree cover. Global changes have occurred many times in the past but anthropogenic CO_2 increase may lead to more rapid changes in the future. Current models propose a doubling of present levels by the middle of the next century, leading to a possible rise in global mean temperature of more than $3°C$, greater in Polar regions, less near the equator. This would have profound effects on precipitation patterns, with increased rainfall generally but summer dryness in some areas. Species limited by temperature, e.g. holly, maple and lime, could colonize regions 100 km further north in the northern hemisphere for each degree rise in mean temperature. Although temperatures may increase, in upland areas strong winds and sudden floods might shorten the life of trees or restrict their growth.

Productivity of C3 plants would increase with elevated CO_2 levels: for example, the rate of photosynthesis of **Populus deltoides** has been found to increase from 16 to 19 μmol CO_2 m^{-2} g^{-1} when CO_2 levels were increased from 300 to 400 ppm (Regehr et al., 1975). There is the possibility that greater productivity might lead to nutrient stress in some tree species: indeed, this has been suggested as one cause of forest dieback. Stomatal conductance is reduced at high CO_2 levels, resulting in less water loss and less absorption of gaseous pollutants. Phenology may change in trees with indeterminate growth as in many introduced species, e.g. horse chestnut, but not where a determinate period of cold is needed before leafing out as in beech. More vigorous growth under favourable conditions may lead to diversion of a tree's resources to vegetative parts rather than reproduction, as is commonly seen in apple orchards. Reproduction may also be affected if trees become out of phase with their insect pollinators. Since species react in different ways, competitive abilities will be altered, causing changes in community structure. Climbing species might become serious competitors as they are in regenerating rainforest. Woodland herbs of the vernal aspect, e.g. bluebell, may be unable to compete with grasses if these started their growth earlier in the spring than they do at present.

11.2   Forest declines and acid rain

There is nothing new about widespread damage to trees, including the progressive deterioration implied by forest decline. Such damage is a natural consequence of climatic extremes or of attacks by pests or pathogens. The idea that atmospheric pollution might be responsible also has a long history. Various authors, commencing with C. E. Moss in 1901, have suggested that pollution has caused decline in plant growth in the industrial Pennines for the past 200 years. There is also support for the view that high levels of SO_2, as well as the exposed nature of the sites, are responsible for the scarcity of Pinus sylvestris in this part of Britain.

What is new, as Innes (1987) shows in his review of air pollution (including acid rain) and forestry, is the unprecedented number of tree species affected and the severity of the damage reported from different parts of the world within a relatively short time during the past two decades; hence neuartige Waldschäden (**new type forest damage**).

Dieback in several high altitude forests in Germany became apparent in the 1970s, and was initially restricted to Abies alba; by 1978 the more commercially important *Picea abies* was also declining. There followed reports of a progressive decline of forest trees throughout Europe, including Scandinavia and Britain, and also in North America. The wide range of species affected includes spruces, pines, firs, larches and yew and, among broadleaves, maples, oaks and beech.

The pattern of decline typically involves a progressive loss of foliage (dieback, crown thinning), with discoloration (yellowing or browning) of remaining needles or leaves. Growth patterns may also be distorted, as in the stork's-nest syndrome in *Abies alba*, recovery shoots in spruce, tinsel shoots in yew and the bunching of twigs in the upper canopy of beech. In some instances there is evidence of death or distortion of roots. Deterioration may eventually result in death of the tree, sometimes hastened by pests or pathogens. Within this general pattern there may also be **regional damage types**, for example five types of spruce damage, with different symptoms and different geographical distributions, are now recognized from Germany (Blank *et al.*, 1988).

In general there has not been the large-scale deforestation foreseen by many environmentalists; typically, dead and severely defoliated trees are scattered among less affected or even healthy trees. Moreover, regional surveys have shown that certain species in certain areas (e.g. fir, spruce and pine in the Black Forest in 1987) are showing signs of improvement, contrary to overall national trends (Blank *et al.*, 1988).

Assessment of forest health

Although attempts have been made to relate tree-ring data to pollution, currently the commonest approach in the European Community involves an annual survey of each country, based on a standardized protocol, which is now co-ordinated by the European Environment Agency. Surveys involve visual examination of trees sampled relative to a (16 km)^2 grid; these are compared with a 'perfect' tree of the same species growing in the same area, in order to estimate the density (i.e. relative thinness, normally described as loss of foliage, which tends to imply damage) and discoloration of the crown. Trees are allocated to the appropriate crown density category (Table 11.1), and may be downgraded if there is extensive discoloration. There is some disagreement about the significance of foliage loss, many accepting that up to 25% (or even 60% in some cases) of the foliage may be lost without necessarily causing damage to the tree; it would therefore seem prudent not automatically to equate loss of foliage with damage (Innes and Boswell, 1987).

Recent results from these surveys indicate that the general health of EC forests is far worse than 10 years ago, and that in 1988 Britain had

Table 11.1   Classification of crown density

Category	Foliage 'loss' (%)	Classification
0	<10	Healthy
1	11–25	Early warning stage (formerly slight damage)
2	26–60	Moderate damage
3	>60	Severe damage
4		Dead

some of the most seriously damaged forests in Europe, in terms of the proportion of sampled trees in categories 2 and 3. However, data from Forestry Commission surveys for 1987–90 have shown no indication of a large-scale decline in forest condition in Britain.

Foliar nutrient indications of vulnerability to forest decline

The use of foliar nutrient levels when assessing the potential advantages of fertilizer applications has already been mentioned (section 9.6). Studies of c. 30–year-old *Picea abies* at 12 sites from south-west Germany to northern Scotland appear to show that nutritional status can also be employed in the evaluation of vulnerability to forest decline (Cape and Freer-Smith, pers. comm.). In German sites where trees appeared to be suffering decline samples were taken from trees with 'good' and with 'poor' crown condition on the basis of crown density and needle yellowing. Samples were obtained of current, 1-year-old and 2-year-old needles. There were wide ranges of nutrient content in current needles of apparently healthy trees and these ranges increased in 2-year-old needles.

At sites with trees having 'poor' crowns, 2-year foliage from 'poor' trees had significantly smaller concentrations of Mg and Ca, and larger concentrations of K, than that from 'good' trees. **Nutrient ratios** were more significant indicators of crown conditions than single nutrient levels alone. 'Poor' crowns had larger ratios of N:Mg, K:Mg, S:Mg, K:Ca and smaller ratios of S:K and N:K. On the basis of nutrient content and nutrient ratios a **risk index** may be employed to identify sites where deterioration in crown condition is likely to occur in the future. It is also advisable to sample several year classes: at sites having trees with 'poor' crowns even apparently healthy trees showed a lack of increase in calcium as needles aged, while nitrogen decreased and magnesium decreased greatly with time.

Possible causes of forest decline

Although **acid deposition** and other forms of atmospheric pollution are most commonly invoked as being responsible for forest decline, there is no shortage of alternative suggestions, some of which may be of local or short-term significance, with the added complication of possible interactions between various factors.

Among **forest management practices**, replacing broadleaves by conifers could lead to increased susceptibility of trees to stress, through soil acidification and compaction. This argument fails to explain either the decline in areas where conifers have grown for centuries or the decline of broadleaves. Studies in West Germany and Switzerland failed to establish a link between management and forest decline.

Pests and diseases are undoubtedly responsible for some of the symptoms recorded in forest health surveys. In Britain, for example, crown thinning can be caused by insects such as spruce aphids and green oak tortrix, or by needle-cast fungi, while leaf miners produce extensive browning in beech (Innes and Boswell, 1987). The severity of attack may be either increased or ameliorated by pollution, depending on interactions between specific pollutants, tree species and pest or pathogen. The only major decline where disease appears to be the primary cause is type 3 spruce decline, where reddening and shedding of older needles is due to needle-cast fungi. On a wider scale, trees in the later stages of decline are particularly susceptible to invasion by secondary pests and diseases, for example the southern pine beetle (*Dendroctonus frontalis*), the larger spruce bark beetle (*Ips typographus*), and root rot fungi. In all cases there is the possibility of complex interactions between biotic agencies and factors such as gaseous pollution, drought and frost.

Long-term climatic change seems to have had little influence on tree health up to now, but **climatic extremes** are strongly implicated, especially in causing stress among trees. The decline of red spruce (*Picea rubens*) in North America has been linked to drought, while beech was particularly badly affected by the European droughts of 1976, 1983 and 1984. Increasing incidence and severity of frosts, as well as sudden fluctuations in temperature, can all lead to damage or even death, especially if frost-hardiness has been reduced, for example by magnesium deficiency. Desiccation, resulting from the combination of frost and strong winds, was particularly damaging to many conifers in Britain early in 1986.

Of the various **gaseous pollutants** known to injure trees, research has largely concentrated on those which are most likely to have effects over large areas, and in particular on SO_2, NO_x and O_3. In trying to establish cause and effect, care must be taken to distinguish between acute and chronic effects, and between long-term average concentrations and episodic extremes. There is also the possibility of synergism between two or more gases. For example, *Picea abies* seedlings are much less sensitive to these three gases separately than when they act together.

Although high levels of SO_2 cause local declines in parts of eastern Europe, mean annual concentrations of SO_2 and NO_x throughout most of western Europe are well below levels known to damage trees. On the other hand, O_3 and other photochemical oxidants are claimed to be the primary cause of decline in California, of ill health in *Pinus strobus* in eastern USA, and of dieback of *Cryptomeria japonica* in Japan. Levels of O_3 which are known to be phytotoxic have been recorded over much of Europe, especially at high altitudes such as in the Bavarian Alps, where some of the earliest instances of forest decline were reported. Although microscopic damage to Norway spruce needles in the Black Forest was similar to O_3 damage, fumigation experiments with O_3 produced a different pattern of damage symptoms. However, fumigation of sensitive clones of *Picea abies* saplings with O_3 during the summer led to their older needles being severely frosted in the following November, an example of an indirect, delayed effect of a pollutant (Brown *et al.*, 1987).

Acid precipitation is unlikely to affect the foliage of mature trees through acidity *per se* at pH values above 3.0; greater acidity than this is most likely in occult deposition, from mist and fog, especially in upland forests. Apart from these effects, and the implications of toxic components such as SO_2 (especially in dry deposition), the major influence of acid rain is on the ionic balance of vegetation, either directly or via the soil. Many of the symptoms of the various types of forest decline resemble, or are associated with, those of nutrient imbalance; possible explanations are provided by the remaining hypotheses.

Soil acidification with aluminium toxicity (the **Ulrich hypothesis**) has been strongly considered as a cause of forest decline, particularly in West Germany (Figure 11.1a). Many forest soils have certainly become more acid since the industrial revolution, but in general it seems that conifers can withstand the levels of aluminium commonly encountered in such soils. However *Fagus sylvatica*, which appears to be favoured by the higher levels of N, P, Ca and Mg on less acidic soils, may be more susceptible, especially in the Solling region, where aluminium levels are exceptionally high. The occurrence of damage to trees on soils well buffered against acidification (e.g. by long-term liming trials) precludes the general applicability of this hypothesis.

The **ozone-acid mist** hypothesis proposes that O_3 damages cell membranes, so increasing rates of leaching of nutrients from foliage by acid rain and acid mists (wet deposition). Adverse effects of O_3 on photosynthesis are claimed to reduce root growth, so that the roots are less able to recycle these nutrients, resulting in foliar deficiencies (Figure 11.1b). Short-term fumigation experiments with *Picea abies* have shown

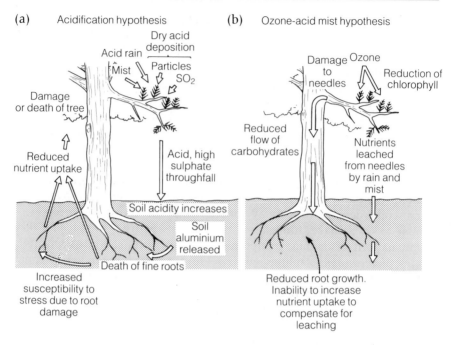

(a)   Acidification hypothesis

(b)   Ozone-acid mist hypothesis

Figure 11.1 (a) Soil acidification hypothesis to explain forest decline. (b) Ozone-acid mist hypothesis to explain forest decline. (From Roberts, 1987; by permission).

that nutrient leaching from needles is indeed accelerated, but that the nutrient content (e.g. Mg) of needles is not reduced. Such fumigation failed to produce chlorosis in older needles, a symptom of type 1 spruce decline, but current needles displayed a distinctive red necrosis.

Magnesium deficiency seems to play a key role in this type 1 decline, which typically occurs in forests above 600 m altitude ('high altitude spruce damage'). There is some evidence that exchangeable magnesium has declined in certain forest soils in Germany, where magnesium nutrition may always have been marginal. This depletion may have been due to soil leaching, driven by acid deposition, and also to tree harvesting, while a succession of dry years possibly exacerbated problems of magnesium uptake, hence the synchronous appearance of symptoms over a wide area. Application of magnesium fertilizer has led to the recovery of some of these trees, but such treatment is unlikely to be relevant to parts of Britain and Norway, where losses are more than offset by atmospheric magnesium input derived from the sea.

Finally, deposition of **excess nutrients** from the atmosphere, especially nitrogen, may be responsible for damage in Sweden, The Netherlands and coastal areas of Germany. Farm slurry is a major source of ammonia, which may lead to nutrient deficiencies in certain soils, as well as causing soil acidification after conversion to nitrate. The fertilizing effect of extra nitrates may be accompanied by reduced uptake of phosphates, and increased susceptibility to fungal attack and cold stress.

Innes (1987) provides an excellent review of the complexities of pollution chemistry and of soil–plant interactions which underlie the phenomena of forest decline. He favours the **multiple stress** hypothesis, whereby decline results from the cumulative effect of several stresses, the pre-eminence of which varies between regions. For example, magnesium deficiency in foliage, perhaps exacerbated by ozone and acid deposition, increases frost susceptibility; severe frosts after long warm summers may trigger the synchronous appearance of decline symptoms over large areas.

In Britain, the Forestry Commission is relating the appearance of trees in its Forest Health Surveys to a wide variety of environmental variables (geographical, topographic, climatic and pollution) but statistical interpretations are limited by inter-correlations. The results of detailed studies of individual sites, and of ambitious long-term experiments such as the SO_2 and O_3 fumigation in Liphook Forest on the Sussex–Hampshire border, are awaited with interest.

Meanwhile, we can only hope that actions such as the reduction of gaseous pollutants will lead to an amelioration of forest decline, as they have in the comparable situation of lake acidification.

11.3   Arboriculture and urban forests

Arboriculture is focused on the well-being of individual trees and involves the replication, selection, establishment, maintenance (including tree surgery) and eventual removal of amenity trees, usually in relatively small areas. Though some of the work involved is similar to that in forestry, arboriculture has evolved its own areas of expertise (Patch, 1987; Bradshaw *et al.*, 1988) and codes of practice, which are in Britain largely the responsibility of the Arboricultural Association. Entrants to the profession receive training in a wide range of technical procedures and especially in the control of safety, whether working on or above the ground, but it takes time for skill and judgement to mature. Tree surgery and the removal of senescent or dangerous trees close to buildings is complex and difficult work; contracts should be carefully framed and awarded to firms with a proven track record and satisfactory insurance cover.

While the skills involved in selecting and placing appropriate trees to produce the best effect, or to hide ugly buildings, are much the same as in the past, others have greatly changed. New species and varieties are becoming available, stock can be grown much more rapidly from seed or by vegetative propagation, and there is a trend towards containerized trees, quite commonly 10 m tall, planted with a good root ball and their own soil. Tree shelters such as Tuley tubes help to reduce casualties in the earlier years. More is known about water relations and irrigation methods, as well as the use of fertilizers in the long-term care of landscape trees. Along with this is the ever present need to be vigilant in detecting and countering outbreaks of disease and pests (sections 6.3 and 6.5) which can spread so easily from one nursery to another. Research has shown that in periods of drought trees can cause shrinkage of soils, particularly those rich in clay, and thus damage foundations of buildings: plantsmen need to be aware of their legal responsibilities in this respect (Biddle, 1987). Identification of the causes of building movement and damage is an important issue; Kelsey (1987) discusses methods of determining whether or not these can be ascribed to the influence of trees.

Root systems lie at the heart of successful planting. Surveys have shown that many early casualties are caused by planting stock whose roots are either inadequate or allowed to dry out after initial lifting. Adequate root systems are essential if the trees are to absorb sufficient mineral nutrients and replace the water lost in transpiration. Time of planting is crucial, broadleaved trees planted in late November (early in the dormant season) often have twice as much root at time of full leaf expansion as those planted in late March. Even in 1678 John Evelyn, who early recognized the utility of growing trees in a nursery and transplanting them to areas where natural regeneration was insufficient, emphasized the need for thorough site preparation before planting. He also advocated the use of supporting systems until the tree's roots were able to hold it steady. Today this support is frequently provided by stakes to which the trees are tied. Sadly, abrasion of amenity trees by stakes or ties, which are often left in position for far too long, is a major cause of damage. Moreover flexing of young trunks in wind helps initiate increased xylem formation, so an excessive period of support should in any case be avoided. On the other hand, support for a brief period tides the tree over the time needed for development of new roots after transplanting. Patch (1987) advocates planting small trees that do not need support. Where trees are over 2 m high support should not extend more than one-third of the total tree height and should be attached only at the top of the stake. The support should be removed at the onset of the second growing season after the tree has survived the winter gales.

Attitudes to the treatment of wounds and the retention of aged or weakened trees at the far end of the tree's life history have greatly changed. Trees in key town or garden situations are so valuable that it is worth going to considerable trouble and expense to retain them, even for a few extra years. In a few instances weakened limbs are merely propped up: cabling is a neater solution. Before it is employed, however, the strength and stability of the tree in relation to loads – mainly wind loads – should be determined so that an appropriate strengthening system can be devised. This new, largely mathematical, approach to the stability of trees is known as **tree statics** (Sinn and Wessolly, 1989).

Tree surgeons formerly applied wound dressings immediately after major pruning or lopping, in the belief that their fungicidal or mechanical sealing effects would prevent the entry of microorganisms. It has now been shown that these sealants have a relatively short-term influence; much greater reliance is now placed on the tree's natural defences, especially its ability to produce internal physical and chemical barriers near to entry points of agents of decay. This system of **compartmentalization** (Shigo, 1984) is very well developed in some clones of yellow birch which also have effective systems for shedding dead branches and healing the remaining wounds. In contrast, aspens are usually poor compartmentalizers and infections from dead branches spread rapidly into adjoining stems. In the long term there is advantage in selecting trees with good natural defences, as well as encouraging good callus formation by making cuts distal to the 'branch bark ridge', avoiding flush cutting which results in larger wounds which stand little chance of healing over. There is still a case, however, for using gel formations of fungicides to treat cankers and fresh wound parasites of fruit trees (Clifford and Gendle, 1987). The fungus *Trichoderma* is quite effective in limiting decay in stem wounds of *Fagus sylvatica* and such biological means of preventing decay may be of value in the future.

Urban forests

Although urban forests differ in their nature and in the way they are conceived, the origin of the term urban forestry is well documented (Jorgensen, 1986). In Britain there is a tendency to employ the idea of urban forests in relation to attempts to 'green the cities' using trees, but the concept adopted in Canada and used by the USDA Forest Service has much wider implications. Jorgensen first coined the term urban forestry in 1965 in relation to municipal tree planting projects in metropolitan Toronto. By 1970 he had defined it very clearly: '**Urban forestry** is a specialised branch of forestry and has as its objective the cultivation and management of trees for their present and potential contribution to the physiological, sociological and economic well-being of urban society. These contributions include the over-all ameliorating effects of trees on their environment, as well as their recreational and general amenity value.' (Jorgensen, 1970).

On this definition urban forestry is concerned with tree management in the whole area utilized and influenced by the population of the town or city. This area involves forested watersheds and outlying recreational areas, as well as trees inside the political boundaries of the municipality. Urban forestry entails multidisciplinary teamwork with major inputs from forestry, arboriculture, horticulture, engineering, landscape architecture and the social sciences. It is one of the two branches of **community, or social forestry**; the other being agroforestry (section 10.7) which also has to be closely tailored to the needs of the population it serves. In both cases the long-term resource is the soil and its productivity, which can only be maintained in perpetuity if the forest is properly managed.

On a narrower definition, as would apply in Britain, 'urban forestry' can be regarded as pertaining rather strictly to highly populated areas; indeed, Miller's (1988) textbook is subtitled 'planning and managing green spaces', i.e. parks and gaps between buildings and roads. The book shows just how complex an operation this is. Species selection, record keeping, programmed maintenance, contract development, task scheduling and budgeting, and economic analyses of various management alternatives, are all involved in what by any standards is very big business that can only be controlled properly with the aid of computer-assisted inventories.

At the same time it is essential to involve the local community in the whole programme of providing and maintaining trees. Unfortunately it is those areas which most need the life and colour trees provide which often have the highest levels of vandalism. Here the arboricultural officer should devote a considerable proportion of his time to educating the public, especially children, so that trees are cherished rather than destroyed.

11.4 Farm forestry

Agricultural surpluses in the EEC have made it necessary to change government policy in Britain towards alternative land use of so-called improved land. This removal of some better quality land from production is by 'set-aside' or by the development of alternative enterprises such as farm woodlands. Terminology in this area has been confused; the following definitions are now accepted: **farm woodlands** are conventional forest plantations on agricultural land: **agroforestry** is the intimate integration of forest and agricultural production with trees planted at wide spacing, while **farm forestry** encompasses both these forms of management as well as short rotation systems which include coppicing. In Britain existing unmanaged farm woodlands occupy 200 000–300 000 ha with, in addition, 200 000 ha of land covered with 'non-woodland' trees, small groups, lines and individual trees. In total they contain approximately 40–60 million m³ of timber.

The Farm Woodland Scheme, initiated in 1987, has as its central feature a system of annual payments to farmers over a period of between 20 and 40 years depending on the type of woodland. These payments are in lieu of the farming income that would come from the land used and are taxable in the same way. Eligibility of land for planting is restricted to arable and grassland up to 10 years old. Old grasslands are rich wildlife habitats and are excluded from planting except on a limited basis in the uplands. To encourage the planting of mixed woodlands with a high proportion of broadleaves, the basic payment time of 20 years is extended to 30 years for woodlands with more than 50% broadleaves and to 40 years for crops containing only oak and beech on the grounds that these species take longer to mature. Very small areas of planting such as hedgerows and field corners are not eligible; the minimum holding is 3 ha (no block less than 1 ha), the maximum holding is 40 ha. Not only is it intended that timber will be grown but also that the landscape will be enhanced and new wildlife habitats created. Where it is suitable there may be recreational use and tourist interest. Farm woodlands are unlikely to decrease the timber deficit, nor will they significantly affect the supplies available to industrial processors, but they should thrive as an adjunct to the traditional forestry industry.

Britain has almost no tradition of farm forestry, unlike most of the rest of Europe, where woodlands are often seen as a natural and integral part of farm management. Some use long-established woods, others have only been involved during the last two decades. In Britain new plantings of trees will vary from small areas that may be difficult to cultivate due to their position, shape or landform, to larger areas where more traditional forestry patterns may be employed.

The central problem of forestry on a farm is the time interval between planting and harvesting the timber crop. The traditional forestry use of discounting techniques is not appropriate to the farm situation. Farmers traditionally do not aim to maximize economic returns in the long term. The annual picture of profit and loss is the most important, the requirement to trade profitably, survival with a satisfactory income rather than a maximized one. This is not true of the new style 'agri-businessmen'.

Although many farmers are already aware of the use of trees and woods in the landscape, most existing farm woodlands are largely unmanaged. They have been neglected or partially felled with little regard for replanting or regeneration. Whilst such neglect may provide excellent conditions for a wide range of birds it is likely to detract from the economic value of the woodland.

The output from farm forestry as cash may be intermittent, as lump sums, or as a sustained flow of income. In economic terms, the farmer is making use of farm resources with low opportunity costs. The timing of most forestry operations can be slotted easily into other farm work on most farms, since many of them are much less dependent on day-to-day weather conditions. Necessary farm labour will have an opportunity cost but this may be zero at certain times of the year. Most farmers are able to provide high inputs of labour for intensive silviculture (pruning, thinning, etc.) with the aim of more specialist markets than extensive forestry. Timber production in Britain is concentrated on a very few species and this has advantages when selling into the timber trade. Because many plantations are concentrated on high ground and poor soils there is a predominance of conifers. On sheltered, fertile, lowland sites a much wider range of species can be grown than on upland and less fertile sites. It is only these better sites that can produce top quality broadleaf timber. Forestry is beginning to 'come down from the hill'.

In many parts of Britain it has been traditional to plant broadleaves in mixtures with conifers. For farm woodlands, one of the main advantages is earlier financial returns from the faster growing species which mature earlier. On most sites, growth of broadleaves in mixture with conifers is superior to that of a pure crop and will often produce taller, straighter stems on the broadleaves. The expected conifer rate of growth should never be more than double that for the broadleaf. Row mixtures can create landscape problems with 'pyjama stripe' effects, avoidable by planting the broadleaves in groups in a matrix of conifers. However, the use of mixtures or a wide range of species may bring management and marketing problems later in the life of the wood.

Timber markets usually handle large quantities of the commodity and this poses a difficulty for the sale of timber from small farm woodlands. High quality timber can be felled and sold in small quantities, even lorry loads. Small-scale harvesting usually means gradual regeneration, maintenance of game and wildlife cover, visual amenity and landscape continuity. Production of the highest quality trees will thus allow management to move towards several other objectives for farm woodlands.

It is important, when planning new farm woodlands, to consider the existing landscape character of the area (Hibberd, 1988, Insley, 1988). The main features to look at are the landform, existing vegetation patterns (especially semi-natural vegetation), land-use patterns (in particular the prominence of hedgerows and hedgerow tree patterns) and the scale of the landscape (for instance whether the landscape is wide and rolling, or small and enclosed, as in a narrow valley). Enhancing the role of woodlands in the landscape should not be seen as an alternative to management for timber production, sport or other use, but as a complementary objective.

When land is converted into woodland the wildlife on that land is transformed. The type of land upon which the Farm Woodland Scheme is focused is of little value in its existing state and the new woodland should be a considerable improvement. Woods can eventually be made wildlife-rich by careful attention to design and choice of species at the outset. Larger woods (5+ ha) tend to contain more structural diversity. The edge between trees and open areas such as rides and glades is usually good for wildlife, especially if it is irregular. These open areas will eventually be used for access, timber storage and machinery turning. Wildlife value is also increased by retaining open areas next to the sides of streams and by not draining and planting wet areas but letting a marsh/bog community develop or by creating ponds. Native tree species are generally richer in wildlife than are exotic species so, if the main priority is conservation, planting ideally should be restricted to native species. Many shrub species colonize new woods naturally once the young trees are large enough to attract birds, but it is beneficial to help this process by planting some hazel, dog rose or similar native shrub species along margins and the edge of glades. Small but connected areas of broadleaved trees arranged between larger blocks of conifers will provide refuges and reservoirs from which many plants and animals can re-invade. This 'string of pearls' arrangement is particularly valuable along stream sides and running through large areas of one tree crop, rather like the 'corridors' provided by hedges in farmland. Valuable work on the conservation of farm woodland and small-scale planting has already been done by Farming and Wildlife Advisory Groups.

Game shooting rents can contribute much more quickly and sometimes contribute a far higher sum to the farm economy than the timber crop of woodland. However, this is only if the design and management of the woods are laid out and conducted with game conservation and shooting in mind. This may result in the sacrifice of 10–25% of the potential timber production area. Game coverts traditionally take 10 years or so before trees and shrubs make sufficient growth to hold pheasants and even then it is only when protective rabbit netting is removed that the pedestrian pheasant uses such plantations to the full. Not all woodlands are suitable, however; cold, draughty woods have never and will never hold game. A new technique, called the 'instant spinney' was developed by the Game Conservancy in the 1970s to assist landowners creating game coverts. Individual tree guards are used against rabbits, hares and, if necessary, deer, while a sheltering game cover crop is grown between the tree rows. The crop makes it possible for the new wood to be used for shooting within one year, whilst the individual guards avoid the need for rabbit netting.

Many farmers close to large urban areas are developing the recreation potential of their farms, including farm woodlands. Mature woodland, or woodland with a diverse structure of tree crops of different ages, open areas and rides, and additional features such as water, is particularly suitable for recreation. Visitor trails, pony trekking or riding circuits and survival and war games are all suitable uses for farm woodlands. Woodland absorbs sound to some extent and can therefore be used as a screen for activities such as clay pigeon shooting or motor cycling. However, if public access to woodland is encouraged it is important to consider all aspects of legal liability.

The management objectives of conservation, recreation, sympathetic landscaping, game management and timber production can all be achieved in the same woodland, but it is wrong to expect that there will never be conflicts.

11.5 Conservation and habitat creation in planted woodlands

After reaching a low point early in the twentieth century the area of woodlands and forests in Britain has risen steadily. Until the 1980s, however, the area under broadleaves continued to fall and losses of ancient semi-natural broadleaved woodland were particularly severe. At this time nature conservation policies became more widespread in public and private woodlands; indeed, in areas such as the Wyre Forest, the Forestry Commission began allocating 10% of total surface area to conservation. Woodland communities are, as we have seen, extremely complex and conservation policies often have to give priority to particular groups, or even species, in some sites or habitats. Most woodland butterflies are found in sunny glades and many are quite demanding in the plants needed to provide foliage for the caterpillar or nectar for the adult (sections 1.2 and 5.7). The Speckled Wood (*Pararge aegeria*) is more of a generalist, tolerating shady rides and having caterpillars which feed on grasses.

Habitat creation may be concerned with increasing the number and diversity of plant and animal species in plantations, some of which may have been created on former agricultural land with residual problems of over-fertility. Here there is a particular need to enhance both the shrub and the herb flora (Cohn and Packham, 1991). This is of value in itself and also provides a greater range of food plants for animals. A much more difficult task is to create a complete species-rich woodland from a bare field. While this is possible on a very small scale, by transplanting large trees and then using essentially horticultural techniques simultaneously to establish the shrubs and herbs, it is only worth doing so if the object is the artificial production of small blocks of particular community types. Larger areas, such as the New National Forest with its projected 30 million trees in the English Midlands, can be created less expensively by planting shrubs and small trees. The nutrient and shade regimes can then be manipulated to favour woodland herb species when these are introduced to future recreational areas at appropriate stages in canopy closure. Are such procedures valid and, if successful, will they encourage further destruction of original native habitats? The answer to the first question is that the public is entitled to the recreational use of attractive species-rich woodland. Moreover, if this need can be met by the use of set-aside land in places close to large centres of population it will relieve the pressure on ancient woodland conserved for scientific purposes.

The second question is more difficult, but there are cases where the conservation interest is completely overruled and habitats are destroyed forever. Here, expertise in habitat creation or transfer can at least save something, as in the case of Biggins Wood at the Channel Tunnel Terminal site at Folkestone, Kent, where topsoil from over a hectare was moved to a new site and the woodland herbs salvaged (Buckley and Knight, 1989). In such construction contracts very large sums of money are involved and with greater financial compensation it would be possible to have far more complete transfers involving plant litter, rotting wood and ancient trees. The latter might not survive but would certainly aid in the transfer of associated invertebrates, lichens, fungi and slime fungi.

Streams, ponds, glades and rides are important features in forest designs. Ponds and streams may be adversely affected by sedimentation if their margins are disturbed by felling, but adequate light encourages aquatic life. In the Wyre Forest care is taken to avoid the invasion of valuable wetland sites by trees, while populations of adders (*Vipera berus*) are favoured by providing sunny clearings for basking in the sun. Coppicing rotations encourage birds such as the nightingale and tree pipit. Many forests now have hides from which deer and birds can be observed. Bird and bat boxes may also be employed.

Glade orientation and width are very important, and calculations have been made for maximum possible sunshine duration between March and September. On flat ground at 52°N, the latitude of Ipswich, the southern aspect of an east–west ride 4.5 m wide surrounded by trees 5 m high (or a ride 18 m wide with trees 20 m high) will receive 10 hours of sunshine centred on midday. In contrast, the northern aspect receives only 1.5 hours before 7 a.m. and a similar amount after 5 p.m. (Carter and Anderson, 1987). North–south rides receive 2.75 hours at equinox and 3.75 hours at midsummer for the same glade width, with the eastern aspect receiving direct light in the morning only and the western in the afternoon; the centre of the ride will receive 4–5 hours continuous direct light centred on midday. In northern Britain, where sun angles are lower, diversification will be enhanced by north–south rides on southerly slopes, particularly when the trees are very tall.

The presence of power or pipe lines sometimes makes the introduc-tion of long straight gaps inevitable and they act as useful firebreaks. However the landscape effect is considerably enhanced if rides are not straight; as in a well-designed garden fresh vistas should unfold as the visitor travels along the track. Much used forest roads are often metalled. Here an appropriate choice of rock type is important, rolling in Wenlock Limestone may encourage the subsequent development of calcicoles in woodlands whose soils are predominantly acid. The damp ditches along the road margins will encourage rushes (*Juncus effusus*, *J. conglomeratus* and *J. articulatus*), creeping buttercup (*Ranunculus repens*) and coltsfoot (*Tussilago farfara*). It is often necessary to cut willows and birches which establish here. A quite different flora, often with woodrushes (*Luzula* spp.), bilberries and common tormentil (*Potentilla erecta*), is found on the well-drained banks above.

Management of woodland edges has received much attention in recent years (Ferris-Kaan, 1991) and Figure 11.2 shows how road and ride margins can be scalloped by cutting opposing or staggered bays to enhance the visual effect and to allow more light to penetrate. The taller the surrounding trees the longer the bays need to be, 7 m is the minimum useful length even when the trees are small. Opposed bays let more light reach the ground. Management should encourage average plant height to increase away from the ride margin whose short-turf plants should be cut annually. Tall grass swards need to be cut every 2–5 years and the shrubs every 3–7 years. Cutting, which prevents dense scrub formation and habitat loss, is best carried out in October with brash being removed from glade margins. It is desirable to bare 1 or 2 m² of ground in each bay and to arrange the maintenance system so that the various bays show different stages in the succession from open ground to scrub.

Habitat restoration is a valid option in ancient woodlands which have been neglected or replanted with conifers. In either case, opening up the wood so that the field layer receives more light is likely to encourage any remaining woodland herbs. Forestry operations usually bare patches of soil, where plants locally common long ago may arise from a persistent seed bank. If there is a possibility of selling coppice products ancient stools may be re-coppiced, if not they can be singled, and diversity encouraged by creating adequate rides and glades. Coppice stools surviving beneath conifers, and even sycamores, can sometimes be revived. Field layers which have gradually succumbed to bracken or bramble can be diversified by the periodic cutting of these potential dominants. Woodland field layers are influenced by many factors and their management is a complex process demanding constant vigilance. Deer populations, for example, provide a great deal of interest but must be controlled by culling or, as now seems a more practical proposition, excluded by fencing if tree regeneration is to continue.

The large-scale creation of diverse woodland communities of high quality provides an opportunity to increase our understanding of how they function, it is also the acid test of our abilities as ecologists.

(a)

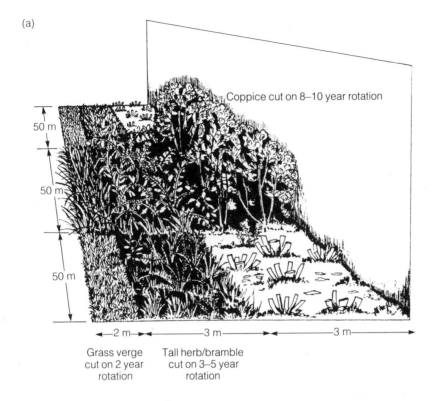

Coppice cut on 8–10 year rotation

50 m

50 m

50 m

←—2 m—→←————3 m————→←—————————3 m—————————→

Grass verge
cut on 2 year
rotation

Tall herb/bramble
cut on 3–5 year
rotation

(b)

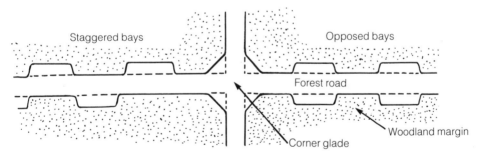

Staggered bays

Opposed bays

Forest road

Corner glade

Woodland margin

Figure 11.2 Modification of ride margins for conservation and visual amenity. (a) Cutting pattern along ride edges (P. R. Hobson). (b) Use of bays and corner glades along woodland rides (modified from Carter and Anderson, 1987).

Epilogue

Much has happened in the ten years since the publication of *Ecology of Woodland Processes*, the predecessor of this book. Although our knowledge of woodlands and forests continues to grow it is not equalled by our ability to prevent their destruction. In the Western world, and particularly in Europe, heightened public awareness through a variety of publications, radio, television, the growth of societies, conservation trusts and forest programmes should mean that management of the world's forests, a marvellous renewable resource, is more likely to be rooted in a sound appreciation of ecology, and natural economics, than hitherto. Let this be seen in effective action to prevent further destruction and desertification, particularly in the Third World.

Pearl-bordered Fritillary (*Bolonia euphryosoyne*) and associated plants (*Viola riviniana* and *Pteridium aquilinum*).

Opening female reproductive bud of hazel (*Corylus avellana*), the most characteristic of coppice shrubs. Note the glandular hairs and also the tuft of wind-
pollinated red stigmas present in early spring.

Burdekin, D. A. (ed.) (1983) Research on Dutch elm disease in Europe. *Forestry Commission Bulletin No. 60*. HMSO, London.

Burnham, C. P. (1970) The regional pattern of soil formation in Great Britain. *Scottish Geographical Magazine*, **86**, 25–34.

Burnham, C. P. and Mackney, D. F. (1964) Soils of Shropshire. *Field Studies*, **2**, 83–113.

Burrows, C. J. (1990) *Processes of Vegetation Change*, Unwin Hyman, London.

Carter, C. L. and Anderson, M. A. (1987) Enhancement of lowland forest ridesides and roadsides to benefit wild plants and butterflies. Research information note 126. Forestry Commission, Farnham.

Caughley, G. (1970) Eruption of ungulate populations, with emphasis on Himalayan thar in New Zealand. *Ecology*, **51**, 53–72.

Cherrett, J. M. (1989) Key concepts: the results of a survey of our members' opinions, in *Ecological Concepts: the contribution of ecology to an understanding of the natural world* (ed. J. M. Cherrett) Blackwell Scientific, Oxford, pp. 1–16.

Clark, D. A. and Clark, D. B. (1984) Spacing dynamics of a tropical rain forest tree: evaluation of the Janzen–Connell model. *The American Naturalist*, **124**, 769–88.

Clark, F. E. (1967) Bacteria in soil, in *Soil Biology* (eds A. Burges and F. Raw), Academic Press, London, pp. 15–49.

Clark, J. (1961) Photosynthesis and respiration in white spruce and balsam fir. New York State University College of Forestry Technical Publication 85.

Clements, F. E. (1916) Plant succession. Analysis of the development of vegetation. Carnegie Institute, Washington, No. 242.

Clements, F. E. (1936) Nature and structure of the climax. *Journal of Ecology*, **24**, 252–84.

Clifford, D. R. and Gendle, P. (1987) Treatment of fresh wound parasites and of cankers, in *Advances in Practical Arboriculture* (ed. D. Patch), HMSO, London, pp. 145–8.

Cohn, E. V. J. and Packham, J. R. (1991) The introduction and manipulation of woodland field layers: seeds, plants, timing and economics, in *Future Forests: lessons from experience* (ed. J. R. Packham), Wolverhampton Polytechnic Woodland Research Group, Wolverhampton, pp. 15–28.

Cole, D. W. and Rapp, M. (1981) Elemental cycling in forest ecosystems, in *Dynamic Principles of Forest Ecosystems* (ed. D. E. Reichle) Cambridge University Press, Cambridge, pp. 341–409.

Coles, J. M. (1989) The world's oldest road. *Scientific American*, **261(5)**, 78–84.

Compton, S. G., Thornton, I. W. B., New, T. R. and Underhill, L. (1988) The colonization of the Krakatau islands by fig wasps and other chalcids (Hymenoptera, Chalcidoidea). *Philosophical Transactions of the Royal Society of London, Series B*, **322**, 459–70.

Connell, J. H. (1979) Diversity in tropical rain forests and coral reefs. *Science*, **199**, 1302–10.

Connell, J. H. and Sousa, W. P. (1983) On the evidence needed to judge ecological stability or persistence. *American Naturalist*, **121**, 789–824.

Conway, G. (1976) Man versus pests, in *Theoretical Ecology, Principles and Applications* (ed. R. M. May), Blackwell Scientific, Oxford, pp. 257–81.

Cooke, A. (1983) The effects of fungi on food selection by *Lumbricus terrestris* L., in *Earthworm Ecology* (ed. J. E. Satchell), Chapman and Hall, London, pp. 365–73.

Cooper, W. S. (1913) The climax forest of the Isle Royale, Lake Superior, and its development. I. *Botanical Gazette*, **55**, 1–44.

Corner, E. J. H. (1964) *The Life of Plants*, Weidenfeld and Nicolson, London.

References

Adamson, R. S. (1912) An ecological study of a Cambridgeshire woodland. *Journal of the Linnean Society (Botany)*, **40**, 339–87.

Alexander, I. (1989) Mycorrhizas in tropical forests, in *Mineral Nutrients in Tropical Forest and Savanna Ecosystems* (ed. J. Proctor), Blackwell Scientific, Oxford, pp. 169–88.

Amman, G. D. (1977) The role of the mountain pine beetle in lodgepole pine ecosystems: impact on succession, in *The Role of Arthropods in Forest Ecosystems* (ed. W. J. Mattson), Springer, New York, pp. 3–18.

Anderson, J. M. (1971) Observations on the vertical distribution of Oribatei (Acarina) in two woodland soils. *IV. Colloquium Pedobiologiae* I.N.R.A., Paris, pp. 257–72.

Anderson, J. M. (1973) The breakdown and decomposition of sweet chestnut (*Castanea sativa* Mill.) and beech (*Fagus sylvatica* L.) leaf litter in two deciduous woodland soils. I and II. *Oecologia (Berlin)*, **12**, 251–74; 275–88.

Anderson, J. M. (1975) Succession, diversity and trophic relationships of some soil animals in decomposing leaf litter. *Journal of Animal Ecology*, **44**, 475–95.

Anderson, J. M. (1978a) Competition between two unrelated species of soil Cryptostigmata (Acari) in experimental microcosms. *Journal of Animal Ecology*, **47**, 787–803.

Anderson, J. M. (1978b) Inter- and intra-habitat relationships between woodland Cryptostigmata species diversity and the diversity of soil and litter microhabitats. *Oecologia (Berlin)*, **32**, 341–8.

Anderson, J. M. and Bignell, D. E. (1980) Bacteria in the food, gut contents and faeces of the litter-feeding millipede *Glomeris marginata* (Villers). *Soil Biology and Biochemistry*, **12**, 251–4.

Anderson, J. M., Huish, S. A., Ineson, P. *et al.* (1985) Interactions of invertebrates, micro-organisms and tree roots in nitrogen and mineral element fluxes in deciduous woodland soils, in *Ecological Interactions in soil* (eds A. H. Fitter, D. Atkinson, D. J. Read and M. B. Usher), Blackwell Scientific, Oxford, pp. 377–92.

Anderson, J. M. and Macfadyen, A. (eds) (1976) *The Role of Terrestrial and Aquatic Organisms in Decomposition Processes*. Blackwell Scientific, Oxford.

Anderson, J. M. and Swift, M. J. (1983) Decomposition in tropical forests, in *Tropical Rain Forest: ecology and management* (eds S. L. Sutton, T. C. Whitmore and A. C. Chadwick), Blackwell Scientific, Oxford, pp. 287–309.

Anderson, J. P. E. and Domsch, K. H. (1980) Quantities of plant nutrients in the microbial biomass of selected soils. *Soil Science*, **130**, 211–16.

Anderson, M. C. (1964a) Studies of the woodland light climate I. The photographic computation of light conditions. *Journal of Ecology*, **52**, 27–41.

Anderson, M. C. (1964b) Studies of the woodland light climate II. Seasonal variation in the light climate. *Journal of Ecology*, **52**, 643–63.

Angevine, M. W. and Handel, S. N. (1986) Invasion of forest floor space, clonal architecture and population growth in the perennial herb *Clintonia borealis*. *Journal of Ecology*, **74**, 547–60.

Archibald, J. F. and Stubbs, A. E. (1980) The effects of Dutch elm disease on wildlife. *Quarterly Journal of Forestry*, **74**, 30–7.

Ashton, P. S. (1977) A contribution of rain forest research to evolutionary theory. *Annals of the Missouri Botanic Garden*, **64**, 649–705.

Aubreville, A. (1938) La foret coloniale: les forets de l`Afrique occidentale francais. *Annales Academie des Sciences coloniales, Paris*, **9**, 1–245.

Ausmus, B. S., Edwards, N. T. and Witkamp, M. (1976) Microbial immobilization of carbon, nitrogen, phosphorus and potassium: implications for forest ecosystem processes, in *The Role of Terrestrial and Aquatic Organisms in Decomposition Processes* (eds J. M. Anderson and A. Macfadyen), Blackwell Scientific, Oxford, pp. 397–416.

Auspurger, C. K. (1983) Offspring recruitment around tropical trees: changes in cohort distance with time. *Oikos*, **40**, 189–96.

Ayensu, E. S. (ed.) (1980) *Jungles*, Jonathan Cape, London.

Baker, H. G. (1973) Evolutionary relationships between flowering plants and animals in American and African tropical forests, in *Tropical Forest Ecosystems in Africa and South America: a comparative review* (eds B. J. Meggers, E. S. Ayensu and W. D. Duckworth), Smithsonian Institution Press, Washington, pp. 145–59.

Baltensweiler, W. (1984) The role of environment and reproduction in the population dynamics of the larch bud moth, *Zeiraphera diniana* Gn. (Lepidoptera, Tortricidae). *Advances in Invertebrate Reproduction*, **3**, 291–302.

Barbour, D. A. (1985) Patterns of population fluctuation in the pine looper moth *Bupalus piniaria* L. in Britain, in *Site Characteristics and Population Dynamics of Lepidopteran and Hymenopteran Forest Pests* (eds D. Bevan and J. T. Stoakley), Forestry Commission, Edinburgh, pp. 8–20.

Barbour, D. A. (1988) The Pine Looper in Britain and Europe, in *Dynamics of Forest Insect Populations* (ed. A. A. Berryman), Plenum, New York, pp. 292–308.

Barbour, D. A. (1990) Synchronous fluctuations in spatially separated populations of cyclic forest insects, in *Population Dynamics of Forest Insects* (eds A. D. Watt, S. R. Leather, M. D. Hunter and N. A. C. Kidd), Intercept, Andover, pp. 339–46.

Begon, M., Harper, J. L. and Townsend, C. R. (1990) *Ecology: individuals, populations and communities*, 2nd edn, Blackwell Scientific, Oxford.

Berryman, A. A. (ed.) (1988) *Dynamics of Forest Insect Populations: patterns, causes, implications*, Plenum Press, New York.

Bevan, D. (1974) Control of forest insects: there is a porpoise close behind us, in *Biology in Pest and Disease Control* (eds D. Price Jones and M. E. Solomon), Blackwell Scientific, Oxford, pp. 302–12.

Biddle, P. G. (1987) Trees and buildings, in *Advances in Practical Arboriculture* (ed. D. Patch), HMSO, London, pp. 121–32.

Binns, W. O., Mayhead, G. J. and Mackenzie, J. M. (1980) *Nutrient Deficiencies of Conifers in British Forests*, HMSO, London.

Birks, H. J. B., Deacon, J. and Peglar, S. (1975) Pollen maps for the British Isles 5000 years ago. *Proceedings of the Royal Society of London, Series B*, **189**, 87–105.

Blackman, G. E. and Rutter, A. J. (1946) Physiological and ecological studies in the analysis of plant environment. I. The light factor and the distribution of the bluebell (*Scilla non-scripta*) in woodland communities. *Annals of Botany, New Series*, **10**, 361–90.

Blackman, G. E. and Rutter, A. J. (1954) *Endymion non-scriptus* (L.) Garcke. Biological flora of the British Isles. *Journal of Ecology*, **42**, 629–38.

Blank, L. W., Roberts, T. M. and Skeffington, R. A. (1988) New forest decline. *Nature*, **336**, 27–30.

Boardman, N. K. (1977) Comparative photosynthesis of sun a: *Annual Review of Plant Physiology*, **28**, 355–77.

Bobek, B., Boyce, M. S. and Kosobucka, M. (1984) Factors a (*Cervus elephus*) population density in south-eastern Po *Applied Ecology*, **21**, 881–90.

Bonan, G. B. and Shugart, H. H. (1989) Environmental facto processes in boreal forests. *Annual Review of Ecology and Syste*

Bond, W. J. (1989) The tortoise and the hare: ecology of angios and gymnosperm persistence. *Biological Journal of the Lin* 227–49.

Bormann, B. T. and Sidle, R. C. (1990) Changes in productivit of nutrients in a chronosequence at Glacier Bay National Par *of Ecology*, **78**, 561–78.

Borschmann, G. (1984) *Greater Daintree. World Heritage Tropica* Australian Conservation Foundation, Hawthorn, Victoria.

Boyd, J. M. (1987) Commercial forests and the nature cons *Forestry*, **60**, 113–34.

Bradshaw, A. D., Biddle, P. G., Patch, D. and Spurway, P. E. the review group on research on arboriculture. *Arboricu* 307–60.

Brasier, C. M. (1983) The future of Dutch elm disease in in *Forestry Commission Bulletin No. 60* (ed. D. A. Burdekin) pp. 96–104.

Brasier, C. M. and Gibbs, J. N. (1973) Origin of the Dutch eln in Britain. *Nature*, **242**, 607–9.

Brasier, C. M. and Gibbs, J. N. (1975) Highly fertile form of tl of *Ceratocystis ulmi*. *Nature*, **257**, 128–31.

Bray, J. R. and Gorham, E. (1964) Litter production in fo *Advances in Ecological Research*, **2**, 101–57.

Brazier, J. D. (1979) Information Paper 12/79. Building Rese Princes Risborough Laboratory.

Brown, A. H. F. (1974) Nutrient cycles in oakwood ec England, in *The British Oak* (eds M. G. Morris and F. H. Per Farringdon, pp. 141–61.

Brown, A. H. F. (1981) The role of buried seed in coppicewoc vation, **21**, 19–38.

Brown, K. A., Roberts, T. M. and Blank, L. W. (1987) Interac and cold sensitivity in Norway spruce: a factor contributing in central Europe? *New Phytologist*, **105**, 149–55.

Brown, R. T. and Curtis, J. T. (1952) The upland conifer- northern Wisconsin. *Ecological Monographs*, **22**, 217–34.

Brown, V. K. and Gange, A. C. (1990) Insect herbivory bel in *Ecological Research*, **20**, 1–58.

Buckley, G. P. (ed.) (1992) *The Ecology and Management* Chapman and Hall, London.

Buckley, G. P. and Knight, D. G. (1989) The feasibility of v tion, in *Biological Habitat Reconstruction* (ed. G. P. Buckl London, pp. 171–88.

Bunce, R. G. H. (1977) The range of variation within the *Pinewoods of Scotland* (eds R. G. H. Bunce and J. N. R. Terrestrial Ecology, Cambridge, pp. 10–25.

Cousens, J. (1974) *An Introduction to Woodland Ecology*, Oliver and Boyd, Edinburgh.

Crawley, M. J. (1983) *Herbivory: the dynamics of animal–plant interactions*, Blackwell Scientific, Oxford.

Crocker, R. L. and Major, J. (1955) Soil development in relation to vegetation and surface age at Glacier Bay, Alaska. *Journal of Ecology*, **43**, 427–48.

Cromer, D. A. N. and Pawsey C. K. (1957) Initial spacing and growth of *Pinus radiata*. *Forestry and Timber Bureau Bulletin* 37, Canberra.

Crompton, E. (1962) Soil formation. *Outlook on Agriculture* **3**, 209–18.

Cutler, D. F. (1991) Tree planting for the future: lessons of the storms of October 1987 and January 1990. *Arboricultural Journal*, **15**, 225–34.

Cutler, D. F., Gasson, P. E. and Farmer, M. C. (1989) The wind blown tree root survey: preliminary results. *Arboricultural Journal*, **13**, 219–36.

Cutler, D. F., Gasson, P. E. and Farmer, M. C. (1990) The wind blown tree survey: analysis of results. *Arboricultural Journal*, **14**, 265–86.

Daniel, T. W., Helms, J. A. and Baker, F. S. (1979) *Principles of Silviculture*, McGraw-Hill, New York.

Darlington, A. (1974) The galls on oak, in *The British Oak* Botanical Society of the British Isles (eds M. G. Morris and F. H. Perring), E. W. Classey, Faringdon, pp. 298–311.

Darwin, C. (1859) *Origin of Species*, John Murray, London.

Daubenmire, R. (1952) Forest vegetation of Northern Idaho and adjacent Washington, and its bearing on concepts of vegetation classification. *Ecological Monographs*, **22**, 301–30.

Daubenmire, R. (1966) Vegetation: identification of typal communities. *Science*, **151**, 291–8.

Davies, R. J. (1987) *Trees and Weeds: weed control for successful tree establishment*, HMSO, London.

Deevey, E. S. (1947) Life tables for natural populations of animals. *Quarterly Review of Biology*, **22**, 283–314.

de Kroon, H. (1986) De vegetaties van Zuidlimburgse hellingbossen in relatie tot het hakhoutbeheer. *Natuurhistorich Maanblad*, **75(10)**, 167–92.

De Silva, B. L. T. (1934) The distribution of 'Calcicole' and 'Calcifuge' species in relation to the content of the soil in calcium carbonate and exchangeable calcium, and to soil reaction. *Journal of Ecology*, **22**, 532–53.

Dempster, J. P. (1975) *Animal Population Ecology*, Academic Press, London.

Dickinson, C. H. and Preece, T. F. (eds) (1976) *Microbiology of Aerial Plant Surfaces*, Academic Press, London.

Dixon, A. F. G. (1970) Quality and availability of food for a sycamore aphid population, in *Animal Populations in Relation to their Food Resources* (ed. A. Watson), Blackwell Scientific, Oxford, pp. 271–87.

Dixon, A. F. G. (1971a) Aphids, in *Methods of Study in Quantitative Soil Ecology* (ed. J. Phillipson), Blackwell Scientific, Oxford, pp. 233–46.

Dixon, A. F. G. (1971b) The role of aphids in wood formation. I and II. *Journal of Applied Ecology*, **8**, 165–79; 393–9.

Dixon, A. F. G. (1977) Aphid ecology: life cycles, polymorphism and population regulation. *Annual Review of Ecology and Systematics*, **8**, 329–53.

Dixon, A. F. G. (1979) Sycamore aphid numbers: the role of weather, host and aphid, in *Population Dynamics* (eds R. M. Anderson, B. D. Turner and L. R. Turner), Blackwell Scientific, Oxford, pp. 105–21).

Dixon, A. F. G. (1990) Population dynamics and abundance of deciduous tree-dwelling aphids, in *Population Dynamics of Forest Insects* (eds A. D. Watt, S. R. Leather, M. D. Hunter and N. A. C. Kidd), Intercept, Andover, pp. 11–23.

Drift, J. van der and Witkamp M. (1959) The significance of the breakdown of oak litter by *Enoicyla pusilla* Burm. *Archives Néerlandaises de Zoologie*, **13**, 486–92.

Dunn, E. (1977) Predation by weasels *Mustela nivalis* on breeding tits (*Parus* spp.) in relation to the density of tits and rodents. *Journal of Animal Ecology* **46**, 633–52.

Duvigneaud, P. (1967) La productivité primaire des écosystèmes terrestres, in *Problèmes de Productivité Biologique* (eds M. Lamotte and F. Bourlière) Masson, Paris, pp. 37–92.

Duvigneaud, P. (ed.) (1971) *Productivity of Forest Ecosystems*, UNESCO, Paris.

Edwards, C. A. (ed.) (1988) Biological interactions in soil. *Agriculture, Ecosystems and Environment*, Special volume, **24**, 377 pp.

Edwards, C. A. and Heath, G. W. (1963) The role of soil animals in breakdown of leaf material, in *Soil Organisms* (eds J. Doeksen and J. van der Drift), North Holland, Amsterdam, pp. 76–84

Edwards, M. V. and Howell, R. S. (1962) Planning an experimental programme for species trials. Paper to *8th British and Commonwealth Forestry Conference East Africa*, Forestry Commission, London.

Edwards, P. J. (1982) Studies of mineral cycling in a montane rain forest in New Guinea. V. Rates of cycling in throughfall and litter fall. *Journal of Ecology*, **70**, 807–27.

Edwards, P. J. (1989) Insect herbivory and plant defence theory, in *Toward a more Exact Ecology* (eds P. J. Grubb and J. B. Whittaker), Blackwell Scientific, Oxford, pp. 275–97.

Egler, F. E. (1954) Vegetation science concepts. I. Initial floristic composition, a factor in old-field vegetation development. *Vegetatio*, **4**, 412–17.

Ellenberg, H. (1988) *Vegetation Ecology of Central Europe*, 4th edn, Cambridge University Press, Cambridge.

Elliston, J. E. (1982) Hypovirulence. *Advances in Plant Pathology*, **1**, 1–33.

Elton, C. S. (1927) *Animal Ecology*, Sidgwick and Jackson, London.

Elton, C. S. (1966) *The Pattern of Animal Communities*, Methuen, London.

Ernst, W. H. O. (1979) Population biology of *Allium ursinum* in northern Germany. *Journal of Ecology*, **67**, 347–62.

Evans, G. C. (1976) A sack of uncut diamonds: the study of ecosystems and the future resources of mankind. *Journal of Animal Ecology*, **45**, 1–39.

Evans, G. C., Bainbridge, R. and Rackham, O. (eds) (1975) *Light as an Ecological Factor*, Vol. II, Blackwell Scientific, Oxford.

Evans, G. C. and Coombe, D. E. (1959) Hemispherical and woodland canopy photography and the light climate. *Journal of Ecology*, **47**, 103–13.

Evans, J. (1984) *Silviculture of Broadleaved Woodland*, HMSO, London.

Evans, J. (1992) in *The Ecology and Management of Coppice Woodland* (ed. G. P. Buckley), Chapman and Hall, London.

Evers, P. W., Donkers, J., Prat, A. and Vermeer, E. (1988) *Micropropagation of Forest Trees through Tissue Culture*, PUDOC, Wageningen.

Falinski, J. B. (1986) *Vegetation Dynamics in Temperate Lowland Primeval Forests: ecological studies in Bialowieza forest*, Dr W. Junk, Lancaster.

Faulkner, R. (1987) Genetics and breeding of Sitka spruce. *Proceedings of the Royal Society of Edinburgh*, **93B**, 41–50.

Feeny, P. (1970) Seasonal changes in oak leaf tannins and nutrients as a cause of spring feeding by winter moth caterpillars. *Ecology*, **51**, 565–81.

Ferris-Kaan, R. (ed.) (1991) *Edge management in Woodlands*, Forestry Commission, Edinburgh.

Flanagan, P. W. and Van Cleve, K. (1977) Microbal biomass, respiration and nutrient cycling in a black spruce taiga ecosystem. *Ecological Bulletin*, **25**, 261–73.

Flenley, J. R. (1979) *The Equatorial Rain Forest: a geological history*, Butterworth, London.

Flowerdew, J. R. (1987) *Mammals: their reproductive biology and population ecology*, Edward Arnold, London.

Flowerdew, J. R. and Gardner, G. (1978) Small rodent populations and food supply in a Derbyshire ashwood. *Journal of Animal Ecology*, **47**, 725–40.

Ford, E. D. and Newbould, P. J. (1977) The biomass of ground vegetation and its relation to tree cover through a deciduous woodland cycle. *Journal of Ecology*, **65**, 201–12.

Foster, J. R. (1988) The potential role of rime ice defoliation in tree mortality of wave-regenerated balsam fir forests. *Journal of Ecology*, **76**, 172–80.

Foster, D. R. and King, G. A. (1986) Vegetation pattern and diversity in S. E. Labrador, Canada: *Betula papyrifera* (birch) forest development in relation to fire history and physiography. *Journal of Ecology*, **74**, 465–83.

Fowler, D., Cape, J. N. and Unsworth, M. H. (1989) Deposition of atmospheric pollutants on forests. *Philosophical Transactions of the Royal Society of London, Series B*, **324**, 247–65.

Fowler, P. J. (1983) *The Farming of Prehistoric Britain*, Cambridge University Press, Cambridge.

Frome, M. (1984) *The Forest Service*, 2nd edn, Westview Press, Boulder, Colorado.

Froment, A., Tanghe, M., Duvigneaud, P. *et al.* (1971) La chênaie mélangée calcicole de Virelles-Blaimont, en haute Belgique, in *Productivity of Forest Ecosystems* (ed. P. Duvigneaud), UNESCO, Paris, pp. 635–65

Fuller, R. J., Stuttard, P. and Ray, C. M. (1989) The distribution of breeding songbirds within mixed coppice woodland in Kent, England, in relation to vegetation age and structure. *Ann. Zool. Fennici*, **26**, 265–75.

Gardner, G. (1977) The reproductive capacity of *Fraxinus excelsior* on the Derbyshire limestone. *Journal of Ecology*, **65**, 107–18.

Garrett, S. D. (1951) Ecological groups of soil fungi; a survey of substrate relationships. *New Phytologist*, **50**, 149–66.

Gasson, P. E. and Cutler, D. F. (1990) Tree root plate morphology. *Arboricultural Journal*, **14**, 193–264.

Gause, G. F. (1934, reprinted 1964) *The Struggle for Existence*, Hafner, New York.

Gholz, H. L. (1987) *Agroforestry: realities, possibilities and potentials*, ICRAF, Lancaster.

Gibbs, J. N. and Greig, B. J. W. (1989) Investigations on tree damage caused by the great gale, in 'After the storm – one year on' *Proceedings of Tree Council Conference*, 4 October 1988, Oxford. Tree Council, London.

Gibbs, J. N., Greig, B. J. W. and Rishbeth, J. (1992) The diseases of Thetford Forest and their influence on its ecology and management, in *Thetford Forest Park: the ecology of a pine forest* (eds P. R. Ratcliffe and S. Harris), HMSO, London.

Gibbs, J. N. and Wainhouse, D. (1986) Spread of forest pests and pathogens in the northern hemisphere. *Forestry*, **59**, 141–53.

Gilbert, N. (1989) *Biometrical Interpretation*, Oxford University Press, Oxford.

Gimingham, C. H. and Birse, E. M. (1957) Ecological studies on growth-form in bryophytes. I. Correlations between growth-form and habitat. *Journal of Ecology*, **45**, 533–45.

Godwin, H. (1975) *The History of the British Flora*, 2nd edn, Cambridge University Press, Cambridge.

Goodland, R. J. A. and Irwin, H. S. (1975) *Amazon Jungle: green hell to red desert?*, Elsevier, Amsterdam.

Gosz, J. R. (1981) Nitrogen cycling in coniferous ecosystems, in *Terrestrial Nitrogen Cycles* (eds F. E. Clark and T. Rosswall), Ecological Bulletin (Stockholm), **33**, 405–26.

Gosz, J. R., Holmes, R. T., Likens, G. E. and Bormann, F. H. (1978) The flow of energy in a forest ecosystem, *Scientific American*, **238(3)**, 92–102.

Grabham, P. N. and Packham, J. R. (1983) A comparative study of the bluebell, *Hyacinthoides non-scripta* (L) Chouard, in two different woodland situations in the West Midlands, England. *Biological Conservation*, **26**, 105–26.

Gradwell, G. R. (1974) The effect of defoliators on tree growth, in *The British Oak* (eds M. G. Morris and F. H. Perring), E. W. Classey, Faringdon, pp. 182–92.

Grainger, A. (1990) *The Threatening Desert*, Earthscan, London.

Grime, J. P. (1966) Shade avoidance and shade tolerance in flowering plants, in *Light as an Ecological Factor* (eds R. Bainbridge, G. C. Evans and O. Rackham), Blackwell Scientific, Oxford, pp. 187–207.

Grime, J. P. (1974) Vegetation classification by reference to strategies. *Nature*, **250**, 26–31.

Grime, J. P. (1979) *Plant Strategies and Vegetation Processes*, Wiley, Chichester.

Grime, J. P., Hodgson, J. G. and Hunt, R. (1988) *Comparative Plant Ecology*, Unwin Hyman, London.

Grime, J. P. and Lloyd, P. S. (1973) *An Ecological Atlas of Grassland Plants*, Edward Arnold, London.

Grubb, P. J. (1977) The maintenance of species-richness in plant communities: the importance of the regeneration niche. *Biological Review*, **52**, 107–45.

Gurnell, J. (1987) *The Natural History of Squirrels*, Christopher Helm, London.

Gysel, L. W. (1971) A 10-year analysis of beechnut production and use in Michigan. *Journal of Wildlife Management*, **35**, 516–19.

Haddock, P. G., Walters, J. and Kozak, A. (1967) Growth of coastal and interior provenances of Douglas fir (*Pseudotsuga menziesii* (Mirb.) Franco) at Vancouver and Haney in British Columbia. Forestry Research Paper No. 79, University of British Columbia.

Hamilton, G. J. (1985) *Forest Mensuration Handbook*, HMSO, London.

Hamilton, G. J. and Christie, J. M. (1971) Forest Management Tables (metric). *Forestry Commission Booklet, No. 34*, HMSO, London.

Hamilton, G. J. and Christie, J. M. (1974) Influence of spacing on crop characteristics and yield. *Forestry Commission Booklet, No. 52*, HMSO, London.

Hanley, T. A. (1984) Habitat patches and their selection by wapiti and black-tailed deer in a coastal montane coniferous forest. *Journal of Applied Ecology*, **21**, 423–36.

Harborne, J. B. (1982) *Introduction to Ecological Biochemistry*, Academic Press, London.

Harding, D. J. L. (1967) Faunal participation in the breakdown of cellophane inserts in the forest floor, in *Progress in Soil Biology* (eds O. Graff and J. E. Satchell), North Holland, Amsterdam, pp. 10–20.

Harding, D. J. L. (ed.) (1987) *Agricultural Surpluses?*, Institute of Biology, London.

Harding, D. J. L. and Easton, S. M. (1984) Development of two species of phthiracarid mites in beech cupules, in *Acarology VI* (eds D. A. Griffiths and C. E. Bowman), Ellis Horwood, Chichester, pp. 860–70.

Harding, D. J. L. and Stuttard, R. A. (1974) Microarthropods, in *Biology of Plant Litter Decomposition*, Vols I and II (eds C. H. Dickinson and G. J. F. Pugh), Academic Press, London, pp. 489–532.

Harper, J. L. (1977) *Population Biology of Plants*, Academic Press, London.

Harper, J. L., Lovell, P. H. and Moore, K. G. (1970) The shapes and sizes of seeds. *Annual Review of Ecology and Systematics*, **1**, 327–56.

Harris, W. F. (1981) Root dynamics, in *Dynamic Properties of Forest Ecosystems* (ed. D. E. Reichle), Cambridge University Press, Cambridge, pp. 318–39.

Harrison, P. (1987) *The Greening of Africa*, Collins, London.

Hart, J. W. (1988) *Light and Plant Growth*, Unwin Hyman, London.

Hassell, M. P. (1980) Foraging strategies, population models and biological control: a case study. *Journal of Animal Ecology*, **49**, 603–28.

Hayes, A. J. (1979) The microbiology of plant litter decomposition. *Science Progress, Oxford*, **66**, 25–42.

Heal, O. W. and MacLean, S. F. (1975) Comparative productivity in ecosystems, in *Unifying Concepts in Ecology* (eds W. H. van Dobben and R. H. Lowe-McConnell), Dr W. Junk, The Hague, pp. 89–108.

Heding, N. (1969) Stem number reduction and diameter development in non-thinned Norway spruce stands with various spacings. *Forstilige Forsogsvaesen i Danmark* **XXXII(2)**, 193–243.

Helliwell, D. R. (1989) Tree roots and the stability of trees. *Arboricultural Journal*, **13**, 243–8.

Hibberd, B. G. (ed.) (1986, 1991) *Forestry Practice*, HMSO, London.

Hibberd, B. G. (1988) *Farm Woodland Practice*, HMSO, London.

Hill, M. O., Bunce, R. G. H. and Shaw, M. W. (1975) Indicator species analysis: a divisive polythetic method of classification and its application to a survey of native pinewoods in Scotland. *Journal of Ecology*, **63**, 597–613.

Hilton, G. M. and Packham, J. R. (1986) Annual and regional variation in English beech mast (*Fagus sylvatica*, L.). *Arboricultural Journal*, **10**, 3–14.

Hilton, G. M., Packham, J. R. and Willis, A. J. (1987) Effects of experimental defoliation on a population of pedunculate oak (*Quercus robur* L.). *New Phytologist*, **107**, 603–12.

Hobson, P. M. (1988) Methven Wood (Almondbank, Perth): the history of its management. *Scottish Forestry*, **44**, 104–12.

Hobson, P. R. and Packham, J. R. (1991) Multipurpose forests: lessons from the Wyre. In *Future Forests: lessons from experience* (ed. J. R. Packham), Wolverhampton Polytechnic Woodland Research Group, Wolverhampton, pp. 43–53.

Hodkinson, I. D. and Hughes, M. K. (1982) *Insect Herbivory*, Chapman and Hall, London.

Hong, S. O. (1975) Vegetative propagation of plant material for seed orchards with special reference to graft-incompatibility problems, in *Seed Orchards* (ed. R. Faulkner), HMSO, London, pp. 38–48.

Hopkins, B. (1965) *Forest and Savanna*, Heinemann, London.

Horn, H. S. (1971) *The Adaptive Geometry of Trees*, Princeton University Press, New Jersey.

Horn, H. S. (1975) Forest succession. *Scientific American*, **232(5)**, 90–8.

Howe, D. A. and Smallwood, J. (1982) Ecology of seed dispersal. *Annual Review of Ecology and Systematics*, **13**, 201–28.

Hubbell, S. P. (1980) Seed predation and the coexistence of tree species in tropical forests. *Oikos*, **34**, 214–29.

Hudson, H. J. (1968) The ecology of fungi on plant remains above the soil. *New Phytologist*, **67**, 837–74.

Hughes, A. P. (1959) Effects of the environment on leaf development in *Impatiens parviflora* DC. *Journal of the Linnean Society (Botany)*, **56**, 161–5.

Hunt, D. J. and Hague, N. G. M. (1974) The distribution and abundance of *Parasitaphelenchus aldhami*, a nematode parasite of *Scolytus scolytus* and S.

multistriatus, the bark beetle vectors of Dutch elm disease. *Plant Pathology*, **23**, 133–5.

Hutchings, M. J. and Barkham, J. P. (1976) An investigation of shoot interactions in *Mercurialis perennis* L., a rhizomatous perennial herb. *Journal of Ecology*, **64**, 723–43.

Hutchings, M. J. and Slade, A. J. (1988) *Plants Today*, **1**, 28–33.

Huxley, P. A. (1983) The tree/crop interface. Working paper 13. ICRAF, Nairobi.

Hytteborn, H. and Packham, J. R. (1985) Left to nature; forest structure and regeneration in Fiby urskog, central Sweden. *Arboricultural Journal*, **9**, 1–11.

Hytteborn, H. and Packham, J. R. (1987) Decay rate of *Picea abies* logs and the storm gap theory: a re-examination of Sernander Plot III Fiby urskog, central Sweden. *Arboricultural Journal*, **11**, 299–311.

Hytteborn, H., Packham, J. R. and Verwijst, T. (1987) Tree population dynamics, stand structure and species composition in the montane virgin forest of Vallibacken, northern Sweden. *Vegetatio*, **72**, 3–19.

Innes, J. L. (1987) *Air Pollution and Forestry*. HMSO, London.

Innes, J. L. and Boswell, R. C. (1987) Forest health surveys 1987, Part 1: Results. *Forestry Commission Bulletin 74*, HMSO, London.

Insley, H. (1988) *Farm Woodland Planning*, HMSO, London.

Iversen, J. (1941) Landnam i Danmarks Stenalder. En pollenanalytisk Undersogelse over det forste Landbrugs Indvirkning paa Vegetationsudviklingen. *Danmarks Geologiske Undersogelse* RII, No. 66.

Jackson, R. M. and Mason, P. A. (1984) *Mycorrhiza*, Edward Arnold, London.

Janzen, D. (1971) Seed predation by animals. *Annual Review of Ecology and Systematics*, **2**, 465–92.

Janzen, D. (1974) Tropical blackwater rivers, animals and mast fruiting by the Dipterocarpaceae. *Biotropica*, **6**, 69–103.

Janzen, D. H. (1975) *Ecology of Plants in the Tropics*, Edward Arnold, London.

Janzen, D. (1976) Why do bamboos wait so long to flower?, in *Tropical Trees: variation, breeding and conservation* (eds J. Burley and B. T. Styles), Academic Press, London, pp. 135–9.

Janzen, D. (1979) How to be a fig. *Annual Review of Ecology and Systematics*, **10**, 13–51.

Jenik, J. (1979) *Pictorial Encyclopedia of Forests*, Hamlyn, London.

Jenni, L. (1987) Mass concentrations of bramblings, *Fringilla montifringilla*, in Europe, 1900–1983: their dependence upon beech mast and the effect of snow cover. *Ornis Scandinavica*, **18**, 84–94.

Jensen, V. (1974) Decomposition of angiosperm tree leaf litter, in *Biology of Plant Litter Decomposition*, Vols I and II, (eds C. H. Dickinson and G. J. F. Pugh), Academic Press, London, pp. 69–104.

Jones, E. W. (1945) The structure and reproduction of the virgin forest of the North Temperate Zone. *New Phytologist*, **44**, 130–48.

Jones, E. W. (1959) *Quercus* L. Biological flora of the British Isles. *Journal of Ecology*, **47**, 169–222.

Jordan, C. F. (1985) *Nutrient Cycling in Tropical Forest Ecosystems*, John Wiley, Chichester.

Jordan, C. F. (1989a) *An Amazonian Rain Forest*, UNESCO, Paris.

Jordan, C. F. (1989b) Are process rates higher in tropical forest ecosystems? in *Mineral Nutrients in Tropical Forest and Savanna Ecosystems* (ed. J. Proctor), Blackwell Scientific, Oxford, pp. 205–15.

Jorgensen, E. (1970) *Urban Forestry in Canada*, The Shade Tree Research Laboratory, Faculty of Forestry, University of Toronto.

Jorgensen, E. (1986) Urban forestry in the rear view mirror. *Arboricultural Journal*, **10**, 177–90.

Kaarik, A. A. (1974) Wood, in *Biology of Plant Litter Decomposition*, Vols I and II (eds C. H. Dickinson and G. J. F. Pugh), Academic Press, London, pp. 129–74.

Karban, R. and Myers, J. H. (1989) Induced plant responses to herbivory. *Annual Review of Ecology and Systematics*, **20**, 331–48.

Kellman, M. (1974) Preliminary seed budgets for two plant communities in coastal British Columbia, *Journal of Biogeography*, **1**, 123–33.

Kelsey, P. J. (1987) Building movement and damage – identifying the causes. *Arboricultural Journal*, **11**, 345–61.

Khosla, P. K., Puri, S. and Kurana, D. K. (1986) *Agroforestry Systems, a New Challenge*, Indian Society of Tree Scientists, Solan, India.

Khurana, D. K. and Kohli, R. K. (eds) (1987) Workshop on agroforestry for rural needs: abstracts. *Indian Society of Tree Scientists*, Solan, India.

King, C. M. (1980) The weasel *Mustela nivalis* and its prey in an English woodland. *Journal of Animal Ecology*, **49**, 127–59.

King, C. (1989) *The Natural History of Weasels and Stoats*, Christopher Helm, London.

Kira, T. (1975) Primary production of forests, in *Photosynthesis and Productivity in Different Environments* (ed. J. P. Cooper), Cambridge University Press, Cambridge, pp. 5–40.

Kirby, K. J. (1988) Changes in the ground flora under plantations on ancient woodland sites. *Forestry*, **61**, 317–38.

Kjøller, A. S. and Struwe, S. (1982) Microfungi in ecosystems: fungal occurrence and activity in litter and soil. *Oikos*, **39**, 389–422.

Klomp, H. (1966) The dynamics of a field population of the pine looper, *Bupalus pinarius* L. (Lep. Geom.). *Advances in Ecological Research*, **3**, 207–305.

Koop, H. (1987) Vegetative reproduction of trees in some European natural forests. *Vegetatio*, **72**, 103–10.

Krebs, C. J. (1985) *Ecology: the experimental analysis of distribution and abundance*, 3rd edn, Harper and Row, New York.

Kurcheva, G. F. (1960) Role of invertebrates in the decomposition of oak litter. *Soviet Soil Science*, **4**, 360–5.

Larcher, W. (1975) *Physiological Plant Ecology*, Springer, Berlin.

La Roi, G. H. (1967) Ecological studies in the boreal spruce–fir forests of the North American taiga. I. Analysis of the vascular flora. *Ecological Monographs*, **37**, 229–53.

Lawton, J. H. (1989) Food webs, in *Ecological Concepts: the contribution of ecology to an understanding of the natural world* (ed. J. M. Cherrett), Blackwell Scientific, Oxford, pp. 43–78.

Lee, K. E. and Wood, T. G. (1971) *Termites and Soils*, Academic Press, London.

Likens, G. E., Bormann, F. H., Johnson, N. M. *et al.* (1970) Effects of forest cutting and herbicide treatment on nutrient budgets in the Hubbard Brook watershed ecosystem. *Ecological Monographs*, **40**, 23–47.

Lindquist, B. (1931) Den Skandinaviska bogskogens biologi (The ecology of the Scandinavian beechwoods). *Svenska Skogsvardsforeningens Tidskrift*, **29** (English digest), 486–520.

Lines, R. (1987) *Choice of Seed Origins for the Main Forest Species in Britain*, HMSO, London.

Livingston, R. B. and Allessio, M. L. (1968) Buried viable seed in successional field and forest stands, Harvard Forest, Massachusetts. *Bulletin of the Torrey Botanical Club*, **95**, 58–69.

Llewellyn, M. (1972) The effects of the lime aphid, *Eucallipterus tiliae* L. (Aphididae) on the growth of the lime *Tilia x vulgaris*. I. *Journal of Applied Ecology*, **9**, 261–82.

Llewellyn, M. (1975) The effects of the lime aphid, *Eucallipterus tiliae* L. (Aphididae) on the growth of the lime *Tilia x vulgaris*. II. *Journal of Applied Ecology*, **12**, 15–23.

Luxton, M. (1972) Studies on the oribatid mites of a Danish beech wood soil. I. Nutritional biology. *Pedobiologia*, **12**, 434–63.

McArdle, B. H., Gaston, K. J. and Lawton, J. H. (1990) Variation in the size of animal populations: patterns, problems and artefacts. *Journal of Animal Ecology*, **59**, 439–54.

McCleery, R. H. and Perrins, C. M. (1985) Territory size, reproductive success and population dynamics in the great tit *Parus major*, in *Behavioural Ecology: ecological consequences of adaptive behaviour* (eds R. M. Sibley and R. H. Smith) Blackwell Scientific, Oxford, pp. 353–73.

McIntosh, R. (1978) Response of Sitka spruce to remedial fertilization in Galloway. *Scottish Forestry*, **32**, 271–82.

McIntosh, R. (1984) *Fertilizer Experiments in Established Conifer Stands*, HMSO, London.

MacArthur, R. H. and Wilson, E. O. (1967) *The Theory of Island Biogeography*, Princeton University Press, New Jersey.

Macfadyen, A. (1961) Metabolism of soil invertebrates in relation to soil fertility. *Annals of Applied Biology*, **49**, 216–19.

Madden, J. L. (1977) Physiological reactions of *Pinus radiata* to attack by woodwasp, *Sirex noctilio* F. (Hymenoptera: Siricidae). *Bulletin of Entomological Research*, **67**, 405–26.

Madge, D. S. (1965) Leaf fall and litter disappearance in a tropical forest. *Pedobiologia*, **5**, 273–88.

Martin, M. H. (1968) Conditions affecting the distribution of *Mercurialis perennis* L. in certain Cambridgeshire woodlands. *Journal of Ecology*, **56**, 777–93.

Martin, R. (1988) An investigation into masting and regeneration in the West Midlands, The Polytechnic, Wolverhampton. (Degree thesis, unpublished.)

Martinez-Ramos, M., Alvarez-Buylla, E., Sarukhán, J. and Piñero, D. (1988). Treefall age determination and gap dynamics in a tropical forest. *Journal of Ecology*, **76**, 700–16.

Matthews, J. D. (1955) The influence of weather on the frequency of beech mast years in England. *Forestry*, **28**, 107–16.

Matthews, J. D. (1989) *Silvicultural Systems*, University Press, Oxford.

Mattson, W. J. (1980) Herbivory in relation to plant nitrogen content. *Annual Review of Ecology and Systematics*, **11**, 119–61.

Mattson, W. J. and Addy, N. D. (1975) Phytophagous insects as regulators of forest primary production. *Science*, **190**, 515–22.

Maurer, E. (1964) Buchen- und Eichensamenjahre in Unterfranken wahrend der letzten 100 Jahre. *Allgemeine Forstzeitschrift*, **31**, 469–70.

May, R. M. (1976) Patterns in multispecies communities, in *Theoretical Ecology, Principles and Applications* (ed. R. M. May), Blackwell Scientific, Oxford, pp. 142–62.

Meidner, H. and Sheriff, D. W. (1976) *Water and Plants*, Blackie, London.

Mellanby, K. (1968) The effects of some mammals and birds on regeneration of oak. *Journal of Applied Ecology*, **5**, 359–66.

Miles, J. (1979) *Vegetation Dynamics*, Chapman and Hall, London.

Miles, J. (1987) Vegetation succession: past and present perceptions, in *Coloni-*

zation, Succession and Stability (eds A. J. Gray, M. J. Crawley and P. J. Edwards), Blackwell Scientific, Oxford, pp. 1–29.

Millar, C. S. (1974) Decomposition of coniferous leaf litter, in *Biology of Plant Litter Decomposition*, Vols I and II (eds C. H. Dickinson and G. J. F. Pugh), Academic Press, London, pp. 105–28.

Miller, H. G. (1979) The nutrient budgets of even-aged forests, in *The Ecology of Even-aged Forest Plantations* (eds E. D. Ford, D. C. Malcolm and J. Atterson), Institute of Terrestrial Ecology, Cambridge, pp. 221–56.

Miller, H. G. (1981a) Forest fertilization: some guiding concepts. *Forestry*, **54**, 157–67.

Miller, H. G. (1981b) Nutrient cycles in forest plantations, their change with age and the consequences for fertilizer practice, in *Proceedings of the Australian Forest Nutrition Workshop, Productivity in Perpetuity*, CSIRO, Canberra, pp. 187–99.

Miller, R. W. (1988) *Urban forestry: planning and managing urban greenspaces*, Prentice Hall, London.

Minot, E. (1978) Interspecific competition in tits. *Nature*, **275**, 463.

Mitchell, F. J. G. and Kirby, K. J. (1990) The impact of large herbivores on the conservation of semi-natural woods in the British uplands. *Forestry*. **63**, 333–53.

Mitchell, P. L. (1992) Growth stages and microclimate in coppice and high forest, in *The Ecology and Management of Coppice Woodland* (ed. G. P. Buckley), Chapman and Hall, London.

Mitchell, P. L. and Kirby, K. J. (1989) *Ecological Effects of Forestry Practices in Long Established Woodland and their Implications for Nature Conservation*, Oxford Forestry Institute, Oxford.

Mitchell, P. L. and Woodward, F. I. (1988) Responses of three woodland herbs to reduced photosynthetically active radiation and low red to far-red ratio in shade. *Journal of Ecology*, **76**, 807–25.

Moore, J. C., Walter, D. E. and Hunt, H. W. (1988) Arthropod regulation of micro- and mesobiota in below-ground detrital food webs. *Annual Review of Entomology*, **33**, 419–39.

Moss, C. E. (1901) Changes in the Halifax flora in the last century and a quarter. *Naturalist*, **26**, 99–107.

Mueller-Dombois, D. and Ellenberg, H. (1974) *Aims and Methods of Vegetation Ecology*, Wiley, London.

Mukerji, S. K. (1936) Contributions to the antecology of *Mercurialis perennis* L. *Journal of Ecology*, **24**, 38–81.

Mutch, R. W. (1970) Wild fires and ecosystems – a hypothesis. *Ecology*, **51**, 1047–51.

Myers, J. M. (1988) Can a general hypothesis explain population cycles of forest Lepidoptera? *Advances in Ecological Research*, **18**, 179–242.

Namkoong, G. (1986) Genetics and the forests of the future. *Unasylva*, **152**, 138(2), 2–18.

Nanson, A. (1968) La valeur des tests précoces dans la selection des arbres forestiers, en particulier au point de vue la croissance. Ph. D. thesis, Fac. Sci. Agron. Etat, Gembloux, Belgium.

Newbould, P. J. (1967) *Methods for Estimating the Primary Production of Forests*, Blackwell Scientific, Oxford.

Nicholson, P. B., Bocock, K. O. and Heal, O. W. (1966) Studies on the decomposition of the faecal pellets of a millipede (*Glomeris marginata* (Villers)). *Journal of Ecology*, **54**, 755–66.

Nielsen, B. Overgaard (1977) Beech seeds as an ecosystem component. *Oikos*, **29**, 268–74.

Nielsen, B. Overgaard (1978) Above ground food resources and herbivory in a beech forest ecosystem. *Oikos*, **31**, 273–9.

Nielsen, P. C. and de Muckadeli, M. S. (1954) Flower observations and controlled pollinations in *Fagus*. *Silvae Genetica*, **3**, 6–17.

Nilsson, I. (1978) The influence of *Dasychira pubibunda* (Lepidoptera) on plant nutrient transports and tree growth in a beech (*Fagus sylvatica*) forest in southern Sweden. *Oikos*, **30**, 133–48.

Nilsson, S. G. and Wastljung, U. (1987) Seed predation and cross-pollination in mast-seeding beech (*Fagus sylvatica*) patches. *Ecology*, **68(2)**, 260–5.

O'Callaghan, D. P., Atkins, P. M. and Fairhurst, C. P. (1984) Behavioural responses of elm bark beetles to baited and unbaited elms killed by cacodylic acid. *Journal of Chemical Ecology*, **10**, 1623–34.

Olmsted, N. W. and Curtis, J. D. (1947) Seeds of the forest floor. *Ecology*, **28**, 49–52.

Ovington, J. D. (1965) *Woodlands*, English Universities Press, London.

Packham, J. R. (1978) *Oxalis acetosella* L. Biological flora of the British Isles. *Journal of Ecology*, **66**, 669–93.

Packham, J. R. (1979) Factors affecting the growth and distribution of the wood sorrel (*Oxalis acetosella*) on the Long Mynd, Shropshire. Caradoc and Severn Valley Field Club, Occasional paper No. 3, pp, 1–14.

Packham, J. R. (1983) *Lamiastrum galeobdolon* (L.) Ehrend, and Polatschek. Biological flora of the British Isles. *Journal of Ecology*, **71**, 975–97.

Packham, J. R. and Cohn, E. V. J. (1990) Ecology of the woodland field layer. *Arboricultural Journal*, **14**, 357–71.

Packham, J. R. and Harding, D. J. L. (1982) *Ecology of Woodland Processes*, Edward Arnold, London.

Packham, J. R. and Willis, A. J. (1976) Aspects of the ecological amplitude of two woodland herbs, *Oxalis acetosella* L. and *Galeobdolon luteum* Huds. *Journal of Ecology*, **64**, 485–510.

Packham, J. R. and Willis, A. J. (1977) The effects of shading on *Oxalis acetosella*. *Journal of Ecology*, **65**, 619–42.

Packham, J. R. and Willis, A. J. (1982) The influence of shading and of soil type on the growth of *Galeobdolon luteum*. *Journal of Ecology*, **70**, 491–512.

Pardé, J. (1980) Forest biomass. *Forestry Abstracts*, **41**, 343–62.

Parker, G. G. (1983) Throughfall and stemflow in the forest nutrient cycle. *Advances in Ecological Research*, **13**, 57–133.

Patch, D. (1987) Trouble at the stake, in *Advances in Practical Arboriculture* (ed. D. Patch) HMSO, London, pp. 77–84.

Payne, C. C. (1988) Prospects for biological control, in *Britain Since Silent Spring* (ed. D. J. L. Harding), Institute of Biology, London, pp. 103–16.

Pennington, W. (1974) *The History of British Vegetation*, 2nd edn, English Universities Press, London.

Perrins, C. M. (1966) The effect of beech crops on great tit populations and movements. *British Birds*, **59**, 419–32.

Perrins, C. M. (1980) The great tit *Parus major*. *Biologist*, **27**, 73–80.

Perrins, C. M. (1991) Tits and their caterpillar food supply. *Ibis*, **133**, Suppl. 1, 49–54.

Peterken, G. F. (1981) *Woodland Conservation and Management*, Chapman and Hall, London.

Peterman, R. M., Clark, W. C. and Holling, C. S. (1979) The dynamics of resilience: shifting stability domains in fish and insect systems, in *Population Dynamics* (eds R. M. Anderson, B. D. Turner and L. R. Taylor), Blackwell Scientific, Oxford, pp. 321–41.

Petersen, H. and Luxton, M. (1982) A comparative analysis of soil fauna populations and their role in decomposition processes. *Oikos*, **39**, 287–388.

Petrusewicz, K. (ed) (1967) *Secondary Productivity in Terrestrial Ecosystems*, Polish Academy of Sciences, Warsaw.

Petrusewicz, K. and Macfadyen, A. (1970) *Productivity of Terrestrial Animals – Principles and Methods*, Blackwell Scientific, Oxford.

Pfister, R. D., Kovalchik, B. L., Arno, S. F. and Presby, R. C. (1977) *Forest Habitat Types of Montana*, United States Department of Agriculture Forest Service General Technical Report INT-34 Intermountain Forest and Range Experimental Station, Ogden, Utah.

Philipson, J. J. (1983) The role of gibberellin $A_{4/7}$, heat and drought in the induction of flowering in Sitka spruce. *Journal of Experimental Botany*, **34**, 291–302.

Phillipson, J. (1960) The food consumption of different instars of *Mitopus morio* (F.) (Phalangiida) under natural conditions. *Journal of Animal Ecology*, **29**, 299–307.

Phillipson, J. (1971) *Methods of Study in Quantitative Soil Ecology: population, production and energy flow*, Blackwell Scientific, Oxford.

Phillipson, J., Putman, R. J., Steel, J. and Woodell, S. R. J. (1975) Litter input, litter decomposition and the evolution of carbon dioxide in a beech woodland – Wytham Woods, Oxford. *Oecologia (Berlin)*, **20**, 203–17.

Pickett, S. T. A. and McDonnell, M. J. (1989) Changing perspectives in community dynamics: a theory of successional forces. *Trends in Ecology and Evolution*, **4**, 241–5.

Pickett, S. T. A. and White, P. S. (eds) (1985) *The Ecology of Natural Disturbance and Patch Dynamics*, Academic Press, London.

Pigott, C. D. (1975) Natural regeneration of *Tilia cordata* in relation to forest-structure in the forest of Bialowieza, Poland. *Philosophical Transactions of the Royal Society, B*, **270**, 151–79.

Pigott, C. D. and Taylor, K. (1964) The distribution of some woodland herbs in relation to the supply of nitrogen and phosphorus in the soil. *Journal of Ecology*, **52** (Supplement), 175–85.

Pimm, S. L. (1982) *Food Webs*, Chapman and Hall, London.

Pisek, A. and Tranquillini, W. (1951) Transpiration und Vasserhaushalt der Fichte (*Picea excelsa*) bei Zunehmerder Luft) – und Bodentrockenheit. *Physiologia Plantarum*, **4**, 1–27.

Podoler, H. and Rogers, D. (1975) A new method for the identification of key factors from life-table data. *Journal of Animal Ecology*, **44**, 85–114.

Pope, D. J. and Lloyd, P. S. (1975) Hemispherical phalography, topography and plant distribution, in *Light as an Ecological Factor II* (eds G. C. Evans, R. Bainbridge and O. Rackham), Blackwell, Oxford, pp. 385–408.

Prentice, I. C. and Leemans, R. (1990) Pattern and process and the dynamics of forest structure: a simulation approach. *Journal of Ecology*, **78**, 340–55.

Proctor, J. (ed.) (1989) *Mineral Nutrients in Tropical Forest and Savanna Ecosystems*, Blackwell Scientific, Oxford.

Proctor, J., Anderson, J. M., Chai, P. and Vallack, H. W. (1983) Ecological studies in four contrasting lowland rain forests in Gunung Mulu National Park, Sarawak. *Journal of Ecology*, **71**, 237–60.

Proctor, M. and Yeo, P. (1973) *The Pollination of Flowers*, Collins, London.

Pugh, G. J. F. (1974) Terrestrial fungi, in *Biology of Plant Litter Decomposition*, Vols 1 and 2, (eds C. H. Dickinson and G. F. G. Pugh), Academic Press, London, pp. 303–36.

Putman, R. J. (1988) *The Natural History of Deer*, Christopher Helm, London.

Rackham, O. (1975) *Hayley Wood: its history and ecology*, Cambridgeshire and Isle of Ely Naturalists' Trust, Cambridge.

Rackham, O. (1976) *Trees and Woodland in the British Landscape*, Dent, London.

Rackham, O. (1980) *Ancient Woodland: its history, vegetation and uses in England*, Edward Arnold, London.

Rackham, O. (1986) *The History of the Countryside*, Dent, London.

Rappaport, R. A. (1971) The flow of energy in an agricultural society. *Scientific American*, **225(3)**, 116–32.

Ratcliffe, D. A. and Oswald, P. H. (1987) *Birds, Bogs and Forestry: the peatlands of Caithness and Sunderland*, NCC, Peterborough.

Ratcliffe, P. R. (1991) The influence of mammals on the regeneration of forest communities, in *Future Forests: Lessons from Experience* (ed. J. R. Packham) Wolverhampton Woodland Research Group, Wolverhampton, pp. 54–70.

Ratcliffe, P. R. (1992) The interaction of deer and vegetation in coppiced woods, in *The Ecology and Management of Coppice Woodland* (ed. G. P. Buckley), Chapman and Hall, London.

Raunkiaer, C. (1934) *The Life Forms of Plants and Statistical Plant Geography*, Clarendon Press, Oxford.

Rayner, A. D. M. and Boddy, L. (1988) Fungal communities in the decay of wood. *Advances in Microbial Ecology*, **10**, 115–66.

Regehr, D. L., Bazzaz, F. A. and Boggess, W. R. (1975) Photosynthesis, transpiration and leaf conductance of *Populus deltoides* in relation to flooding and drought. *Photosynthetica*, **9(1)**, 52–61.

Reichle, D. E. (1971) Energy and nutrient metabolism of soil and litter invertebrates, in *Productivity of Forest Ecosystems* (ed. P. Duvigneaud), UNESCO, Paris, pp. 465–77.

Reichle, D. E. (ed.) (1981) *Dynamic Properties of Forest Ecosystems*, Cambridge University Press, Cambridge.

Reyes, V. G. and Tiedje, J. M. (1976) Ecology of the gut microbiota of *Tracheoniscus rathkei* (Crustacea, Isopoda). *Pedobiologia*, **16**, 67–74.

Richards, W. N. (1985) Problems of water management and water quality arising from forestry activities, in *Woodlands, Weather and Water* (eds D. J. L. Harding and J. K. Fawell), Institute of Biology, London, pp. 67–85.

Rishbeth, J. (1963) Stump protection against *Fomes annosus*. III. Inoculation with *Peniophora gigantea*. *Annals of Applied Biology*, **52**, 63–77.

Rishbeth, J. (1976) Chemical treatment and inoculation of hardwood stumps for control of *Armillaria mellea*. *Annals of Applied Biology*, **82**, 57–70.

Roberts, T. M. (1987) Effects of air pollution on agriculture and forestry. *Acid Rain: CEGB Research*, **20**, 39–52.

Robertson, A. (1987) The centroid of tree crowns as an indicator of abiotic processes in a balsam fir wave forest. *Canadian Journal of Forest Research*, **17(7)**, 746–55.

Rodin, L. E. and Bazilevich, N. I. (1967) *Production and Mineral Cycling in Terrestrial Vegetation*, Oliver and Boyd, Edinburgh.

Rohmeder, E. (1967) Beziehungen zwischen Frucht-bzw. Samenerzeugung und Holzerzeugung der Waldbaunme. *Allgemeine Forstzeitschrifte*, **22**, 33–9.

Rollinson, T. J. D. (1985) *Thinning Control*, HMSO, London.

Romme, W. H. and Despain, D. G. (1989) The Yellowstone Fires. *Scientific American*, **261(5)**, 20–9.

Rosenthal, G. A. and Janzen, D. H. (eds) (1979) *Herbivores: their interaction with secondary plant metabolites*, Academic Press, New York.

Roulund, H. (1981) Problems of clonal forestry in spruce and their influence on breeding strategy. *Forestry Abstracts*, **42**, 457–71.

Russell, R. S. (1977) *Plant Root Systems; their function and interaction with the soil*, McGraw-Hill, London.

Sale, J. S. P., Tabbush, P. M. and Lane, P. B. (1986) *The Use of Herbicides in the Forest*, 2nd edn, Forestry Commission, Edinburgh.

Salisbury, E. J. (1916a) The oak-hornbeam woods of Hertfordshire. I and II. *Journal of Ecology*, **4**, 83–117.

Salisbury, E. J. (1916b) The emergence of aerial organs in woodland plants. *Journal of Ecology*, **4**, 121–8.

Salisbury, E. J. (1920) The significance of the calcicolous habit. *Journal of Ecology*, **8**, 202–15.

Salisbury, E. J. (1924) The effects of coppicing as illustrated by the woods of Hertfordshire. *Transactions of the Hertfordshire Natural History Society and Field Club*, **18**, 1–21.

Salisbury, E. J. (1925) The vegetation of the forest of Wyre: a preliminary account. *Journal of Ecology*, **13**, 314–21.

Salisbury, E. J. (1942) *The Reproductive Capacity of Plants*, Bell, London.

Salisbury, E. J. (1961) *Weeds and Aliens*, Collins, London.

Salmon, J. T. (1986) *A Field Guide to the Native Trees of New Zealand*, Reed Methuen, Auckland.

Satchell, J. E. (1967) Lumbricidae, in *Soil Biology* (eds A. Burges and F. Raw), Academic Press, London, pp. 259–322.

Satchell, J. E. and Lowe, D. G. (1967) Selection of leaf litter by *Lumbricus terrestris*, in *Progress in Soil Biology* (eds O. Graff and J. E. Satchell), North Holland, Amsterdam, pp. 102–19.

Schmithüsen, J. (1976) *Atlas zur Biogeographie* Geographisck-Kartographishes Institut Meyer, Mannheim.

Schowalter, T. D., Hargrove, W. W. and Crossley, D. A. (1986) Herbivory in forested ecosystems. *Annual Review of Entomology*, **31**, 177–96.

Schulze, E. D. (1970) Der CO_2-Gaswechsel de Buche (*Fagus sylvatica* L.) in Abhangigkiet von den Klimafaktoren im Freiland. *Flora, Jena*, **159**, 177–232.

Schulze, E. D. (1972) Die Wirkung von Licht und Temperatur auf den CO_2-Gaswechsel verscheidener Lebensformen aus der Krautschicht eines montanen Buchenwaldes. *Oecologia (Berlin)*, **9**, 235–58.

Schulze, E. D., Fuchs, M. I. and Fuchs, M. (1977a) Spatial distribution of photosynthetic capacity and performance in a mountain spruce forest of Northern Germany. I. Biomass distribution and daily CO_2 uptake in different crown layers. *Oecologia (Berlin)*, **29**, 43–61.

Schulze, E. D., Fuchs, M. I. and Fuchs, M. (1977b) Spatial distribution of photosynthetic capacity and performance in a mountain spruce forest of Northern Germany. III. The significance of the evergreen habit. *Oecologia (Berlin)*, **30**, 239–48.

Schwintzer, C. R. and Tjepkema, J. D. (1990) *The Biology of Frankia and Actinorhizal Plants*, Academic Press, London.

Scott, S. (1985) The regeneration of hedgerow elm. *Arboricultural Journal*, **9**, 285–92.

Seastedt, T. R. (1984) The role of microarthropods in decomposition and mineralization processes. *Annual Review of Entomology*, **29**, 25–46.

Seeger, M. (1930) Erfahrungen Uber die Eiche in der Rhinehene bei Emmendingen (Baden). *Allgemeine Forst-u. Jagdzeitung*, **106**, 201–19.

Sernander, R. (1936) The primitive forests of Granskar and Fiby: a study of the part played by storm-gaps and dwarf trees in the regeneration of the Swedish spruce forest. *Acta Phytogeographica Suecica*, **8**, 1–232. (English summary, pp. 220–7.)

Shaw, M. W. (1974) The reproductive characteristics of oak, in *The British Oak* (eds M. G. Morris and F. H. Perring) E. W. Classey, Faringdon, pp. 161–81.

Shigo, A. L. (1984) Compartmentalisation: a conceptual framework for understanding how trees grow and defend themselves. *Annual Review of Phytopathology*, **22**, 189–214.

Shimwell, D. W. (1971) *The Description and Classification of Vegetation*, Sidgwick and Jackson, London.

Shugart, H. H. (1984) A Theory of Forest Dynamics, Springer, New York.

Shugart, H. H. (1987) Dynamic ecosystem consequences of tree birth and death patterns. *Bioscience*, **37**, 596–602.

Shugart, H. H., Emanuel, W. R. and de Angelis, D. L. (1980) Environmental gradients in a simulation model of a beech-yellow poplar stand. *Mathematical Bioscience*, **50**, 163–70.

Shugart, H. H. and Urban, D. L. (1989) Factors affecting the relative abundances of forest trees, in *Toward a More Exact Ecology* (eds P. J. Grubb and J. B. Whittaker) Blackwell Scientific, Oxford, pp. 249–73.

Silvertown, J. W. (1980) The evolutionary ecology of mast seeding in trees. *Biological Journal of the Linnean Society*, **14**, 235–50.

Sinclair, A. R. E. (1989) Animal population regulation, in *Ecological Concepts* (ed. J. M. Cherrett) Blackwell Scientific, Oxford, pp. 197–241.

Sinn, G. and Wessolly, L. (1989) A contribution to the proper assessment of the strength and stability of trees. *Arboricultural Journal*, **13**, 45–65.

Slade, A. J. and Hutchings, M. J. (1987a) The effects of nutrient availability on foraging in the clonal herb *Glechoma hederacea*. *Journal of Ecology*, **75**, 95–112.

Slade, A. J. and Hutchings, M. J. (1987b) The effects of light intensity on foraging in the clonal herb *Glechoma hederacea*. *Journal of Ecology*, 75, 639–50.

Slade, A. J. and Hutchings, M. J. (1987c) Clonal integration and plasticity in foraging behaviour in *Glechoma hederacea*. *Journal of Ecology*, **75**, 1023–36.

Smith, P. H. (1972) The energy relations of defoliating insects in a hazel coppice. *Journal of Animal Ecology*, **41**, 567–87.

Solomon, M. (1976) *Population Dynamics*, 2nd edn, Edward Arnold, London.

Southern, H. N. (1970) The natural control of a population of tawny owls (*Strix aluco*). *Journal of Zoology, London*, **162**, 197–285.

Southwood, T. R. E. (1973) The insect/plant relationship – an evolutionary perspective, in *Insect/Plant Relationships* (ed. H. F. van Emden), Blackwell Scientific, Oxford, pp. 3–30.

Southwood, T. R. E. (1976) Bionomic strategies and population parameters, in *Theoretical Ecology, Principles and Applications* (ed. R. M. May), Blackwell Scientific, Oxford, pp. 26–48.

Southwood, T. R. E. (1977) Habitat, the templet for ecological studies? *Journal of Animal Ecology*, **24**, 337–65.

Speight, M. R. and Wainhouse, D. (1989) *Ecology and Management of Forest Insects*, Oxford Science Publications, Oxford.

Spencer, D. A. (1964) Porcupine population fluctuations in past centuries revealed by dendrochronology. *Journal of Applied Ecology*, **1**, 127–49.

Sprugel, D. G. (1976) Dynamic structure of wave-regenerated *Abies balsamea* forests in the north-eastern United States. *Journal of Ecology*, **64**, 889–911.

Spurr, S. H. and Barnes, B. V. (1980) *Forest Ecology*, 3rd edn, Wiley, New York.

Stark, N. M. and Jordan, C. F. (1978) Nutrient retention by the root mat of an Amazonian rain forest. *Ecology*, **59**, 434–7.

Steele, R. C. (1972) *Wildlife Conservation in Woodlands*, HMSO, London.

Stern, K. and Roche, L. (1974) *Genetics of Forest Ecosystems*, Chapman and Hall, London.

Swift, M. J. (1976) Species diversity and the structure of microbial communities in terrestrial habitats, in *The Role of Terrestrial and Aquatic Organisms in Decomposition Processes*, (eds J. M. Anderson and A. Macfadyen) Blackwell Scientific, Oxford, pp. 185–222.

Swift, M. J. (1977a) The ecology of wood decomposition. *Science Progress, Oxford*, **64**, 179–203.

Swift, M. J. (1977b) The role of fungi and animals in the immobilisation and release of nutrient elements from decomposing branch-wood, in *Soil Organisms as Components of Ecosystems* (eds U. Lohm and T. Persson) *Ecological Bulletins (Stockholm)*, **25**, 193–202.

Swift, M. J., Heal, O. W. and Anderson, J. M. (1979) *Decomposition in Terrestrial Ecosystems*, Blackwell Scientific, Oxford.

Sydes, C. and Grime, J. P. (1981) Effects of tree leaf litter on herbaceous vegetation in deciduous woodland. I and II. *Journal of Ecology*, **69**, 237–48; 249–63.

Sykes, J. M. and Bunce, R. G. H. (1970) Fluctuations in litter-fall in a mixed deciduous woodland over a three-year period. *Oikos*, **21**, 326–9.

Tansley, A. G. (1935) The use and abuse of vegetational concepts and terms. *Ecology*, **16**, 284–307.

Tansley, A. G. (1939) *The British Islands and their Vegetation*, Cambridge University Press, Cambridge.

Tanton, M. T. (1965) Acorn destruction potential of small mammals and birds in British woodlands. *Quarterly Journal of Forestry*, **59**, 230–4.

Teasdale, J. B. (1986) Fire protection, in *Forestry Practice* (ed. B. G. Hibberd), HMSO, London.

Thompson, K. and Grime, J. P. (1979) Seasonal variations in the seed banks of herbaceous species in ten contrasting habitats. *Journal of Ecology*, **67**, 893–921.

Tittensor, R. M. (1980) Ecological history of yew *Taxus baccata* L. in southern England. *Biological Conservation*, **17**, 243–65.

Tranquillini, W. (1979) *Physiological Ecology of the Alpine Timberline*, Springer, Berlin.

Trewavas, A. (1987) Sensitivity and sensory adaptation in growth substance responses, in *Hormone Action in Plant Development* (eds G. V. Hoad, J. R. Lenton, M. B. Jackson and R. K. Atkin) Butterworths, London, pp. 19–38.

Tribe, H. T. (1957) Ecology of microorganisms in soil as observed during their development upon buried cellulose film, in *Microbial Ecology* (eds R. E. O. Williams and C. C. Spicer), Cambridge University Press, Cambridge, pp. 287–98.

Tuley, G. (1983) Shelters improve the growth of young trees in the forest. *Quarterly Journal of Forestry*, **77**, 77–87.

UNESCO (1978) *Tropical Forest Ecosystems*, UNESCO, Paris.

Usher, M. B. and Parr, T. W. (1977) Are there successional changes in arthropod decomposer communities? *Journal of Environmental Management*, **5**, 151–60.

Varley, G. C. (1967) The estimation of secondary production in species with an annual life-cycle, in *Secondary Productivity of Terrestrial Ecosystems* (ed. K. Petrusewicz), Polish Academy of Sciences, Warsaw, pp. 447–57.

Varley, G. C. (1970) The concept of energy flow applied to a woodland community, in *Animal Populations in Relation to their Food Resources* (ed. A. Watson), Blackwell Scientific, Oxford, pp. 389–405.

Varley, G. C. and Gradwell, G. R. (1962) The effect of partial defoliation by caterpillars on the timber production of oak trees in England. *Proceedings of the 11th International Congress of Entomology*, Vienna 1960, Vol 2, 211–14.

Varley, G. C., Gradwell, G. R. and Hassell, M. P. (1973) *Insect Population Ecology, An Analytical Approach*, Blackwell Scientific, Oxford.

Veblen, T. T., Ashton, D. H. and Schlegel, F. M. (1979) Tree regeneration strategies in a lowland *Nothofagus*-dominated forest in south-central Chile. *Journal of Biogeography*, **6**, 329–40.

Visser, S. (1985) Role of soil invertebrates in determining the composition of soil microbial communities, in *Ecological Interactions in Soil* (eds A. H. Fitter, D. Atkinson, D. J. Read and M. B. Usher), Blackwell Scientific, Oxford, pp. 297–317.

Vitousek, P. M. and Sanford, R. L. (1986) Nutrient cycling in moist tropical forest. *Annual Review of Ecology and Systematics*, **17**, 137–67.

Vogl, R. J. (1970) Fire and the northern Wisconsin pine barrens. *Proc. 10th Ann. Tall Timbers Fire Ecol. Conf.*, pp. 175–209.

Vogt, K. A., Grier, C. C. and Vogt, D. J. (1986) Production, turnover and nutrient dynamics of above- and below-ground detritus of world forests. *Advances in Ecological Research* **15**, 303–77.

Wallwork, J. A. (1970) *Ecology of Soil Animals*, McGraw-Hill, London.

Wallwork, J. A. (1983) Oribatids in forest ecosystems. *Annual Review of Entomology*, **28**, 109–30.

Walter, H. (1973) *Vegetation of the Earth in Relation to Climate and the Eco-physiological Conditions*, EUP-Springer, London.

Waring, R. H. (1989) Ecosystems: fluxes of matter and energy, in *Ecological Concepts* (ed. J. M. Cherrett) Blackwell Scientific, Oxford, pp. 17–41.

Waring, R. H. and Schlesinger, W. H. (1985) *Forest Ecosystems: concepts and management*, Academic Press, London.

Watkins, C. (1990) *Woodland Management and Conservation*, David and Charles, London.

Watt, A. D. (1990) The consequences of natural, stress-induced and damage-induced differences in tree foliage on the population dynamics of the pine beauty moth, in *Population Dynamics of Forest Insects* (eds A. D. Watt, S. R. Leather, M. D. Hunter and N. A. C. Kidd), Intercept, Andover, pp. 157–68.

Watt, A. D. and Leather, S. R. (1987) Pine beauty moth population dynamics: synthesis, simulation and prediction, in *Population Biology and Control of the Pine Beauty Moth* (eds S. R. Leather, J. T. Stoakley and H. F. Evans), Forestry Commission Bulletin 67, HMSO, London, pp. 41–5.

Watt, A. D., Leather, S. R. Hunter, M. D. and Kidd, N. A. C. (eds) (1990) *Population Dynamics of Forest Insects*, Intercept, Andover.

Watt, A. S. (1919) On the causes of failure of natural regeneration in British oakwoods. *Journal of Ecology*, **7**, 173–203.

Watt, A. S. (1923) On the ecology of British beechwoods with special reference to their regeneration. Part I. *Journal of Ecology*, **11**, 1–48.

Watt, A. S. (1925) On the ecology of British beechwoods with special reference to their regeneration. Part II, Sections II and III. *Journal of Ecology*, **13**, 27–73.

Watt, A. S. (1926) Yew communities on the South Downs. *Journal of Ecology*, **14**, 282–316.

Watt, A. S. (1947) Pattern and process in the plant community. *Journal of Ecology*, **35**, 1–22.

Way, M. J. and Bevan, D. (1977) Dilemmas in forest pest and disease management, in *Ecological Effects of Pesticides* (eds F. H. Perring and K. Mellanby), Academic Press, New York, pp. 95–110.

Webb, D. P. (1977) Regulation of deciduous forest litter decomposition by soil arthropod feces, in *The Role of Arthropods in Forest Ecosystems* (ed. W. J. Mattson), Springer Verlag, New York, pp. 57–69.

Webber, J. (1981) A natural biological control of Dutch elm disease. *Nature*, **292**, 449–51.

Webber, J. and Brasier, C. M. (1984) The transmission of Dutch elm disease: a study of the processes involved, in *Invertebrate–Microbial Interactions (British Mycological Society Symposium 6)* (eds J. Anderson, A. D. M. Rayner and D. Walton), Cambridge University Press, Cambridge, pp. 271–306.

Weir, R. J. and Zobel, B. J. (1975) Advanced generation seed orchards, in *Seed Orchards* (ed. R. Faulkner), HMSO, London, pp. 118–27.

Whale, D. M. (1984) Habitat requirements in *Primula* species. *New Phytologist*, **97**, 665–79.

Whitmore, T. C. (1975, 2nd edn, 1984) *Tropical Rain Forests of the Far East*, Clarendon, Oxford.

Whitmore, T. C. (1982) On pattern and process in forests, in *The Plant Community as a Working Mechanism* (ed. E. I. Newman), Special publication of the British Ecological Society 1, Blackwell Scientific, Oxford, pp. 45–59.

Whitmore, T. C. (1989) Tropical forest nutrients, where do we stand? A *tour de horizon*, in *Mineral Nutrients in Tropical Forest and Savanna Ecosystems* (ed. J. Proctor), Blackwell Scientific, Oxford, pp. 1–14.

Whitmore, T. C. (1990) *An Introduction to Tropical Rain Forests*, Clarendon Press, Oxford.

Whitney, G. G. (1990) The history and status of hemlock-hardwood forests of the Allegheny plateau. *Journal of Ecology*, **78**, 443–58.

Whittaker, R. H. (1975) *Communities and Ecosystems*, 2nd edn, Collier Macmillan, London.

Whittaker, R. H. and Marks, P. L. (1975) Methods of assessing terrestrial vegetation, in *Productivity of the Biosphere* (eds H. Lieth and R. H. Whittaker), Springer, New York, pp. 55–118.

Whittaker, R. J., Asquith, N. M., Bush, M. B. and Partomihardjo, T. (eds) (1990) *Krakatau Research Project 1989 Expedition Report*, School of Geography, University of Oxford.

Whittaker, R. J., Bush, M. B. and Richards, K. (1989) Plant recolonization and vegetation succession on the Krakatau Islands, Indonesia. *Ecological Monographs*, **59(2)**, 59–123.

Willis, A. J. (1973) *Introduction to Plant Ecology*, Allen and Unwin, London.

Willis, K. G. and Benson, J. F. (1989) Recreational values of forests. *Forestry*, **62**, 93–110.

Wise, J. R. (1883) *The New Forest: its history and scenery*, Henry Sotheran and Co, London.

Witkamp, M. and Ausmus, B. S. (1976) Processes in decomposition and nutrient transfer in forest systems, in *The Role of Terrestrial and Aquatic Organisms in Decomposition Processes* (eds J. M. Anderson and A. Macfadyen), Blackwell Scientific, Oxford, pp. 375–96.

Witkamp, M. and Crossley, D. A. (1966) The role of arthropods and microflora in breakdown of white oak litter. *Pedobiologia*, **6**, 293–303.

Wong, M., Wright, S. J., Hubbell, S. P. and Foster, R. B. (1990) The spatial and reproductive consequences of outbreak defoliation in *Quararibea asterolepis*, a tropical tree. *Journal of Ecology*, **78**, 579–88.

Wood, M. (1989) *Soil Biology*, Blackie, London.

Wood, R. F. (1974) *Fifty Years of Forestry Research*, HMSO, London.

Wood, T. G. (1971) The effects of soil fauna on the decomposition of *Eucalyptus* leaf litter in the Snowy Mountains, Australia. *IV. Colloquium Pedobiologiae*, INRA, Paris, pp. 349–58.

Woodward, I. (1989) Plants in the greenhouse world. *New Scientist: inside science*, No. 21.

Wright, J. W. (1976) *Introduction to Forest Genetics*, Academic Press, London.

Young, J. E. (1975) Effects of spectral composition of light sources on the growth of a higher plant, in *Light as an Ecological Factor*, Vol II (eds G. C. Evans, R. Bainbridge and O. Rackham), Blackwell Scientific, Oxford, pp. 135–60.

Zlotin, R. I. and Khodashova, K. S. (1980) *The Role of Animals in Biological Cycling of Forest-Steppe Ecosystems*, Dowden, Hutchinson and Ross, Stroudsburg.

Further reading

Begon, M. and Mortimer, M. (1986) *Population Ecology, a Unified Study of Animals and Plants*, 2nd edn, Blackwell Scientific, Oxford.

Burrows, C. J. (1990) *Processes of Vegetation Change*, Unwin Hyman, London.

Cannell, M. G. R., Malcom, D. C. and Robertson, P. A. (eds) (1992) *The Ecology of Mixed-species Stands of Trees*, Blackwell Scientific, Oxford.

Cherrett, J. M. (ed.) (1989) *Ecological Concepts: the contribution of ecology to an understanding of the natural world*, Blackwell Scientific, Oxford.

Clapham, A. R., Tutin, T. G. and Moore, D. M. (1987) *Flora of the British Isles*, 2nd edn, Cambridge University Press, Cambridge.

Crawley, M. J. (1983) *Herbivory: the dynamics of animal–plant interactions*, Blackwell Scientific, Oxford.

Dickinson, C. H. and Pugh, G. J. F. (1974) *Biology of Plant Litter Decomposition*, Vols 1 and 2, Academic Press, London.

Ellenberg, H. (1988) *Vegetation Ecology of Central Europe*, 4th edn, Cambridge University Press, Cambridge.

Elton, C. S. (1966) *The Pattern of Animal Communities*, Methuen, London.

Evans, G. C., Bainbridge, R. and Rackham, O. (eds) (1975) *Light as an Ecological Factor*, Vol II, Blackwell Scientific, Oxford.

Evans, J. (1984) *Silviculture of Broadleaved Woodland*, HMSO, London.

Evans, J. (1992) *Plantation Forestry in the Tropics*, 2nd edn, Clarendon Press, Oxford.

Gray, A. J., Crawley, M. J. and Edwards, P. J. (eds) (1987) *Colonization, Succession and Stability*, Blackwell Scientific, Oxford.

Grime, J. P., Hodgson, J. G. and Hunt, R. (1988) *Comparative Plant Ecology: a functional approach to British species*, Unwin Hyman, London.

Hart, J. W. (1988) *Light and Plant Growth*, Unwin Hyman, London.

James, N. D. G. (1982) *The Forester's Companion*, 3rd edn, Basil Blackwell, Oxford.

Krebs, C. J. (1985) *Ecology: the experimental analysis of distribution and abundance*, 3rd edn, Harper and Row, New York.

Matthews, J. D. (1989) *Silvicultural Systems*, University Press, Oxford.

Mitchell, A. (1974) *A Field Guide to the Trees of Britain and Northern Europe*, Collins, London.

Mitchell, P. L. and Kirby, K. J. (1989) *Ecological Effects of Forestry Practices in Long Established Woodland and their Implications for Nature Conservation*, Oxford Forestry Institute, Oxford.

Peterken, G. F. (1981) *Woodland Conservation and Management*, Chapman and Hall, London.

Pickett, S. T. A. and White, P. S. (eds) (1985) *The Ecology of Natural Disturbance and Patch Dynamics*, Academic Press, London.

Rackham, O. (1980) *Ancient Woodland: its history, vegetation and uses in England*, Edward Arnold, London.

Rackham, O. (1986) *The History of the Countryside*, J. M. Dent, London.

Ricklefs, R. E. (1990) *Ecology*, 3rd edn, W. H. Freeman, New York.

Rodwell, J. S. (ed.) (1991) *British Plant Communities*, Vol 1. *Woodlands and Scrub*, Cambridge University Press, Cambridge.

Speight, M. R. and Wainhouse, D. (1989) *Ecology and Management of Forest Insects*, Clarendon, Oxford.

Spurr, S. H. and Barnes, B. V. (1980) *Forest Ecology*, 3rd edn, Wiley, New York.

Stern, K. and Roche, L. (1974) *Genetics of Forest Ecosystems*, Chapman and Hall, London.

Swift, M. J., Heal, O. W. and Anderson, J. M. (1979) *Decomposition in Terrestrial Ecosystems*, Blackwell Scientific, Oxford.

Waring, R. H. and Schlesinger, W. H. (1985) *Forest Ecosystems: concepts and management*, Academic Press, London.

Whitmore, T. C. (1984) *Tropical Rain Forests of the Far East*, 2nd edn, Clarendon, Oxford.

Wood, R. F. (1974) *Fifty Years of Forestry Research*, HMSO, London.

The Forestry Commission publishes, through HMSO, a wide range of booklets, books and bulletins and reports on all aspects of forestry. Its annual reports on forestry research are a useful guide to current trends as are the publications of the US Forest Service.

Index

Bold numbers refer to pages containing a definition of a key term

English and scientific names of some common woodland organisms

Alder, common *Alnus glutinosa*
Ash, common *Fraxinus excelsior*
Aspen, *Populus tremula*
Bank vole *Clethrionomys glareolus*
Beech, common *Fagus sylvatica*
Big tree, Wellingtonia *Sequoiadendron giganteum*
Bilberry *Vaccinium myrtillus*
Birch *Betula*
Birch-bracket *Piptoporus (= Polyporus) betulinus*
Bluebell *Hyacinthoides non-scripta (= Endymion non-scriptus)*
Bracken *Pteridium aquilinum*
Bryony, black *Tamus communis*
  white *Bryonia dioica*
Bugle *Ajuga reptans*
Campion, red *Silene dioica*
Cherry, bird *Prunus padus*
  wild, gean *Prunus avium*
Chestnut, horse *Aesculus hippocastanum*
  sweet *Castanea sativa*
Coastal redwood *Sequoia sempervirens*
Cow-wheat *Melampyrum pratense*
Creeping soft-grass *Holcus mollis*
Dog's mercury *Mercurialis perennis*
Douglas fir *Pseudotsuga menziesii*
Dutch elm disease *Ophiostoma (= Ceratocystis) ulmi*
Elder *Sambucus nigra*
Elm, wych *Ulmus glabra*
Enchanter's nightshade *Circaea lutetiana*
Fir, grand *Abies grandis*
  silver *A. alba*
Fireweed *Chamaenerion angustifolium (= Epilobium angustifolium)*
Fly agaric *Amanita muscaria*
Foxglove *Digitalis purpurea*
Goosegrass, cleavers *Galium aparine*
Great tit *Parus major*
Green oak roller *Tortrix viridana*
Gypsy moth *Porthetria dispar (= Lymantria dispar)*
Hart's tongue fern *Phyllitis scolopendrium*
Hazel *Corylus avellana*
Hawthorn *Crataegus monogyna*
Heather, ling *Calluna vulgaris*
Hemlock, eastern *Tsuga canadensis*
  western *T. heterophylla*
Holly *Ilex aquifolium*
Honey fungus *Armillaria mellea*
Honeysuckle *Lonicera periclymenum*
Hornbeam *Carpinus betulus*
Ivy *Hedera helix*

Juniper *Juniperus communis*
Knopper gall *Andricus quercus-calicis*
Larch, European *Larix decidua*
Lesser celandine *Ranunculus ficaria*
Lime, linden, hybrid *Tilia x europaea*
  large-leaved *T. platyphyllos*
  small-leaved *T. cordata*
Lime aphid *Eucallipterus tiliae*
Maple, field *Acer campestre*
  sugar *A. saccharum*
Oak, English (pedunculate) *Quercus robur*
  sessile (durmast) *Q. petraea*
Old man's beard, traveller's joy *Clematis vitalba*
Oribatid mites Cryptostigmata
Oxlip *Primula elatior*
pill millipede *Glomeris marginata*
Pine, Austrian, Corsican *Pinus nigra*
  subspecies *laricio*
  lodgepole *P. contorta*
  Monterey *P. radiata*
  Scots *P. sylvestris*
Pine beauty *Panolis flammea*
Pine looper *Bupalus piniaria*
Pine shoot beetle *Tomicus piniperda*
Primrose *Primula vulgaris*
Ramsons, wild garlic *Allium ursinum*
Rowan, mountain ash *Sorbus aucuparia*
Sedge, pendulous *Carex pendula*
  wood *C. sylvatica*
Southern beech *Nothofagus*
Spindle *Euonymus europaeus*
Spruce, Norway *Picea abies*
  Sitka *P. sitchensis*
Spruce budworm, eastern *Choristoneura fumiferana*
Spurge, wood *Euphorbia amygdaloides*
Stinging nettle *Urtica dioica*
Sycamore *Acer pseudoplatanus*
Sycamore aphid *Drepanosiphum platanoidis*
Tawny owl *Strix aluco*
Tufted hair-grass *Deschampsia cespitosa*
Tulip tree, tulip poplar *Liriodendron tulipifera*
Weasel *Mustela nivalis*
Willow *Salix*
Winter moth *Operophtera brumata*
Wood anemone *Anemone nemorosa*
Wood mouse *Apodemus sylvaticus*
Wood sorrel *Oxalis acetosella*
Yellow archangel *Lamiastrum galeobdolon (= Galeobdolon luteum)*
Yew *Taxus baccata*